Advances in pig nutrition

It is widely recognised that agriculture is a significant contributor to global warming and climate change. Agriculture needs to reduce its environmental impact and adapt to current climate change whilst still feeding a growing population, i.e. become more 'climate-smart'. Burleigh Dodds Science Publishing is playing its part in achieving this by bringing together key research on making the production of the world's most important crops and livestock products more sustainable.

Based on extensive research, our publications specifically target the challenge of climate-smart agriculture. In this way we are using 'smart publishing' to help achieve climate-smart agriculture.

Burleigh Dodds Science Publishing is an independent and innovative publisher delivering high quality customer-focused agricultural science content in both print and online formats for the academic and research communities. Our aim is to build a foundation of knowledge on which researchers can build to meet the challenge of climate-smart agriculture.

For more information about Burleigh Dodds Science Publishing simply call us on +44 (0) 1223 839365, email info@bdspublishing.com or alternatively please visit our website at www.bdspublishing.com.

Related titles:

Achieving sustainable production of pig meat Volume 2
Print (ISBN 978-1-78676-092-0); Online (ISBN 978-1-78676-094-4, 978-1-78676-095-1)

Developing animal feed products
Print (ISBN 978-1-78676-463-8); Online (ISBN 978-1-78676-465-2, 978-1-78676-466-9)

Optimising pig herd health and production
Print (ISBN 978-1-78676-883-4); Online (ISBN 978-1-78676-885-8, 978-1-78676-886-5)

Seaweed and microalgae as alternative sources of protein
Print (ISBN 978-1-78676-620-5); Online (ISBN 978-1-78676-622-9, 978-1-78676-623-6)

Understanding gut microbiomes as targets for improving pig gut health
Print (ISBN 978-1-78676-487-4); Online (ISBN 978-1-78676-489-8, 978-1-78676-490-4)

Chapters are available individually from our online bookshop:
https://shop.bdspublishing.com

climate•SMART•publishing
IN AGRICULTURAL SCIENCE

BURLEIGH DODDS SERIES IN AGRICULTURAL SCIENCE
NUMBER 155

Advances in pig nutrition

Edited by Emeritus Professor Julian Wiseman, University of Nottingham, UK

burleigh dodds
SCIENCE PUBLISHING

Published by Burleigh Dodds Science Publishing Limited
82 High Street, Sawston, Cambridge CB22 3HJ, UK
www.bdspublishing.com

Burleigh Dodds Science Publishing, 1518 Walnut Street, Suite 900, Philadelphia, PA 19102-3406, USA

First published 2024 by Burleigh Dodds Science Publishing Limited

Library of Congress Control Number: 2024932216

British Library Cataloguing in Publication Data
A catalogue record for this book is available from the British Library

ISBN 978-1-80146-694-3 (print)
ISBN 978-1-80146-696-7 (PDF)
ISBN 978-1-80146-695-0 (ePub)
ISSN 2059-6936 (print)
ISSN 2059-6944 (online)

DOI: 10.19103/AS.2024.0140

Typeset by Deanta Global Publishing Services, Dublin, Ireland

Contents

Part 2 Feed formulation

Series list

Introduction

Pig nutrition faces many challenges, including the need to meet the changing requirements of animals as they grow while minimising environmentally and economically-damaging nutrient losses. Additionally, with growing crops for feed seen as a significant contributor to climate change, there is an emerging consensus that the sector must seek alternative, more sustainable feed sources which have a reduced impact on the environment.

This volume provides a comprehensive overview of the range of research addressing these challenges. The chapters are split into four parts: Part 1 chapters explore recent advances in understanding feed intake, highlighting pig nutritional requirements and digestive physiology. Part 2 chapters cover feed formulation, focussing on advances in understanding pig nutritional guidelines and ensuring feed safety. Chapters in Part 3 describe the role of feed additives in optimising pig nutrition, with discussion of amino acids, probiotics and exogenous enzymes. Finally, Part 4 chapters review alternative feed sources, including their development, the use of insects as an alternative protein source, green feed in organic pig farming, and corn fermented protein from dry grind ethanol.

Part 1 Feed intake

Chapter 1 focuses on advances in understanding pig nutritional requirements and metabolism. It begins by outlining optimising sow nutrition throughout gestation and lactation, covering the management of gilt weight on herd entry, nutrition to prevent sow lameness, minimising variation in nutrient intake during gestation, the transition from gestation to lactation, maintaining body weight during lactation, and nutritional interventions to improve productivity of the gilt, sow and litter. The chapter then discusses reducing variation in pig production systems, and the strategic use of metabolic modifiers and other modes of growth promotion. Next, challenges in matching nutrient requirements to diet specifications are examined, covering the rapid measurement of nutritional quality, updated information on nutritional requirements, and water requirements for pigs. This is followed by a description of optimising the utilisation of co-products, as well as gut health, nutrient utilisation capacity and health interactions.

The following chapter discusses the pivotal developmental changes within a pig's gastrointestinal system (GIT) across various life stages, emphasizing their importance for achieving growth, health, and feed efficiency. Chapter 2 starts by describing how, during the prenatal phase, the GIT experiences substantial growth, particularly in intestinal mass. It then moves on to examine how after birth, dynamic GIT changes occur, notably how the establishment of the pig's intestinal microbiota begins shortly after birth. Next, discussion explores the challenges that weaning presents, including reduced feed intake, stress, and gut microbiota shifts affecting GIT pH regulation, enzyme production, and intestinal barrier function. A section outlining how dietary strategies can address these issues is also included, covering the benefits of dairy products, fermentable carbohydrates, exogenous enzyme supplementation, the challenges of high protein content, and the negative effects of leguminous plant proteins. Finally, digestion and absorption of nutrients are considered, in particular the manipulation of intestinal nutrient transporters through nutritional interventions.

Part 2 Feed formulation

Part 2 begins by highlighting the development of nutritional guidelines for pigs. Chapter 3 opens with an outline of pig feeding and nutrition from birth to market weight, covering: breed and genetic traits affecting nutrient energy requirements, production goals, growth stage and weight, nutrient interactions, feed ingredients and quality, and feed cost and availability. Next, formulating diets to maximise feed efficiency and performance but minimise nutrient excretion is explored. The basis of diet formulation opens this section, followed by discussion covering diets for nursery pigs, growing-finishing pigs, boars, and gilts and sows. Following this, key challenges in formulating nutritional guidelines and possible solutions are described, including heat stress, the presence of toxigenic fungi and mycotoxins in feed ingredients, post-weaning diarrhoea, slowing down the growth of pigs, and the potential detrimental effect of high dietary leucine concentration.

The subject of chapter 4 is modelling feed requirements for pigs to optimise feed efficiency. To begin, the chapter defines feed efficiency, factors, and strategies to improve feed efficiency. Next, estimating feed requirement is explored, bringing into focus modelling growing pigs and gestating sows, data collection and uncertainties, and estimating for groups and individuals. A section on the improvements of models of gestating sows is also included, covering behaviour and health and environmental factors. The chapter ends with a case study describing the characteristics of gestating sows and their farrowing performance, and how they can be used to predict their nutrient requirements during gestation.

Chapter 5 concerns new approaches for determining the nutritional value of pig feed. The chapter starts by reviewing methodologies and studies to determine energy, nutrient content, and digestibility in feed ingredients. Next, models and equations that dynamically predict energy, nutrient content, and digestibility in feed ingredients are summarised. This is followed by an exploration of anti-nutritional factors, including: trypsin inhibitors and urease activity, glucosinolates, protein and lipid oxidation, biogenic amines, and co-occurring, emerging, and masked mycotoxins. An overview of functional nutrients is also included. Finally, the environmental impact indicators of feed ingredients are highlighted, with a focus on the findings of Life Cycle Assessments.

Chapter 6 covers the safety of feed for pigs. The chapter begins with an overview of the pathogens of concern for feed-based transmission, including prions and parasites, bacteria, and viruses. Next, feed contamination and its prevention are considered, covering: ingredient sourcing, receiving and loadout practices, zones in the mill, and employee training. A section on mitigation of pathogens in feed is also included, covering point-in-time mitigation strategies and those with residual effects. Finally, feed mill decontamination is discussed.

Part 3 The role of feed additives in optimising pig nutrition

The first chapter of Part 3 discusses the principles and practical aspects of meeting amino acid requirements and the use of functional amino acids in pig nutrition. Chapter 7 first explores the metabolism of protein and amino acids, followed by the principles of amino acid needs, including maintenance, growth and production. Next, the estimation of amino acid requirements is described, covering empirical and mechanistic approaches, as well as the availability and use of supplemental amino acids. Following this, the concept and application of functional amino acids, such as arginine, methionine, tryptophan, phenylalanine, and glutamine, are outlined. The chapters ends with a section discussing the practical aspects of feed formulation meeting amino acid needs.

Chapter 8 examines the use of prebiotics in the modulation of the gastrointestinal (GI) microbiota and the consequences this may have for pig health. The prebiotics discussed are oligosaccharides, inulin, polysaccharides and non-carbohydrate prebiotics. The chapter reviews recent research on prebiotic modes of action as well as their effects in different dietary formulations.

The focus of Chapter 9 is understanding and optimising the use of exogenous enzymes in pig nutrition. To begin, augmenting neonate and mature pig enzymatic capacities are reviewed, covering: protease, amylase, the absorptive capacity of the intestine, and market considerations. Next, a section detailing the hydrolysis of phytic acid and lower esters, their anti-nutritive effects, and optimising phytate hydrolysis. Following this, the chapter examines the benefits of extraphosphoric activities, including the reduction of endogenous losses, increases in micromineral absorption, reduction in the incidence and severity of post-weaning diarrhoea, and impacts on the parturition of sows. The chapter then considers how fibre degrading enzymes function in pigs and, finally, constraints on enzyme functionality.

Part 4 Alternative feed sources

Chapter 10 provides an overview of co-products and other alternative feedstuffs for sustainable pig production. The chapter opens with a discussion of feed evaluation, animal growth and carcass characteristics. Focus then moves to meat quality and animal health. After this, the chapter highlights the connection between sustainability and feed, in particular the greenhouse gas emissions of pigs, and the human-edible protein balance of swine and amount of land required to produce it. The chapter ends with a case study of the use of canola co-products as feedstuffs for pigs.

The following chapter concerns the use of insects as a source of protein in pig nutrition. Chapter 11 initially addresses the question of why use insects in feed, then describes the current state of insect production, potential candidates for feed and their current production volumes. This is followed by sections on production and the health responses in pigs to insect protein, competition for resources, and rearing insects on manure. Next, current research is reviewed, covering insects as antimicrobials, and energy use and life cycle analysis, followed by future trends in research, including alternative insect species, insect welfare, the impact of COVID-19 on global feed production and sources, and workplace labour constraints. Finally, barriers to uptake and challenges to production are outlined.

The subject of Chapter 12 is green feed in organic pig farming with an emphasis on studies conducted under the conditions of Denmark and Northern Europe. To begin, the chapter presents findings from recent research on the energy and protein (amino acid) requirements of organic pigs, covering locomotor activity, thermoregulation, and milk production. This is followed by an exploration of the feeding value, daily intakes and performance impact of direct foraging, as well as alternative forage cops and below-ground contributions. The feeding value, daily intake and performance impact of roughage is then explored. Finally, the feeding value, daily intake and performance impact of green protein is discussed.

The final chapter of the book explores high protein corn fermentation products for swine derived from corn ethanol production. Chapter 13 starts by describing the ethanol production by-product, distillers dried grains with solubles, and its use as pig feed. Next, the production process and value of corn fermented protein (CFP) are covered, followed by the challenges to its production. Focus then turns to four case studies. These are: the standardised ileal digestibility of CFP for pigs, concentrations of digestible and metabolisable energy in CFP products fed to pigs, the effect of an addition of phytase to fermentation on the availability of calcium and phosphorous in corn fermented products, and the effects on performance of pigs of inclusion of CFP in the diet.

Part 1

Feed intake

Chapter 1

Advances in understanding pig nutritional requirements and metabolism

Robert J. van Barneveld, Robert J. E. Hewitt and Darryl N. D'Souza, SunPork Group, Australia

1 Introduction

Nutritional requirements defined for different classes of pig vary depending on the desired production outcomes. Historically, requirements have mainly been specified for the maximum productivity of the breeding herd and optimal feed conversion efficiency of the progeny. In most cases, this approach will also contribute to the optimal sustainability of the production system. Yet, with increasing demand for feed inputs, greater pressure on land use, an enhanced focus on greenhouse gas emissions, and greater credence given to social imperatives for farming systems, combined with the pressure to remain profitable, defining nutrient requirements of pigs in modern production systems and our approach to meeting them must be broader than ever before.

Pork has evolved as the most consumed meat on the planet on the enduring premise that the fundamental role of the pig is to accumulate resources, otherwise discarded as waste, and to act as a hedge against oscillating availability and the price of grain (Fairlie, 2010). Increasing demand for food by an ever-growing population in the modern world means grain is now the

http://dx.doi.org/10.19103/AS.2024.0140.01

primary input of pork production systems, exposing producers to significant fluctuations in price with no option to capitalise on grain overproduction, further confounded by global trade, an increased number of grain accumulation points, futures trading, freight availability and more variable weather patterns. In addition, recycling, waste reduction targets, and renewable energy systems and mandates are placing pressure on the availability and price of traditional waste streams used as nutrient sources for pigs (rendered products, tallows, oils and milling co-products).

For the above reasons, much of the world's pig production has shifted away from the banking of otherwise wasted nutrients towards specialisation, with prescriptive diets becoming dependent on a reduced number of energy and nutrient concentrated ingredients. Consequently, we need to question whether this production is competing directly with human consumption for inputs, and we need to understand the environmental impacts of these production systems. As the demand for meat increases (meat consumption expected to increase by 50% by 2050 (FAO, 2018a)), we need to continue to advance our pork production systems to be sustainable and further ensure outcomes that are not only good for the consumer but also for the pig and the environment.

Strategies to increase sustainability in livestock generally fall into three categories – efficiency, sufficiency or consistency (Schader et al., 2015). Sufficiency strategies centre around reducing the demand for less sustainable animal products, with changes in dietary patterns – less meat or substituting grain-fed ruminant products with non-ruminant products. Improving productivity so that demand is met whilst reducing environmental impacts is the key focus of efficiency strategies. Consistency strategies focus on the net contribution of livestock to the human food supply, exploiting those resources that cannot be otherwise used for food production.

As a first step in demonstrating that pork production can contribute to Consistency strategies, van Barneveld et al. (2023) demonstrated that the net protein contribution (NPC) from an intensive Australian pork supply chain was 3.26 (Table 1). The NPC was calculated using (a) actual, published or estimated values for human-edible fractions of feedstuffs, (b) the percentage of protein available within raw materials considered to be human-edible, (c) recommended amino acid scoring patterns for infants, adolescents and adults, (d) published and calculated from standard reference nutrient databases, digestible indispensable amino acid scores, (e) carcase yields and carcase composition from published studies and (f) actual feed formulations, feed volumes and production data. Key to generating a positive NPC from this Australian value chain was the efficiency of the system driven by the annual production of carcase and offal relative to human-edible protein consumed, the leanness of the carcase, and the utilisation of waste streams, co-products

Table 1 The net protein contribution from an intensive Australian pork supply chain

	Pork
Total annual carcase weight (kg)	88 080 878
Total annual offal weight (kg)	3 951 213
Carcase protein content (g/kg)	140
Offal protein content (g/kg)	180
Human-edible protein produced (HePp[1], g)	13 042 541 328
Human-edible protein fed (HePf, g)	18 653 955 545
Human-edible protein conversion efficiency (HePCE[2], g/g)	0.70
DIAAS pork (Ertl et al., 2016)	113.9
DIAAS diet (from DIAAS Calculation sheets (FAO, 2011))	24.4
Protein quality ratio (DIAAS pork/DIAAS diet)	4.66
Net protein contribution (HePCE × PQR)	3.26

[1]HePp, sum of total annual carcase weight multiplied by carcase protein content and total annual offal weight multiplied by offal protein content.
[2]HePCE, human-edible protein produced divided by human edible protein fed.
Source: Adopted from: van Barneveld et al. (2023).

and human-inedible feedstuffs. In relation to the three categories of livestock sustainability strategies described above, if we optimise efficiency through closely matched diet specifications to requirements for a lean carcase, maximise productivity of the breeding herd (including minimising sow wastage), make better use of lower-quality ingredients and promote consumption of pork, pig production will be highly sustainable.

In relation to advances in pig nutritional requirements and metabolism, these sustainability strategies can be achieved through:

1. Optimising sow nutrition throughout gestation and lactation, including a focus on nutritional strategies that increase litter size;
2. Reducing variation in pork production systems, with particular focus on nutrition that increases viability of piglets and their birthweight;
3. Strategic use of metabolic modifiers, if the market accepts their use;
4. Closer matching of nutrient requirements to feed composition;
5. Optimising the utilisation of co-products not suitable for human consumption;
6. Optimising gut health and the capacity for nutrient utilisation; and
7. Enhanced understanding of nutrition and health interactions.

This chapter aims to consider these facets, focussing on advances made in nutritional requirements and metabolism and how these advancements contribute to the sustainable production of pig meat.

2 Optimising sow nutrition throughout gestation and lactation

In pig production systems, improved resource use efficiency can be achieved by optimising nutritional resources directed towards the breeding herd relative to a target output. Traditionally, sow nutrition focussed on just two production stages - gestation and lactation. However, changes to the sow as a result of genetic selection programmes for efficient, lean growth and a greater understanding of the physiology of the sow and the developing progeny have resulted in more targeted nutritional interventions.

However, with the above in mind, the apex key performance indicator for efficient use of resources in a pig production system continues to be whole herd feed conversion. This accounts for all nutrients consumed by both the breeding herd and progeny per kilogram of pork produced. Minimisation of sow replacement through optimal gilt development, prevention of lameness and optimisation of reproductive efficiency will contribute significantly to reduced whole herd feed conversion efficiency and we advocate that a primary driver in the attainment of these objectives is maintenance of adequate sow body condition and minimisation in variation in this condition across the reproductive cycle.

2.1 Management of gilt weight on herd entry

Improper management of the sow at any stage, from her selection into the herd through her subsequent parities, has the potential to affect her lifetime productivity and thus her impact on the efficiency of the herd. Nutritionally, the developing gilt represents a challenge with the focus on fast, lean, efficient production in progeny. When gilt management programmes are effective, they will result in improved lifetime performance from a smaller gilt pool. Non-negotiable aspects of gilt development have been well established and still hold true (Beltranena et al., 2005; Williams et al., 2005), that includes recognising the importance of a dedicated gilt development area, ensuring that you have enough gilts in the pool to meet requirements, breed to weight not age and on the second parity and avoid slow-growing gilts. Despite the significant genetic improvements observed, there has been a relatively low shift in weight-for-age targets, Close and Cole (2000) recommended a weight of 130–140 kg at 220-230 days, whilst Theil et al. (2022b) reported current Danish recommendations as 140–160 kg at 220–240 days. Although genetic gains in leanness have led to suggested target backfat thicknesses falling from 18–20 mm to 13–15 mm.

Moturi et al. (2022) reinforced these principles when they studied gilts that were mated early (222 days of age), middle (244 days) or late (264 days), with respective weights of 141, 153 and 165 kg. The late-mated gilts had higher backfat thickness at farrowing compared to either the early- or middle-mated groups, that resulted in lower feed intake in the first lactation, and a subsequently higher loss of backfat. This resultant loss of condition did not impact on subsequent reproductive performance, whilst no significant reduction in piglet growth was observed. Interestingly, low (13.0 MJ ME/kg) or high (13.9 MJ ME/kg) energy diets fed from 150 days of age had no influence on performance.

In many production systems, the management of the gilt begins when she is selected from the grower herd at 20 weeks of age or later. However, there is considerable evidence that sow lifetime performance is affected by management decisions that are made from birth. When low birth weight pigs are selected as breeding stock, they produce fewer piglets over their lifetime, and these piglets also tend to have lower birth weights (Magnabosco et al., 2016). Whilst adequate colostrum intake is linked to heavier nursery exit weights and increased odds of farrowing a litter (Vallet et al., 2015), with similar results when pigs are weaned at older ages (Knauer, 2016b). Key aspects of managing and feeding gilts to improve sow lifetime performance have been summarised by Faccin et al. (2022) and are outlined in Table 2. Consequently, feeding programmes during gilt development should be designed with these principles and targets in mind.

2.2 Nutrition to prevent sow lameness

As discussed in the previous edition of this chapter (van Barneveld et al., 2017), the main nutrients affecting foot health in pigs include the amino acids cysteine and methionine; the minerals such as calcium, zinc, copper, selenium and manganese (van Barneveld, 2010) and vitamins A, D, E and biotin (Kornegay, 1986). Skeletal health is also vital to reduce lameness. Osteochondrosis has been suggested as a major cause of leg weakness in growing pigs (Nakano et al., 1987) and has been linked to the premature culling of breeder stock (Gresham, 2003). The primary development of osteochondrosis in pigs occurs at a very young age; however, clinical signs of lameness are not observed until much later, up to 18 months of age (Ytrehus et al., 2004). Vitamin D plays an integral role in mineral nutrition and is responsible for the intestinal absorption of minerals. Direct dietary supplementation with 25-hydroxycholecalciferol, the activated form of vitamin D that normally occurs in the liver of the pig, has been shown to promote normal endochondral ossification and inhibit the progression of osteochondrosis (Sugiyama et al., 2013). Further development of nutritional strategies to improve sow foot health have not yielded any significant insights into this issue.

2.3 Minimising variation in nutrient intake during gestation

Nutrition during gestation is focussed on meeting the requirements of the sow for maintenance and the development of maternal tissues and the conceptus (National Research Council, 2012); however, it is only in the last third of gestation

Table 2 Key aspects of managing and feeding gilts to improve lifetime performance and longevity, and produce high-survival pigs

Area	Action	Reason
Birth weight	Avoid selecting gilts weighing less than 1 kg at birth.	Gilts born <1 kg of birth weight produce 4.5 fewer piglets over three parities than the other birth weight classes.
Litter of origin	At the multiplier, avoid selecting gilts born from sows farrowing litters with average birth weight <1.15 kg in two or more farrows.	Gilts born to low birthweight phenotype sows can pass this feature to their progeny.
Colostrum intake/ cross-fostering	Sows with higher milk production (e.g. multi-parity sows) should nurse future replacement gilts.	Gilts consuming at least 250 g of colostrum will be heavier at 42 days of age, and gilts heavier at weaning have higher survivability and odd to farrow a litter.
Weaning age	Replacement gilts should be at least 24 days old at weaning.	Gut health, postweaning survivability and lifetime productivity are greater for pigs when weaned at 24 days of age or older.
Vitamins and minerals	Add more phosphorous than terminal lines, consider organic trace mineral sources and add extra choline, pyridoxine, folic acid and biotin.	Phosphorus recommendation levels for bone mineralisation are 8% higher than for growth performance. Organic trace mineral supplementation can reduce osteochondrosis incidence. These vitamins are involved in reproductive functions and often not included in finishing pig premixes.
Mammary development	From 90 days of age to puberty, do not restrict energy intake too much; avoid overfeeding gilts in late gestation.	Appropriate mammary gland development can be reached without maximising energy intake, but low intake also reduces mammary DNA. Overfeeding gilts in late gestation will deposit more fat in mammary glands, reducing colostrum and milk production.
Boar exposure	Provide adequate boar exposure with boars with at least 10-months of age. Promote as much physical interaction with gilts as possible.	Having more boars is better to avoid fatigue. Mature boars have more libido and are larger enough to stimulate oestrous without being submissive to the gilts in pens.

(*Continued*)

Table 2 (*Continued*)

Area	Action	Reason
Breeding targets	Breed on second oestrus and 135–160 kg body weight (BW).	Inseminating in the first oestrous reduces farrowing rate and litter size in the first farrowing. Breeding overweight gilts reduces their longevity in the herd.
Flush feed	Only flush feed gilts that might not reach BW target at breeding.	Flush feeding only improves reproductive outputs when gilts are below the target weight for breeding.
Gestation feeding	Early: do not feed gilts below maintenance and growth requirements and avoid feeding more than 31.4 MJ of net energy (NE) per day. Late: unless body condition is low, avoid bump feeding.	Over-conditioned gilts can have lower litter size and have reduced feed intake and milk production in lactation. Bump feeding improves birth weight slightly but increases the stillborn rate and lowers lactation feed intake and colostrum and milk production.
First lactation	Avoid having unsuckled mammary glands.	Mammary glands not suckled during first lactation produce less colostrum and milk in the subsequent farrowing.

Source: Adopted from: Faccin et al. (2022)

that sows devote a substantial proportion of energy and nutrients to the growth of the placenta, foetus and later mammary growth (Theil et al., 2022b). Whilst a higher level of nutrition in early gestation has no impact on embryo survival, it does result in an increase in weight and back fat (Hoving et al., 2012). There appears to be a role for specific amino acids (leucine, arginine and glutamine), monosaccharides (glucose and fructose) and vitamins (riboflavin and folic acid) in early lactation (Langendijk, 2021), but further research is required. Despite the increased energy and lysine requirements of sows as gestation progresses, a higher feeding level is not associated with any increased reproductive response (Theil et al., 2022b). However, it is linked to sow weight gain (Knauer, 2016a). Supplying nutrients above requirements during gestation benefits only the sow and not the developing piglets, and it further results in a reduction in voluntary feed intake during lactation (Weldon et al., 1994).

2.4 The transition from gestation to lactation

The requirements of the sow within the transition period around parturition, where sows shift from late gestation to early lactation, have become more apparent in recent years (Feyera and Theil, 2017). This period involves rapid foetal growth, mammary development, farrowing, colostrum and milk production, leading to profound metabolic changes that often result in sows entering a catabolic state. In addition to these physiological changes, sows are isolated into farrowing pens, and the inability to express maternal behaviours, such as nesting, adds further stress. This transition period also covers the first few days of the piglet's life, where its survival is most challenged by the sow's farrowing performance and colostrum yield. These factors are of prime importance for minimising stillborn rates and early piglet mortality (Theil et al., 2022b).

Piglets that experience an extended farrowing duration have lower vitality and an increased risk of being stillborn (Peltoniemi et al., 2014; Feyera et al., 2018), with both the energy status (plasma glucose) and time from last meal before farrowing identified as key factors increasing farrowing duration (Feyera et al., 2018). In contrast to energy, crude protein levels above requirements, when metabolised for energy, seem to be detrimental to the farrowing process (Tydlitát et al., 2008; Pedersen et al., 2020), resulting in an increased stillborn rate. It is thought that this impact of high crude protein is related to the energetically unfavourable oxidation of protein (Theil et al., 2020).

The potential sow milk yield is impacted by the number of mammary cells present at farrowing, which can be influenced by nutrition when mammary development is already underway (Theil et al., 2022a). Rapid accretion of mammary tissue occurs at three stages in a pig's life – from 90 days of age to puberty, from day 90 of gestation to farrowing, and during lactation. During this transition period, milk-synthesising tissue replaces adipose and stromal tissues (Ji et al., 2006). However, recommendations on how to influence mammary development during this period are equivocal.

Excess energy results in obese gilts with reduced mammary cell division (Head and Williams, 1991). However, feeding at 1.58- or 1.82-times maintenance resulted in increased backfat thickness (17–19 mm, and 21–26 mm, respectively) and an increased mass of parenchymal tissue (Farmer et al., 2017). These levels of backfat contrast with recommendations for the modern sow (Hojgaard and Brunn, 2021), where 'over-conditioning' is associated with decreased lactation feed intake (Eissen et al., 2000). Increased lysine in late gestation resulted in improved piglet growth in lactation (Che et al., 2020), which could be due to enhanced mammary development that accounts for one-sixth of lysine requirements in the last 12 days of gestation (Feyera and Theil, 2017). Similarly, Farmer et al. (2022) saw a 44% increase in total mammary parenchyma when a high-lysine diet (26.0 g/d lysine) was fed in comparison to a control diet (18.6 g/d), yet there was no evidence in this study that piglet growth was enhanced as a result.

Feed quality cannot be overlooked, with mycotoxins having the potential to decrease the circulating concentrations of hormones such as prolactin, which is essential for mammogenesis (Theil et al., 2022a).

Whilst dietary energy levels can improve colostrum yield when fed above maintenance (Decaluwé et al., 2014), excess energy does not result in improved yield. Colostrum yield is maximised at intake levels lower than those delivering maximum farrowing performance (Feyera et al., 2021). The greatest impact on the composition of colostrum appears to be increasing the dietary fibre content in the transition period and supplementation of branched-chain amino acids. High-fibre diets during the last 2 weeks of gestation increased fat content by 49%, with an increased mammary uptake of acetate and butyrate (Feyera et al., 2019), although it is dependent on the fibre source (Theil et al., 2022). Specific amino acids would also appear to be of benefit, with branched-chain amino acids (Wijesiriwardana et al., 2021) and valine, in particular (Zhao et al., 2019; Che et al., 2020), showing a linear response in the fat and crude protein content to increasing dietary valine.

2.5 Maintaining body weight during lactation

A reduction in feed intake during lactation is likely to lead to the mobilisation of body reserves, as the sow, in contrast to gestation, devotes all her efforts to ensuring the growth of the piglets (Bergsma et al., 2009). The body condition upon exit from gestation can impact this process. Lean sows tend to have higher protein reserves and a higher voluntary feed intake compared to fat sows. This can be explained by high levels of exogenous protein that may stimulate milk production and therefore voluntary feed intake (Revell et al., 1998a). Conversely, fat sows have lower protein reserves, thus less ability to mobilise them for milk output when the supply is limited, especially in early lactation (Revell et al., 1998b). One strategy that has been employed to prevent impaired feed intake in lactation is to restrict feed intake in the immediate period both prior to and post parturition (Koketsu et al., 1996); however, more recent research would suggest that *ad libitum* feeding in this period can lower mobilisation of body reserves and increase piglet performance (Cools et al., 2014).

The use of high-fibre feeds during gestation is an approach that has been taken to increase voluntary feed intake in lactation. The bulk of high-fibre diets allows the sow's satiety to be met whilst restricting her energy intake, a factor known to influence voluntary feed intake in lactation (Eissen et al., 2000). The underlying mechanism of this effect is not fully understood; however, high-fibre diets appear to modulate the gut microbiota, supporting the abundance of beneficial *Lactobacilli* and reducing *Proteobacteria,* resulting in a decrease in proinflammatory factors (Li et al., 2023). Odakura et al. (2023) also showed that eubiotic fibre supplementation in the third trimester resulted in increased piglet weaning weight, associated with increased lactation feed intake.

Whilst body tissue catabolism is an immediate issue for the sow and her suckling piglets, the impact on subsequent reproductive performance is considerable. Clowes et al. (2003) investigated the impact of increasing losses of protein on the reproductive performance of first-parity sows. Ovarian function was suppressed in sows with the greatest protein loss; they had fewer medium-sized follicles, and those follicles contained less follicular fluid, with lower oestradiol and IGF-1 concentrations, translating to a lower number of piglets in the subsequent litter (De Bettio et al., 2016). Feeding carbohydrate-rich diets during the follicular development phase improves follicle and oocyte quality, with dextrose, in particular, resulting in improved subsequent litter size (van den Brand et al., 2009; Plush et al., 2019).

The recovery of body tissue losses is best achieved in the first 30 days of gestation (Dourmad et al., 1996), however, it has been traditional to keep the feeding levels low in early gestation. The modern sow though does not benefit from restricting feed intake (Leal et al., 2019), with a higher feed intake supporting the formation of luteal tissues and progesterone secretion by the ovaries (Langendijk, 2021), whilst post implantation, a high feed intake seems to improve embryo survival and the maintenance of pregnancy. There is still some suggestion that increased feed intake benefits the sow to a greater degree (weight gain, body condition and backfat thickness) than the piglet, with higher mortality observed, linked to an increase in the number of low-birth-weight pigs (Carrión-López et al., 2022).

2.6 Nutritional interventions to improve productivity of the gilt and sow and litter performance

The productivity of gilts, sows and litter performance plays a crucial role in the overall success of pig production systems. As nutrition plays a pivotal role in optimising these outcomes, this section explores various nutritional interventions that have been employed to enhance the productivity of gilts, sows and their litters. These recent research outcomes listed below provide insights into the key nutritional strategies that could be adopted to promote improved gilts, sows and litter performance and ultimately contribute to the profitability and sustainability of pig operations.

The main circulating metabolite of vitamin D, 25-hydroxyvitamin D (25(OH)D), plays an important role in the establishment and maintenance of pregnancy and the regulation of foetal growth, with temporal changes in maternal plasma concentrations as gestation progresses (Stenhouse et al., 2022). At day 60 of gestation, maternal 25(OH)D concentration was positively associated with prenatal survival. Stenhouse et al. (2022) also showed a novel association, with the concentration of 25(OH)D at day 90 of gestation being inversely associated with the percentage of males in the litter.

Choline is a lipotropic amine that carries out various functions in the body and is important in the formation of the neurotransmitter acetylcholine. It also plays a role in foetal development and fertility, with certain fertility issues in humans associated with decreased concentrations of choline. Zhan et al. (2021) utilised these findings as the basis for an investigation that showed that choline supplementation promoted ovarian development with an increased number of corpus lutea being present in choline-supplemented groups, meaning more follicles had ovulated. Sow's milk contains high concentrations of free and bound glutamine plus glutamate (Wu et al., 2011), with most of this glutamine synthesised by mammary tissue from branched-chain amino acid metabolism rather than absorbed from circulating reserves. Monosodium glutamate (MSG) is a readily available source of glutamate. When included in sow diets at 1% or 2%, it was found that sows lost less weight during lactation, yet piglets grew faster and pre-weaning mortality was significantly lower (Rezaei et al., 2022). The supplementation with MSG resulted in changes in the amino acid profile of sow plasma and milk samples, which corresponded to a changed piglet plasma profile. In diets that may be deficient in glutamate, such as typical corn-soy diets, supplementation with MSG would appear effective in improving sow lactational performance, therefore, piglet growth.

The leucine metabolite, β-hydroxy β-methyl butyrate (HMB), has been shown to influence protein turnover through both increasing protein synthesis and reducing protein degradation. When HMB was included at 15 mg/kg of body weight from day 100 of gestation piglet birth weights were consistently increased, and piglet growth rate to weaning was also enhanced (Davis et al., 2021). An additional dose-response study (Davis et al., 2022) showed a quadratic effect with piglet weights deteriorating at HMB levels above 15 mg/kg BW. Mortality also tended to follow the same quadratic pattern, with reduced 24-h, week 1, and pre-weaning mortality at 15 mg/kg BW compared to all other inclusion rates. Interestingly, IgG concentration in colostrum was linear, continuing to increase when fed above the turning point for growth and mortality.

Increasing the intake of methionine in late gestation has been shown to increase the antioxidant status of sows and newborn piglets, whilst interest has also been shown in the effect of increasing the methionine-to-lysine ratio to increase the endogenous production of hydrogen sulfide to enable normal placental vascular development (Bruun et al., 2023). When assessed, increasing the dietary SID methionine:lysine from 0.30 to 0.44 in the period from mating until day 50 of gestation failed to result in any improvements in the number of live or still born piglet, nor their birth weight or variation in birth weight (Bruun et al., 2023).

Including fats and oils in lactation diets is an effective strategy to increase energy concentration and can also provide essential fatty acids (EFA) that cannot be synthesised by the sow. Linoleic (LA) and α-linoleic acid (ALA) are two such EFAs that can be elongated to form other polyunsaturated fatty acids such as EPA and DHA. Holen et al. (2022) fed diets with low or high LA and ALA concentrations to sows from day 112 of gestation. Sows receiving high levels of EFAs produced litters that grew faster in lactation, resulting in higher weaning weights, but no difference in survivability was observed. This improved piglet performance did not result from any alteration to the crude composition of colostrum or milk; however, the LA and ALA content in colostrum and milk increased in response to the diet levels of EFAs.

The supply of energy to the lactating sow is primarily supplied by a combination of starch from cereals and fatty acids from fats or oils. Dietary fat is able to improve the growth rate of nursing piglets, but high dietary fat can exacerbate oxidative stress and increase the inflammatory response. This increase in inflammation and the diet composition itself can modify the gut microbiota, preferencing those microbes that are associated with fatty acid metabolism. Therefore, a balanced diet is required. Wang et al. (2023) investigated three starch-to-fat ratios and showed that a high starch-to-fat ratio diet fed from day 85 of gestation resulted in increased piglet birth weight, likely a result of improved diet digestibility and reduced immune challenge. Dietary high starch intake in the lactating sow could ameliorate lipid metabolism disorders, inflammation and insulin resistance of sows and improve the birth weight of piglets.

Short-chain fatty acids (SCFA), in particular acetate and butyrate, have been recognised as substrates for milk synthesis within ruminants for a long time. The circulating level of SCFAs in pigs is much lower than ruminants, yet when high-fibre diets are fed during gestation, serum acetic acid levels are increased, resulting in an increase in the fat content of colostrum and milk (Jensen et al., 2012). As an alternative to feeding high-fibre diets, direct supplementation of diets with 0.1% sodium acetate from day 85 of gestation was undertaken (Qi et al., 2023). Supplementation with sodium acetate resulted in increased colostrum and milk yield, whilst also increasing the nutrient concentration of colostrum and milk fat content at day 21 of lactation, resulting in heavier piglet weaning weights.

In summary, the key to a long reproductive life is to optimise the entry of the gilt into the breeding herd through meeting key weight targets, maintaining an adequate level of nutrition throughout gestation to support maintenance of body tissues and the growth of maternal and foetal tissues, supporting the sow during the transition period, and ensuring a high level of feed intake during lactation to reduce losses of body reserves and maintain good ovarian function.

3 Reducing variation in pig production systems

Variation is an often overlooked contributor to inefficient use of resources, and the nutritional interventions that can be applied to reduce variation are poorly understood or limited.

Variation is inherent in any biological system and is a challenge to manage in modern pork production businesses. Within the growing herd, inherent variation within a population of pigs represents a significant cost, because of the need to select on farm to meet market specifications, poor matching of diet specifications to nutrient requirements, grading losses, higher pre-weaning mortality and challenges associated with health management. Consequently, any management practice that can be applied to reduce variation at the point of sale has the potential to improve the profitability and overall efficiency of a pig enterprise.

There are many factors that can influence birthweight and variability in birthweight; hence, manipulation by nutritional means alone is unlikely to resolve this. It is, however, relevant that most nutrient specifications used in diet formulations today have been derived from far less prolific sows; hence, it is possible that both the sow and foetuses are compromised, given both have an increase in requirements as gestation progresses (Ball et al., 2008). Lawlor et al. (2007) reported that birthweight, weaning weight and within-litter variation were unaffected when five different dietary digestible energy (DE) levels were provided during different gestation phases. In contrast, Wu et al. (2010) demonstrated that supplementation of gestating sow diets with specific amino acids, free arginine (to bring the lysine to arginine ratio to 2.64) and glutamine, significantly influenced litter birthweight variation. These amino acids were targeted as arginine is extensively catabolised by arginase in the small intestine, and only 60% of dietary arginine enters the portal circulation of gestating gilts (Wu et al., 2007), whilst the uptake of glutamine in the uterus of gestating gilts is the greatest amongst all amino acids (Wu et al., 1999), and it is abundant in placental and foetal fluids. When Wu et al. (2010) added 0.6% glutamine and 0.4% arginine to corn-soybean meal-based diets, glutamine concentrations in gilt plasma did not decline, but a marked reduction in ammonia and urea in maternal plasma and a decline of variation in birthweights amongst all piglets born and pigs born alive were observed. Foxcroft (2007) suggested that the smallest pigs within the litter respond the greatest to maternal nutrition interventions, resulting in a reduction in variation in birthweight as evidenced by the significant impact of exogenous somatotropin during early gestation on small pigs (Rehfeldt et al., 2001).

The important role of the placenta and its development has been highlighted in recent years. Arginine, amongst other amino acids, is essential for the proper development of the placenta. They regulate angiogenesis and influence vascular development, and therefore may allow for a greater supply of nutrients and oxygen to the foetus. The average piglet birth weight was heavier in sows supplemented with 1% L-arginine, with a greater percentage of piglets weighing over 1.8 kg at birth, although the percentage of pigs born at lower weights did not change (da Silva Fonseca et al., 2022). Increasing the ratio of valine to lysine has also been shown to increase the total placental area (Wang et al., 2023), which was associated with a decrease in the number of stillborn piglets. Whilst melatonin supplementation during late gestation was shown to reduce oxidative stress in the placenta (Wang et al., 2023), leading to increased birth weight with resultant impacts on survival and weaning weights.

In addition to nutritional interventions that influence variation in birthweight per se, it is possible that a considerable amount of the variation in growth performance after birth may be largely determined, and essentially pre-programmed, during foetal development in the uterus (Foxcroft and Town, 2004). Foxcroft (2007) reported that the early period of myogenesis, involving the differentiation of primary muscle fibres, is generally considered resistant to nutritional manipulation, whereas the nutritional effects on differentiation and hyperplasia of secondary fibres have been demonstrated between day 25 and day 90 of gestation.

As discussed previously, whilst the greatest effects of lactation feeding will be on postnatal piglet performance and variation, nutrition of the sow during lactation can also influence variation in litter birthweights. Increased catabolism during the last week of lactation in primiparous sows is known to reduce embryonic survival and development up to day 30 of gestation in the subsequent litter (Foxcroft, 1997). Quesnel et al. (2008) found significant relationships between sow's body condition and the coefficient of variation (CV) of birthweight. Litter heterogeneity increased with body weight at the beginning and end of gestation. The CV for birthweight was not linked with body weight gain during gestation, but was with back fat gain, which was likely a reflection of changes in feed allowance during gestation because of body reserve mobilisation during the previous lactation (Quesnel et al., 2008).

Regardless of the variation in the birthweight of any litter, all pork production systems should focus on optimising sow milk production and piglet colostrum and milk intake to maximise weaning weights for age. Targets of 75–85 kg of litter weight at 21 days of lactation should be a key focus for production staff, with this target representing the best insight into the effectiveness of any lactation feeding programme.

Douglas et al. (2014) demonstrated that when low-birthweight piglets are grouped and offered supplementary milk, they drink significantly ($P < 0.001$) more than those of mixed group piglets. Further, Quiniou et al. (2002) reported that small birthweights translated into small weaning weights, but this was not connected with reduced growth potential before weaning. Donovan and

Dritz (2000) removed larger pigs from the dam for a short period within 24 h after birth to allow smaller pigs in the litter uninhibited access to the dam as a means of reducing variation in weight at weaning. Whilst no differences in average daily gain or weaning weight were detected, there were significant linear improvements in standard deviation (SD) and CV of the average daily gain, but only for litters with nine piglets or more born alive. Given that smaller piglets respond well whilst suckling, provision of supplementary milk during the lactation period should increase overall piglet viability and weaning weight and reduce SD and CV. Wolter et al. (2002) offered supplemental milk replacer from day 3 of lactation to weaning at 21 days of age. Supplemental milk significantly ($P < 0.001$) enhanced weaning weights (6.60 kg vs 5.69 kg) and also reduced the CV of weaning weights (6.62% vs 7.04%), although this effect was not significant ($P > 0.05$). Recently, Douglas et al. (2014) reported the converse. These authors were unable to demonstrate any benefit of supplementary milk on the performance of lightweight piglets before or after weaning, but it did affect their drinking behaviour and, consequently, the variation in bodyweight of lightweight piglets in mixed litters to slaughter was reduced.

Even when those factors in gestation and pre-weaning that we have already shown to influence variation are addressed, there is still likely to be a level of variation in a population of piglets at weaning, and variation from batch to batch. With the SD of the population increasing with age (Patience et al., 2004), we must also focus efforts to reduce this immediately post-weaning.

There is limited ability to influence the performance of lightweight weaners through nutritional intervention in the post-weaning period (Hewitt et al., 2015). Lightweight weaned pigs (3.8 kg at 19 days of age) were compared with heavier pigs (5.2 kg at same weaning age) and a control group that was reflective of the normal population of weaners (4.4 kg). The lightweight pigs remained on the starter diet (15.0 MJ DE/kg, 0.8 g standardised ileal digestible lysine (SID Lys)/MJ DE) until they reached 30 kg liveweight, with the aim of lifting their performance to that of the heavier-weight group. Despite this significant intervention, lightweight pigs ate less and grew slower, taking an extra 13.5 days to reach market weight. Feed conversion did not differ amongst treatments, suggesting that boosting feed intake should lead to improvements in growth rates. Treating light- or heavier-weight pigs as a distinct group also resulted in an increase in the CV over the growth period, when compared with the control population. This has been reflected in other studies (O'Connell et al., 2005) where uniform groups at weaning became more varied over life, whilst the control population became less varied, possibly due to the natural variation in growth response, differing responses to weaning and increased levels of aggression in even groups where there is no obvious hierarchy. Segregation of small pigs may allow the opportunity to utilise the known growth improvement effects of additional fat in finisher pigs, shifting the lower end of the population

(Hastad et al., 2020), if infrastructure and economics allow. Experimental work to reduce variation continues to produce inconsistent results (see meta-analysis by Tolosa et al., 2021), with managing variation still being the best approach to manage profitability at marketing.

4 Strategic use of metabolic modifiers and other modes of growth promotion

The impact of technologies such as metabolic modifiers on pork production is well documented (Dunshea, 2005; Dunshea et al., 2016). These technologies that increase feed efficiency and lean tissue deposition whilst decreasing fat deposition have been developed to improve the efficiency of animal production. Some of these technologies: somatotropin, immunisation against gonadotropin-releasing factor and orally active dietary additives such as ractopamine, chromium and betaine were discussed comprehensively in the prior edition of this publication (van Barneveld et al., 2017b). The use of many of these technologies is still subject to poor market acceptance due to regulatory bans, despite the significant ability to reduce the industry's environmental footprint.

Moves within the EU to limit the surgical castration of male pigs continues; however, approximately two-thirds of male pigs are still castrated under local anaesthesia. The uptake of immunocastration continues to be low, despite the wealth of evidence supporting its impact on performance and the environment. Comparative meta-analyses of up to 21 studies show that during the finisher phase, immunisation against gonadotropin-releasing factor increases feed intake and average daily gain and decreases feed conversion rate compared with contemporary barrows (Dunshea et al., 2013). Whilst from an environmental point of view, the effect of immunocastration on greenhouse gas emissions was reduced by 3.7% compared with a surgical castration production system (De Moraes et al., 2013). There is also increased interest in the use of immunisation against gonadotropin-releasing factor in female pigs to control puberty development in extensive production systems (Hernández-García et al., 2013), to improve carcase characteristics (Poulsen Nautrup et al., 2020), and to increase backfat thickness in modern lean genotypes to make them more suitable to produce high-quality dry-cured ham (Pérez-Ciria et al., 2021).

Fats are energy-concentrated, containing between 2 and 2.5 times the energy of cereal grains. The inclusion of added fat in finisher pig diets has consistently been shown to increase growth rate through increased feed efficiency of reduced daily feed intake (Liu et al., 2018a; Hastad et al., 2020; Bromm et al., 2022). The growth response to fat appears independent of the source of the fat, although there is some evidence of better growth performance in pigs that receive more saturated fats (Liu et al., 2018a). Fat quality can be an important factor in carcase quality. Fats and oils derived from plants are generally higher in polyunsaturated fatty acids than those derived from animal sources, and lead to a higher iodine value in carcase fat (Bromm et al., 2022), which is detrimental for some processing markets (Soladoye et al., 2014). Therefore, the needs of the market should also be considered when considering fat sources and their inclusion rates.

5 Matching nutrient requirements to diet specifications

The ability to routinely, accurately and cost-effectively measure variation in the nutritional quality of feed ingredients prior to diet formulation represents a fundamental pillar of sustainable pork production worldwide. It is the most effective means of matching nutrient requirements to diet specifications. Central to this is the capacity to rapidly measure nutritional quality of feed ingredients and maintain an up-to-date definition of nutrient requirements.

5.1 Rapid measurement of nutritional quality

Accepting the premise that measurement of 'available' nutrients will provide the greatest capacity to define the comparative nutritional value of ingredients, there is a need to prioritise which available nutrients are measured in which ingredients. This is due to the considerable cost and effort associated with establishing these analyses and the fact that it is unlikely a single analysis will be suitable for all ingredients. Apart from proximate analysis of all ingredients using either chemical or rapid techniques such as near-infrared spectrophotometry (NIRS), available energy in cereals and available lysine in oilseed meals are the highest priority for analysis of available nutrients, based on their relative proportions in pig diets and their associated variation (van Barneveld et al., 2017a).

Significant debate has ensued over the suitability of DE and metabolisable energy (ME) versus net energy (NE) systems as the most appropriate measures of available energy in feed ingredients, particularly cereal grains. The latter has been touted as a better alternative on the basis that DE and ME are poor predictors of animal performance (de Lange and Birkett, 2005), but unsupported by other research (Cadogan, 2009). The digestibility of a feed is affected little by animal factors, except for young piglets and sows (Noblet and Shi, 1993; Nitrayova et al., 2006). However, NE values for a feed depend on characteristics of the feed, nutrients absorbed, and the metabolic status of the animal, with different NE values derived for the same feed when consumed by animals of different genotypes, stages of growth, physiological states and climatic conditions (Kil et al., 2013). The debate about the superiority of

NE is irrelevant because it is impossible to measure NE at the point of feed manufacture (Shurson et al., 2021). Furthermore, NE calculations depend on information on the DE content of a feed. Consider an NE equation reported by Noblet (2006):

$$NE\ (MJ/kg, DM) = 0.703 \times DE\ (MJ/kg, DM) - 0.0041 \times CP + 0.0066 \times EE - 0.0041 \times CF + 0.0020 \times ST$$

where CP is crude protein, EE is ether extract, CF is crude fibre, and ST is starch (all g/kg DM). The equation shows that more than 70% of the NE contribution from an ingredient is derived from DE. Rapid analysis of DE combined with additional chemical components does allow use of NE as an estimate of available energy for an animal, but without the DE component the NE calculation would be impossible.

Further justification for the development of rapid measures of DE in cereals for pigs was provided by van Barneveld (1999). Differences of up to 3.7 MJ/kg dry matter (DM) in DE content were observed following a review of data for more than 70 cultivars of wheat. Similarly, an analysis of data for more than 125 cultivars of barley revealed a range in DE estimates from 11.7 to 16.0 MJ/kg DM. The cultivar itself was shown to have a minimal effect on the availability of energy in cereals, whilst the cultivation conditions and agronomic practices (e.g. fertiliser rate) were reported to have a greater influence than the growing region or the growing year. Many physical and chemical factors were shown to influence DE in cereals, including protein source and type, starch characteristics, fat source and type, non-starch polysaccharide components and anti-nutritional factors. van Barneveld (1999) concluded that the availability of energy to the animal will depend on the particular combination of these components in a grain and how they behave in the presence of nutrients from other feed ingredients. Hence, a rapid measurement of DE is best based on data derived from *in vivo* measurements.

In feed ingredients, such as soybean meal and canola meal, that have undergone processing, heat damage can result in an overestimation of nutritional quality due to reactions between the ε-amino group of lysine with other compounds such as carbohydrates (van Barneveld et al., 1999). It appears that most processors are aware of the potential for heat damage during oil extraction, but quantification of these effects occurs rarely. Rutherfurd et al. (1997) developed the digestible reactive lysine assay as a means of assessing heat damage in feed ingredients. van Barneveld et al. (1999) applied this technique to assess the degree of heat damage in cold-pressed, expeller-extracted and solvent-extracted canola meal. Despite similar total lysine concentrations in the canola meals, true ileal reactive lysine digestibility was significantly higher in the cold-pressed meal. Reactive lysine content was shown to be a good indicator of true ileal digestible reactive lysine content (and thus the degree of heat damage). Given this relationship, the application of rapid assays for total lysine and reactive lysine would allow more accurate assessment of the nutritional quality of ingredients such as soybean meal and canola meal, and account for this at the time of diet formulation and manufacture.

The need for rapid analysis of reactive lysine in oilseed meals as a routine assessment was demonstrated by Kim and Mullan (2012). A total of 209 samples collected from the major soybean meal-producing countries revealed a 27% variation in reactive lysine content (21.9 g-30.1 g/kg 'as is' basis) across all samples and a 13% variation within shipment samples. This finding suggests that there is wide variation in reactive lysine content in soybean meal samples, and the use of spot sampling to analyse reactive lysine content can lead to large inaccuracies. Data analysis showed the proportion of reactive lysine relative to total lysine was not different ($P > 0.05$) due to country of origin, whilst reactive lysine content was significantly different between countries of origin ($P < 0.001$). Kim and Mullan (2012) concluded that the variation in reactive lysine content caused by heat processing is similar between countries, but that the total amount of lysine and reducing sugars present in the raw soybeans prior to heat processing are a major source of variation. Nevertheless, the linear relationship between total and reactive lysine was weak (slope: 6.00, intercept: 0.7152, RSD: 0.913, $R^2 = 0.52$) and cannot be used for robust prediction of reactive lysine from analysed total lysine content.

van Barneveld et al. (1998) developed the first NIRS calibration for the measurement of DE in cereals for growing pigs, using more than 150 samples from across the world that had been subjected to *in vivo* DE analysis (Table 3). Innovative calibration development techniques were employed so that samples from various laboratories could be meaningfully included in a single calibration. This research demonstrated that DE in whole grain cereals can be measured by NIRS to an accuracy of at least ±0.38 MJ/kg, which at the time was comparable to the accuracy obtained when using *in vivo* reference

Table 3 Summary of calibration statistics for digestible energy (MJ/kg) in whole and milled cereal grain samples for pigs

Sample set	SECV (MJ/kg)	RER	SECV/STD
Whole grain	0.518	8.8	0.49
Whole grain (IV)	0.377	12.1	0.36
Milled grain	0.596	7.7	0.53
Milled grain (IV)	0.428	10.7	0.38

SECV, standard error of cross validation; RER, range error ratio; STD, standard deviation; IV, indicator variables.
Source: Adopted from: van Barneveld et al. (1998).

methods. No significant differences in accuracy were observed between milled and whole grain calibrations or between local and global calibrations.

Further to the development of DE calibrations for cereals in pigs, van Barneveld and Ru (2003) also demonstrated that NIRS could be successfully employed to measure the total and reactive lysine content of canola meal, with obvious potential to extend this to soybean meal.

Significant investment has ensued in Australia in the past 25 years to build upon these base calibrations, and we now have a practical means of rapid, online assessment of DE content in cereals used in pig diets, together with the option for routine assessment of batches of heat-treated oilseed meals at the point of receival (Table 4). The extension of NIRS to assess finished pig diets quickly and inexpensively has resulted in acceptable results, with further refinement work continuing (Noel et al., 2022; Paternostre et al., 2021)

The development of NIRS calibrations for the measurement of digestible nutrients in ingredients for pigs, together with more robust NIRS equipment for online applications in feed mills, has led to the collection of data that should concern all commercial nutritionists and pig producers. Online measurement of the DE and crude protein contents of wheat in a large Australian feed mill demonstrated why real-time measurement and the capacity to account for variation should exceed all other priorities in the application of feed evaluation systems (Fig. 1). Over the course of 1 h of production, the crude protein content of the wheat varied by 4% units and the DE content by 1.2 MJ/kg as received. Furthermore, the mean DE content was approximately 0.6 MJ/kg lower than the reported book values for wheat (which would be commonly used in the diet formulation). In this particular mill, production rates were approximately 20 tonnes/h, and it was common for large batches of single diets to be produced over this period. The online data suggests that the first 3 tonnes of feed utilised wheat with a DE content between 12.9 and 13.4 MJ/kg, the next 12 tonnes were fairly consistent at 13.4 MJ/kg, but the last 5-6

Table 4 Summary of calibration statistics for AusScan NIRS calibrations for ileal and faecal digestible energy (MJ/kg, as received in cereals, excluding oats and maize) for pigs, and total and reactive lysine in canola meal (g/kg, as received)

Measurement	*n*	Mean	SD	SECV	1-VR	RPD
Ileal DE (MJ/kg)	236	11.5	1.03	0.48	0.80	2.17
Faecal DE (MJ/kg)	288	13.7	0.69	0.26	0.86	2.65
Total lysine (g/kg)	113	20.19	1.37	0.65	0.78	2.11
Reactive lysine (g/kg)	113	17.97	1.75	0.92	0.73	1.91

DE, digestible energy; *n*, number of samples used in the calibration; SD, standard deviation of values in the population; SECV, standard error of cross validation; VR, variance ratio; RPD, ratio of prediction to deviation (SD/SECV).
Source: Adopted from: Black et al. (2013).

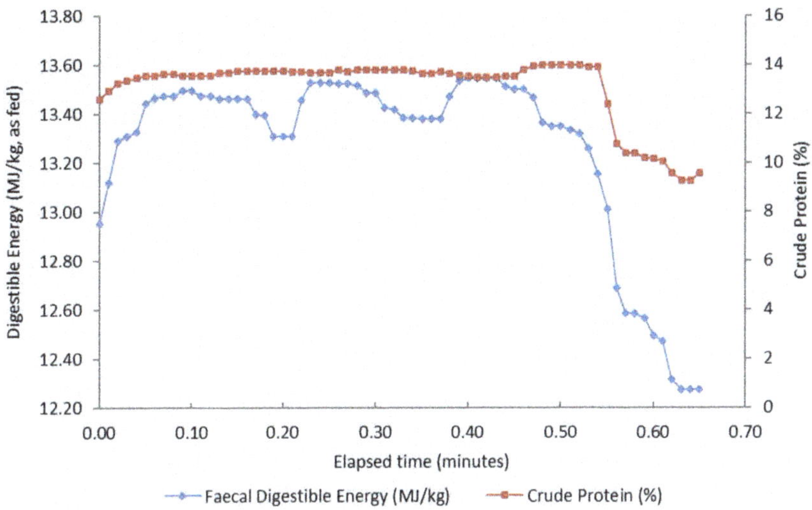

Figure 1 Real-time NIRS analysis of faecal digestible energy (MJ/kg, as fed) and crude protein (%) contents of wheat included in pig diets. Source: Adopted from: van Barneveld et al. (2018).

tonnes varied between 13.2 and 12.2 MJ/kg as fed. If diets were formulated as a starter feed, some piglets would receive a diet equating to 14.0 MJ DE/kg, whilst others would receive a diet equivalent to 14.7 MJ DE/kg, and the nutritionist would be expecting delivery of feed equating to 15.0 MJ DE/kg to these pigs. This difference in energy intake would suppress production efficiency and represent a major source of variation in pig growth, which would increase as the pigs grow.

As we become more reliant on NIRS calibrations for the measurement of DE in cereals to improve the accuracy of diet formulations, it is also important we have practical methods for the assessment of calibration outliers, given that not all samples can be subjected to *in vivo* analysis. van Barneveld et al. (2018) used the faecal DE values that comprise the NIRS calibrations described by Black et al. (2013) and correlated them with 87 chemical and physical components. As a result of this analysis, two models were developed for faecal DE on a dry matter (DM) basis that appear to have equivalent potential and could be applied immediately to assess outliers and candidate grains. Models for ileal DE had limited value. The best-fit models for faecal DE were based on specific weight (SPW; kg/hL), lignin (LIG; % DM) and measures of insoluble fibre including insoluble xylose (INX; % DM) and xylose and arabinose residues (INAX; % DM). Correlation equations using the two respective models for barley, sorghum, triticale and wheat are presented below:

Model 1

Barley DE (MJ/kg DM) = 14.934 – (0.212 × INX) – (0.021 × SPW) – (0.339 × LIG);
Sorghum DE (MJ/kg DM) = 15.650 – (0.212 × INX) – (0.021 × SPW) – (0.339 × LIG);
Triticale DE (MJ/kg DM) = 14.852 – (0.212 × INX) – (0.021 × SPW) – (0.339 × LIG);
Wheat DE (MJ/kg DM) = 15.442 – (0.212 × INX) – (0.021 × SPW) – (0.339 × LIG).

Model 2

Barley DE (MJ/kg DM) = 14.854 – (0.104 × INAX) – (0.337 × LIG) + (0.027 × SPW);
Sorghum DE (MJ/kg DM) = 15.833 – (0.104 × INAX) – (0.337 × LIG) + (0.027 × SPW);
Triticale DE (MJ/kg DM) = 14.863 – (0.104 × INAX) – (0.337 × LIG) + (0.027 × SPW);
Wheat DE (MJ/kg DM) = 15.413 – (0.104 × INAX) – (0.337 × LIG) + (0.027 × SPW).

5.2 Updated information on nutritional requirements

The 11th revised edition of Nutrient Requirements of Swine from the National Research Council (NRC) that was released in 2012 continues to be the fundamental resource when reviewing the nutrient requirements of modern genotypes. As in the tenth edition, the main emphasis of this publication was to present estimates for the nutrient requirements of pigs and on available nutrient contents in feed ingredients. In addition, information was included on the pigs' need for water, co-products from the corn and soybean industries, estimation of nutrient digestibility, non-nutritive feed additives, feed contaminants, feed processing, means to minimise nutrient excretion and research needs. New chapters in this latest edition also included information regarding the nutritional importance of lipids and carbohydrates.

Whilst there are various publications that have sought to update the nutrient requirements of the modern genotype, they are specific to the genotype being supplied (e.g. DanBred Nutrient Specifications, PIC Nutrition and Feeding Guidelines) or published by supplement companies (e.g. OVN Optimal Vitamin Nutrition Guidelines - DSM). It has been some time since the nutritional requirements of the pig were sourced from actual experimental studies to elucidate the nutritional requirement of the growing pig, lactating sow and gestating sows. Instead, mathematical models to estimate nutrient requirements are commonly used (National Research Council, 2012). As summarised by de Lange (2013), models are now used to dynamically represent the partitioning of energy intake and to then estimate amino acid, phosphorus and calcium requirements of pigs in a relatively disease- and stress-free environment. Similarly, for the interpretation of nutrient requirements studies, dietary nutrient levels were recalculated, and, when needed, estimates of nutrient requirements were established based on a reinterpretation of the data sets. Empirical estimates of requirements for vitamins, fatty acids and minerals other than calcium and phosphorus have been updated from National Research Council (1998; de Lange, 2013).

5.3 Water requirements for pigs

Water is the major component of the pig's body and has an important role in most functions of the animal, yet its role in nutrition and metabolism is often forgotten (Thulin and Brumm, 1991). The volume of water consumed is influenced by multiple factors such as environmental conditions, water quality, accessibility, physiological state, health status and animal behaviour (Edwards and Crabb, 2021). However, whether due to its relative cost or a lack of infrastructure to understand its usage, if water is present and accessible, then the pigs' needs are assumed to be met. Water quality standards are generally based on water for human consumption, with limited research to establish pig drinking water quality standards (Edwards, 2018). The consumption of water by pigs in the immediate post-weaning period and sows during lactation are two areas where research has been conducted to improve performance.

Low feed intake immediately post-weaning is common (Arnaud et al., 2023), and with water and feed intake being closely associated (Dybkjær et al., 2006), research has looked at how water is delivered and the subsequent impacts on weaner pig water intake. However, there is little difference in post-weaning feed disappearance, daily gain or feed efficiency between the two major types of drinkers, nipples or cups/bowls, but water disappearance is greater in nipple drinkers (Torrey and Widowski, 2006; Vande Pol et al., 2022a), even under heat stress conditions (Vande Pol et al., 2022b).

Piglet behaviour is the major influencer of water disappearance. When bowl drinkers were placed side-by-side rather than spaced apart, swapping between drinkers more than doubled water disappearance, although this was still less than bite drinkers (Magowan et al., 2007). Hygiene also influences water usage, with bowls that were not cleaned having an approximately 8% lower water disappearance (Phillips and Phillips, 1999). Wastage is an important consideration given the increased administration of treatments and additives in water, whilst it may lead to poor interpretation of actual water consumption. Estimates of wastage range from 57% in nipple drinkers (Phillips and Phillips, 1999) to 15–19% in bowls (Wang et al., 2017; Phillips and Phillips, 1999, respectively). Similarly, no difference in growth performance between drinker types has been observed in grow-finisher pigs, with similar increased wastage in nipple compared with cup/bowl drinkers (Li et al., 2005; Vande Pol et al., 2022b).

Water is a major component of milk, and the average daily water intake during lactation has been measured to be in the range of 17.2–20.2 litres per day (Peng et al., 2007; Oliviero et al., 2009, respectively). During peak lactation, water intakes of 27.0 litres per day have been observed (Quiniou et al., 2000). Water intake is influenced by factors such as feed composition, the amount of feed intake and ambient temperature. Kruse et al. (2011) showed a positive correlation between water and feed intake during lactation with piglet weaning weight and decreased sow body weight loss. Sows within the Kruse et al. (2011) study were fed on a step-up basis until around day 9 of lactation; however, the water-to-feed ratio would suggest that higher feed intake was possible, as is the norm in many production systems today where the sow is *ad libitum* fed from early lactation. The water-to-feed ratio fell from 7 L of water per kg of feed at the start of lactation to 4 L of water per kg of feed after day 9, with parity 2 sows consuming 31.2 L of water per day across the 26 days of lactation, 811 L of water during lactation in farrowing facilities maintained between 20°C and 22°C (Kruse et al., 2011).

Heat stress during lactation reduces a sow's daily feed intake and milk production (Quiniou and Noblet, 1999), impacting piglet growth (Renaudeau and Noblet, 2001), but also impacts fertility with decreased conception rates and an extension of the wean-to-oestrus interval (Prunier et al., 1997). Treatments such as floor cooling have been applied to reduce heat stress, and whilst they have resulted in a reduction in the respiration rates and temperature of the sow, they did not lead to improvements in litter gain (Cabezón et al., 2017). When water was chilled during a period of heat stress, from an ambient temperature of 22°C–10°C or 15°C, feed intake was maintained (5.4 kg per day c.f. 3.8 kg per day), water consumption rose to 38 L per day and piglet weaning weights were 0.5 kg heavier (Jeon et al., 2006), despite accommodation remaining at 29°C, with similar performance gains observed by Willis and Collman (2007).

Other performance improvements in lactation associated with water include the use of self-fed wet/dry feeders, where the sow is able to determine the amount of both feed and water being dropped into a common trough, which resulted in a 0.5 kg improvement in piglet weaning weight (Peng et al., 2007), whilst inadequate flow rates reduced sow feed intake, impacting weaning weights during summer (Leibbrandt et al., 2001). Although there has been some greater interest in the role of water in pig nutrition in recent times, in general, it still appears to be the forgotten nutrient.

6 Optimising utilisation of co-products

In modern intensive production systems focussed on providing people with proteins, the animal can be used to 'value-add' raw materials that would otherwise be difficult to include in a human diet. The pig excels in this context and can make use of co-products and grains that have been downgraded from human grain markets. Characterising the nutritional quality of these raw materials often demonstrates their nutritional value exceeds the physical characteristics that may have compromised their suitability as a human food product, making them comparatively valuable components of pig diets.

The FAO estimates that each year, approximately one-third of all food produced for human consumption in the world is lost or wasted (FAO, 2013). This food wastage represents a missed opportunity to improve global food security but also to mitigate environmental impacts and resources use from food chains. Information regarding this food waste being reused by other industries such as the animal industry is scarce, but one can hazard a guess that this is more likely to be low.

Food waste has historically been recycled as livestock feed, particularly for pigs, where cooked food waste (swill) was fed to pigs. As a non-ruminant species, the pig's digestive system is well adapted for the conversion of food waste into animal protein (Westendorf, 2000). Swill can be a high-quality animal feed that requires no additional land to be brought into production, and hence has minimal or even positive environmental impact (food waste otherwise disposed). However, the use of swill is prohibited in many countries due to biosecurity and food safety concerns/regulation. In addition, collection and processing costs, nutrient value and even implications on product quality are issues that further hinder the use of food waste as an animal feed ingredient. Hence, for the most part, food waste remains a largely untapped resource for the pig industry.

The role of a variety of insects that are part of a human diet in many countries is quickly becoming a mainstream but 'micro-niche' protein for human consumption. Edible insects, such as black soldier fly (BSF, *Hermetia illucens*) and yellow mealworm (*Tenebrio molitor*), continue to be investigated as potential protein sources for pigs (DiGiacomo and Leury, 2019; Hong and Kim, 2022). Additional to their nutritional contribution, BSF larvae are high in monounsaturated fatty acids and chitin, which have been shown to increase intestinal health and performance of weaned pigs (Boontiam et al., 2022). Monounsaturated fats have been shown to increase nutrient absorption and have antimicrobial action, and chitin promotes hindgut fermentation of short-chain fatty acids and an immunomodulatory effect on immune function. However, the ability to use large, consistent quantities of insect meal is still to be achieved. Limitations on the larval feedstock that would allow them to be fed to livestock are currently in place with concerns about transmissible spongiform encephalopathies (Veldkamp and Vernooij, 2021) as well as the accumulation of heavy metals and other contaminants (Schrögel and Wätjen, 2019). It is likely, though, that insect meals may well be able to provide an opportunity to utilise certain food wastes to generate novel pig feed ingredients.

7 Optimising gut health, nutrient utilisation capacity and health interactions

At a time where antimicrobial resistance (AMR) poses a very real threat to human health, the animal industries, whilst unfairly blamed by media as the sole cause of these issues, do contribute to some of the risk around AMR in pigs, farm workers and the general community. The use of antibiotics as growth promoters was extensive in most pig production systems; however, there has been a concerted effort to minimise their use globally, although, some countries have been more proactive than others. The corresponding and significant increase in therapeutic use of antibiotics suggests that perhaps such bans may be less than effective and may take longer for industries to adhere to. Needless to say, the issue of antibiotic use and AMR has caused an increase in both the use and development of alternative nutritional strategies to antibiotic feed additives to improve pig growth performance.

'Gut health' is a term that we define as describing a generalised condition of homeostasis in the gastrointestinal tract (GIT) of the pig (Pluske et al., 2007). As can be expected, the factors and conditions involved in 'gut health' are multifactorial, complex and in many instances poorly described and sometimes incorrectly interpreted, although it is evident that perturbations of the GIT that cause disease, even sub-clinical disease, are a disturbance of this homeostasis. In addition to enteric disease, other factors including the period immediately after weaning, changes in diet and disruptions to meal patterns that disrupt the flow of nutrients through the GIT, also affect gut health (Pluske et al., 2007). Gut health can be viewed as an outcome (positive, negative and the status quo) of the complex interactions occurring in the GIT between nutrition (e.g. feed type, feed composition, presence or absence of antibiotic feed additives), the mucosa of the GIT (e.g. type of receptors, mucin, inflammatory activity, cytokine production and barrier function) and the microbiota (e.g. density, location, pathogenicity and extent of colonisation; Pluske et al., 2007). Whilst much of our research focus has been directed to the newly weaned pig, it is important to consider that growing-finishing pigs and sows also suffer from GIT disruptions causing decreased growth performance, morbidity and sometimes death.

The microbiota of the pig GIT is of undeniable importance. It contains the largest number and greatest diversity of microorganisms of all body systems and is comprised of bacteria, archea, fungi and viruses as well as their genomes and extra-chromosomal elements (Nowland et al., 2022). Demonstrated links between the microbiota and irritable bowel syndrome, obesity, autoimmunity, asthma and mental health have been established in humans, so it is not unreasonable to think that the pig GIT microbiome is likely to impact the growth and efficiency of pigs. The linkages between microbiota diversity and performance differences in pigs have primarily been established by comparing populations with very diverse performance (Gardiner et al., 2020), however, high bacterial diversity is generally seen as favourable. The review of Gardiner et al. (2020) showed that environmental factors have an enormous impact on the bacterial taxa, but certain taxa are consistently linked with better production phenotypes - *Treponema* and *Methanobrevibacter* in the small and large intestines, *Lactobacillus* in the large intestine, whilst others such as *Bacteroides* and butyrate producers, *Roseburia* and *Ruminococcus* appear to improve feed efficiency, with some conflicting results.

Given the apparent importance of the microbiome in achieving efficient growth in the pig, and the known impact of the environment, there has been considerable interest in being able to manipulate or influence its establishment. The exact timing of initial colonisation has yet to be determined. *In utero* colonisation has yet to be observed in pigs (Nowland et al., 2022), although the presence of bacteria in the colon of stillborn pigs suggests colonisation in the immediate prenatal period (Nowland et al., 2021b), suggesting that the sow's microbiota is likely to influence the microbiota of her offspring. Ensuring piglets receive optimal colostrum and milk in early life is essential, with its importance well documented (Le Dividich et al., 2005), especially in comparison to formula feeding (Inman et al., 2010; Iozzo and Sanguinetti, 2018). Whilst the rearing environment and its hygiene were also able to influence colonisation, with the removal of sow faeces from the farrowing crate for the first 10 days or life altering colonisation and improving piglet growth and survival (Nowland et al., 2021a). Evidently there is an opportunity for controlling the exposure of the piglet.

The small intestine is the site of absorption for most minerals; however, inorganic minerals, such as zinc oxide, are dissociated under acidic pH in the stomach. Zinc oxide has been used to control post-weaning diarrhoea but has traditionally been included in weaner diets at supraphysiological levels (i.e. 2000–3000 mg/kg or ppm) such that significant levels can pass through the stomach to be active in the small intestine, and the dissociated zinc is then excreted with the potential to cause environmental harm. Encapsulation technology has been developed that has resulted in significantly lower levels of zinc being required to achieve the same biological outcome (Kim et al., 2011; Jang et al., 2014; Shen et al., 2014). Encapsulation sees the entrapment of the active agent within another substance, such as plant gums, or more commonly, lipid materials that survive the low pH of the stomach (Nedovic et al., 2011). This technology has been applied to improve the delivery of many bioactive molecules such as antioxidants, vitamins and minerals, with other phytomolecules now being commonly fed to pigs.

The movement away from the use of antimicrobial growth promoters has resulted in an increased range of feed additives that were described in the prior edition (van Barneveld et al., 2017b) and reviewed more recently by Liu et al. (2018b) and Bonetti et al. (2021).

8 Conclusion and future trends

Our understanding of the nutritional requirements and metabolism of the pig has significantly advanced over the last 20 years - from the way we formulate diets, our understanding of the feed ingredients and the nutritional technologies that have enabled us to improve the performance of the pig. The magnitude of progress is evident when we compare the performance of pigs today to that of late 1980s, whole-of-life average daily gain has increased from 450 g/day to >720 g/day, whilst herd feed conversion has decreased from >5.5 to 3.6 kg feed/kg carcase. During this time, significant technologies such as pST and ractopamine that have been key contributors to this improved performance are no longer viable technologies for use in the pork industry. Some of these technologies, whilst still effective, are no longer viable due to shifting consumer perceptions. Despite the removal of these technologies, we have not seen any decline in pig performance. The risk associated with AMR in humans and pigs has seen the use of in-feed antibiotics becoming increasingly unsustainable. This has given rise to a range of new technologies that have positively contributed to both growth performance and animal health. The interaction between nutrition and health is especially apparent in the young pig where technologies to alleviate the growth check is seen as a consequence of weaning.

As we look forward, pig production systems are more than likely to undergo significant changes, if the last 10–20 years are anything to go by. Changes in genetics, housing systems, animal health care, as well as systems shaped by changing environmental regulation and consumer attitudes are then likely to affect the nutrient requirements of the modern pig. Further advances in nutrigenomics could help us better understand the genetic mechanisms behind nutrient metabolism in pigs, and how we can optimise diets to meet their specific nutritional needs required to achieve high performance. Our ability to implement precision nutrition strategies to tailor diets to individual animals based on their specific nutritional requirements will further optimise feed efficiency and reduce waste, whilst also improving pig health and performance. The gut microbiome plays a crucial role in nutrient metabolism and overall pig health. Significant research has shed some light on the specific roles and impacts of the microbial communities within the pig gut; however, we are yet to implement nutritional strategies that optimise the gut microbiome to enhance their influence on nutrient absorption and metabolism. As the demand for animal protein continues to grow, there is increasing interest in developing alternative protein sources (insects, algae and single-cell proteins) that can be used in pig diets and provide new ways to meet the nutritional requirements of pigs in a sustainable way.

Our reliance on theoretical data, rather than actual nutrient requirement data, puts us at a particular risk of 'assuming' that we are feeding pigs to grow as closely as possible to their genetic potential. The same applies to our estimation of the nutrient specification of feed ingredients. Technologies such as NIR and other 'precision farming' technologies must be better utilised to ensure that the diets we are feeding our pigs actually meet their requirement. Whilst not discussed, feeder design and innovation to enable us to tailor diets to meet individual pig's requirements, whilst still economically a distant possibility, do present us with opportunities to advance our understanding of the pigs' nutritional requirements and metabolism.

9 Where to look for further information

This chapter has highlighted the areas within pig nutritional requirements and metabolism that have the potential to impact the sustainability of pig meat production. When considering these areas, it is imperative to focus on the primary drivers of sustainability as outlined by groups like the FAO in documents such as Building a Common Vision for Sustainable Food and Agriculture - Principles and Approaches (FAO, 2014) and Transforming the livestock sector through the Sustainable Development Goals (FAO, 2018b).

Our knowledge of nutritional requirements and metabolism continues to evolve, aided by advances in technology and methodology. Therefore, the most up-to-date information will continue to be found in the scientific journals. A range of references were reviewed in the writing of this chapter, and these papers would be an ideal starting point to further understand this topic.

In particular, 'Gilt development to improve offspring performance and survivability' (Faccin et al., 2022) comprehensively reviews the key aspects of managing and feeding gilts to improve lifetime performance and longevity. 'Feeding the modern sow to sustain high productivity' (Theil et al., 2022b) addresses the nutritional requirements of the modern hyper-prolific sows, whilst the impact of metabolic modifiers on livestock production, their role in sustainable production, and the issues affecting their wider use are reviewed in 'Metabolic modifiers as performance-enhancing technologies for livestock production' (Dunshea et al., 2016).

'Advances in Pork Production' is the proceedings of the Banff Pork Seminar and has provided several references used in this chapter, as has 'Manipulating Pig Production', the proceedings of the Australasian Pig Science Association Conference. Covering a range of topics, the invited papers and abstracts are, generally, of high value and have added considerably to our understanding of pork production.

10 References

Arnaud, E. A., Gardiner, G. E. and Lawlor, P. G. (2023). Selected nutrition and management strategies in suckling pigs to improve post-weaning outcomes. *Animals* 13(12), 1998.

Ball, R. O., Samuel, R. S. and Moehn, S. (2008). Nutrient requirements of prolific sows. *Adv. Pork Prod.* 19, 223–236.

Beltranena, E., Patterson, J., Foxcroft, G. and Pettitt, M. (2005). 12 non-negotiable aspects of gilt development. *West. Hog J.* 2004, 45–50.

Bergsma, R., Kanis, E., Verstegen, M. W. A., van der Peet-Schwering, C. M. C. and Knol, E. F. (2009). Lactation efficiency as a result of body composition dynamics and feed intake in sows. *Livest. Sci.* 125(2–3), 208–222.

Black, J. L., Flinn, P. C., Diffey, S. and Tredrea, M. (2013). An update on near infrared reflectance analysis of cereal grains to estimate digestible energy content for pigs. In J. R. Pluske and J. M. Pluske (Eds.), *Manipulating Pig Production XIV*. Australasian Pig Science Association, Werribee, p. 54.

Bonetti, A., Tugnoli, B., Piva, A. and Grilli, E. (2021). Towards zero zinc oxide: Feeding strategies to manage post-weaning diarrhea in piglets. *Animals* 11(3), 642.

Boontiam, W., Phaengphairee, P., Hong, J. and Kim, Y. Y. (2022). Full-fatted *Hermetia illucens* larva as a protein alternative: Effects on weaning pig growth performance, gut health, and antioxidant status under poor sanitary conditions. *J. Appl. Anim. Res.* 50(1), 732–739.

Bromm, J. J., DeRouchey, J. M., Tokach, M. D., Berg, K. M., De Jong, J. A., Pohlen, C. L., Woodworth, J. C., Goodband, R. D. and Gebhardt, J. T. (2022). Effects of fat source and level on growth performance, carcass characteristics, carcass iodine value and economics of finishing pigs in a commercial environment. *Kans. Agric. Exp. Stn. Res. Rep.* 8(10), 10.

Bruun, T. S., Østrup, E., Lyderdik, K. K., Dall, J. and Strathe, A. V. (2023). Effect of increasing dietary methionine-to-lysine ratio during early gestation on fetal development and piglet birth weight. *Ani. Rep. Sci.* 252, 107251.

Cabezón, F. A., Schinckel, A. P., Marchant-Forde, J. N., Johnson, J. S. and Stwalley, R. M. (2017). Effect of floor cooling on late lactation sows under acute heat stress. *Livest. Sci.* 206, 113–120.

Cadogan, D. (2009). Net energy defines lean and fat deposition better than digestible energy. Final Report, Project 2G-104. The Co-operative Research Centre for High Integrity Australian Pork, Roseworthy, SA.

Carrión-López, M. J., Madrid, J., Martínez, S., Hernández, F. and Orengo, J. (2022). Effects of the feeding level in early gestation on body reserves and the productive and reproductive performance of primiparous and multiparous sows. *Res. Vet. Sci.* 148, 42-51.

Che, L., Xu, M., Gao, K., Wang, L., Yang, X., Wen, X., Xiao, H. and Jiang, Z. (2020). Effects of dietary valine supplementation during late gestation on the reproductive performance and mammary gland development of gilts. *J. Anim. Sci. Biotechnol.* 11, 15.

Close, W. H. and Cole, D. J. A. (2000). The pre-breeding gilt. In W. H. Close and D. J. A. Cole (Eds.), *Nutrition of Sows and Boars*. Nottingham University Press, Thrumpton.

Clowes, E. J., Aherne, F. X., Foxcroft, G. R. and Baracos, V. E. (2003). Selective protein loss in lactating sows is associated with reduced litter growth and ovarian function. *J. Anim. Sci.* 81(3), 753–764.

Cools, A., Maes, D., Decaulwe, R., Buyse, J., van Kempen, T. A. T. G., Liesegang, A. and Janssens, G. P. J. (2014). Ad libitum feeding during the peripartal period affects body condition, reproduction results and metabolism of sows. *Anim. Reprod. Sci.* 145(3–4), 130–140.

da Silva Fonseca, L., Lanferdini, E., Moreira, R. H. R., Chaves, R. F., Perazolli, P. H., de Paula, Y. H., Renno, L. N., Garbossa, C. A. P., de Souza Cantarelli, V. and de Abreu, M. L. T. (2022). Arginine supplementation in the feed of gestating sows. *Livest. Sci.* 263, 104999.

Davis, H. E., Jagger, S., Toplis, P. and Miller, H. M. (2021). Feeding β-hydroxy β-methyl butyrate to sows in late gestation improves litter and piglet performance to weaning and colostrum immunoglobulin concentrations. *Anim. Feed Sci. Technol.* 275, 114889.

Davis, H. E., Jagger, S., Toplis, P. and Miller, H. M. (2022). Dietary β-hydroxy β-methyl butyrate supplementto sows in late gestation improves litter and piglet performance to weaning and colostrum immunoglobulin concentrations. *Anim. Feed Sci. Technol.* 275, 114889.

De Bettio, S., Majorka, A., Barrilli, L. N. E., Bergsma, R. and Silva, B. A. N. (2016). Impact of feed restriction on the performance of highly prolific lactating sows and its effect on the subsequent lactation. *Animal* 10(3), 396–402.

De Lange, C. F. M. (2013). New NRC (2012) nutrient requirements of swine. *Adv. Pork Prod.* 24, 17–28.

De Lange, C. F. M. and Birkett, H. (2005). Characterization of useful energy contents in swine and poultry feed ingredients. *Can. J. Anim. Sci.* 85(3), 269–280.

De Moraes, P. J. U., Allison, J., Robinson, J. A., Baldo, G. L., Boeri, F. and Borla, P. (2013). Life cycle assessment (LCA) and environmental product declaration (EPD) of an immunological product for boar taint control in male pigs. *J. Environ. Assess. Policy Manag.* 15(1), 1350001-1350026.

Decaluwé, R., Maes, D., Cools, A., Wuyts, B. and De Smet, S. (2014). Effect of peripartal feeding strategy on colostrum yield and composition in sows. *J. Anim. Sci.* 92(8), 3557-3567.

DiGiacomo, K. and Leury, B. J. (2019). Review: Insect meal: A future source of protein feed for pigs? *Animal* 13(12), 3022–3030.

Donovan, T. S. and Dritz, S. S. (2000). Effect of split nursing on variation in pig growth from birth to weaning. *J. Am. Vet. Med. Assoc.* 217(1), 79-81.

Douglas, S. L., Edwards, S. A. and Kyriazakis, I. (2014). Management strategies to improve the performance of low birth weight pigs to weaning and their long-term consequences. *J. Anim. Sci.* 92(5), 2280–2288.

Dourmad, J. Y., Etienne, M. and Noblet, J. (1996). Reconstitution of body reserves in multiparous sows during pregnancy: Effect of energy intake during pregnancy and mobilization during the previous lactation. *J. Anim. Sci.* 74(9), 2211-2219.

Dunshea, F. R. (2005). Sex and porcine somatotropin impact on variation in growth performance and back fat thickness. *Aust. J. Exp. Agr.* 45(6), 677–682.

Dunshea, F. R., Allison, J. R. D., Bertram, M., Boler, D. D., Brossard, L., Campbell, R., Crane, J. P., Hennessy, D. P., Huber, L., de Lange, C., Ferguson, N., Matzat, P., McKeity, F., Moraes, P. J. U., Mullan, B. P., Noblet, J., Quiniou, N. and Tokach, M. (2013). The effect

of immunization against GnRF on nutrient requirements of male pigs: A review. *Animal* 7(11), 1769-1778.
Dunshea, F. R., D'Souza, D. N. and Channon, H. A. (2016). Metabolic modifiers as performance-enhancing technologies for livestock production. *Anim. Front.* 6(4), 6–14.
Dybkjær, L., Jacobsen, A. P., Tøgersen, F. A. and Poulsen, H. D. (2006). Eating and drinking activity of newly-weaned piglets: Effects of individual characteristics, social mixing, and addition of extra zinc in the feed. *J. Anim. Sci.* 84(3), 702–711.
Edwards, L. (2018). Drinking water quality and its impact on the health and performance of pigs. Report prepared for the CRC for High Integrity Australian Pork (2A-118), porkcrc.com.au/wp-content/uploads/2018/08/2A-118-Drinking-Water-Quality-Final-Report.pdf (accessed 6 July 2023).
Edwards, L. and Crabb, H. (2021). Water quality and management in the Australian pig industry. *Anim. Prod. Sci.* 61(7), 637–644.
Eissen, J. J., Kanis, E. and Kemp, B. (2000). Sow factors affecting voluntary feed intake during lactation. *Livest. Prod. Sci.* 64(2–3), 147–165.
Faccin, J. E. G., Tokach, M. D., Goodband, R. D., DeRouchey, J. M., Woodworth, J. C. and Gebhardt, J. T. (2022). Gilt development to improve offspring performance and survivability. *J. Anim. Sci.* 100(6), 1–10.
Fairlie, S. (2010). *Meat: A Benign Extravagance*. Permanent Publications, East Meon.
FAO. (2013). *Food Wastage Footprint – Impacts on Natural Resources*. Food and Agriculture Organization of the United Nations, Rome.
FAO. (2014). *Building a Common Vision for Sustainable Food and Agriculture – Principles and Approaches*. Food and Agriculture Organization of the United Nations, Rome.
FAO. (2018a). *The Future of Food and Agriculture – Alternative Pathways to 2050*. Food and Agriculture Organization of the United Nations, Rome.
FAO. (2018b). *Transforming the Livestock Sector through the Sustainable Development Goals*. Food and Agriculture Organization of the United Nations, Rome.
Farmer, C., Martineau, J. P., Méthot, S. and Bussières, D. (2017). Comparative study on the relations between backfat thickness in late-pregnant gilts, mammary development and piglet growth. *Transl. Anim. Sci.* 1(2), 154–159.
Farmer, C., Palin, M. F., Hovey, R. C., Falt, T. D. and Huber, L. A. (2022). Dietary supplementation with lysine (protein) stimulates mammary development in late pregnant gilts. *J. Anim. Sci.* 100(5), skac051.
Feyera, T. and Theil, P. K. (2017). Energy and lysine requirements and balances of sows during transition and lactation: A factorial approach. *Livest. Sci.* 201, 50–57.
Feyera, T., Pedersen, T. F., Krogh, U., Foldager, L. and Theil, P. K. (2018). Impact of sow energy status during farrowing on farrowing kinetics, frequency of stillborn piglets and farrowing assistance. *J. Anim. Sci.* 96(6), 2320–2331.
Feyera, T., Skovmose, S. J. W., Nielsen, S. E., Vodolazska, D., Bruun, T. S. and Theil, P. K. (2021). Optimal feed level during the transition period to achieve faster farrowing and high colostrum yield in sows. *J. Anim. Sci.* 99(2), skab040.
Foxcroft, G. R. (1997). Mechanisms mediating nutritional effects on embryonic survival in pigs. *J. Reprod. Fertil.* 52, 47–61.
Foxcroft, G. R. (2007). Pre-natal programming of variation in post-natal performance: How and when? *Adv. Pork Prod.* 18, 167–189.
Foxcroft, G. R. and Town, S. (2004). Prenatal programming of postnatal performance: The unseen cause of variance. *Adv. Pork Prod.* 15, 269–279.

Gardiner, G. E., Metzler-Zebeli, B. U. and Lawlor, P. G. (2020). Impact of intestinal microbiota on growth and feed efficiency in pigs: A review. *Microorganisms* 8(12), 1886.

Gresham, A. (2003). Infectious reproductive disease in pigs. *In Pract.* 25(8), 466–473.

Hastad, C. W., Tokach, M. D., Dritz, S. S., Goodband, R. D., DeRouchey, J. M. and Wu, F. (2020). Effects of added fats on growth performance of finishing pigs sorted by initial weight. *Transl. Anim. Sci.* 4(1), 307-315.

Head, R. H. and Williams, I. H. (1991). Mammogenesis is influenced by pregnancy nutrition. In E. S. Batterham (Ed.), *Manipulating Pig Production III*. Australasian Pig Science Association, Werribee, p. 33.

Hernández-García, F., Duarte, J. L., Pérez, M. A., Raboso, C., Rosario, A. I. and Izquierdo, M. (2013). Successful long-term pre-pubertal immunocastration of purebred Iberian gilts reared in extensive systems. *Acta Agric. Slov. Suppl.* 4, 123-126.

Hewitt, R. J. E., Corso, A. and van Barneveld, R. J. (2015). Reducing variation in finisher growth performance through early post-weaning dietary intervention. *Anim. Prod. Sci.* 55(12), 1573.

Holen, J. P., Woodworth, J. C., Tokach, M. D., Goodband, R. D., DeRouchey, J. M., Gebhardt, J. T., DeDecker, A. E. and Martinez, X. (2022). Evaluation of essential fatty acids in lactating sow diets on sow reproductive performance, colostrum and milk composition, and piglet survivability. *J. Anim. Sci.* 100(6), 1-11.

Hong, J. and Kim, Y. Y. (2022). Insect as feed ingredients for pigs. *Anim. Biosci.* 35(2), 347–355.

Hoving, L. L., Soede, N. M., Feitsma, H. and Kemp, B. (2012). Embryo survival, progesterone profiles and metabolic responses to an increased feeding level during second gestation in sows. *Theriogenology* 77(8), 1557–1569.

Inman, C. F., Haverson, K., Konstantinov, S. R., Jones, P. H., Harris, C., Smidt, H., Miller, B., Bailey, M. and Stokes, C. (2010). Rearing environment affects development of the immune system in neonates. *Clin. Exp. Immunol.* 160(3), 431–439.

Iozzo, P. and Sanguinetti, E. (2018). Early dietary patterns and microbiota development: Still a way to go from descriptive interactions to health-relevant solutions. *Front. Nutr.* 5, 5.

Jang, I., Kwon, C. H., Ha, D. M., Jung, D. Y., Kang, S. Y., Park, M. J., Han, J. H., Park, B. C. and Lee, C. Y. (2014). Effects of a lipid-encapsulated zinc oxide supplement on growth performance and intestinal morphology and digestive enzyme activities in weanling pigs. *J. Anim. Sci. Technol.* 56, 29.

Jensen, M. B., Pedersen, L. J., Theil, P. K., Yde, C. C. and Bach Knudsen, K. E. (2012). Feeding motivation and plasma metabolites in pregnant sows fed diets rich in dietary fiber either once or twice daily. *J. Anim. Sci.* 90(6), 1910–1919.

Jeon, J. H., Yeon, S. C., Choi, Y. H., Min, W., Kim, S., Kim, P. J. and Chang, H. H. (2006). Effects of chilled drinking water on the performance of lactating sows and their litters during high ambient temperatures under farm conditions. *Livest. Sci.* 105(1-3), 86-93.

Ji, F., Hurley, W. L. and Kim, S. W. (2006). Characterization of mammary gland development in pregnant gilts. *J. Anim. Sci.* 84(3), 579-587.

Kil, D. Y., Kim, B. G. and Stein, H. H. (2013). Feed energy evaluation for growing pigs. *Asian-Australas. J. Anim. Sci.* 26(9), 1205–1217.

Kim, J. C. and Mullan, B. P. (2012). Quantification of the variability in the amino acid and reactive lysine content of soybean meal and development of a NIR calibration for rapid prediction of reactive lysine content. Final Report, Project 4B-106. The Co-operative Research Centre for High Integrity Australian Pork, Roseworthy, SA.

Kim, J. C., Mullan, B. P. and Pluske, J. R. (2011). Micro-encapsulated zinc oxide as a means of controlling post-weaning diarrhoea in pigs. In R. J. van Barneveld (Ed.), *Manipulating Pig Production XIII*. Australasian Pig Science Association, Werribee, p. 95.

Knauer, M. T. (2016a). Effect of increasing sow feeding level in late gestation on piglet quality and sow body condition. *J. Anim. Sci.* 94(Suppl. 2), 97.

Knauer, M. (2016b). *Effects of Preweaning Factors on Sow Lifetime Productivity*. National Pork Board Research Report #11-146.

Koketsu, Y., Dial, G. D., Pettigrew, J. E., Marsh, W. E. and King, V. L. (1996). Characterization of feed intake patterns during lactation in commercial swine herds. *J. Anim. Sci.* 74(6), 1202–1210.

Kornegay, E. T. (1986). Biotin in swine production: A review. *Livest. Prod. Sci.* 14(1), 65–89.

Kruse, S., Traulsen, I. and Krieter, J. (2011). Analysis of water, feed intake and performance of lactating sows. *Livest. Sci.* 135(2–3), 177–183.

Langendijk, P. (2021). Latest advances in sow nutrition during early gestation. *Animals* 11(6), 1720.

Lawlor, P. G., Lynch, P. B., O'Connell, M. K., McNamara, L., Reid, P. and Stickland, N. C. (2007). The influence of over feeding sows during gestation on reproductive performance and pig growth to slaughter. *Arch. Anim. Breed.* 50, 82–91.

Leal, D. F., Muro, B. B. D., Nichi, M., Almond, G. W., Viana, C. H. C., Vioti, G., Carnevale, R. F. and Garbossa, C. A. P. (2019). Effects of post-insemination energy content of feed on embryonic survival in pigs: A systematic review. *Anim. Reprod. Sci.* 205, 70–77.

Le Dividich, J. L., Rooke, J. A. and Herpin, P. (2005). Nutritional and immunological importance of colostrum for the new-born pig. *J. Agric. Sci.* 143(6), 469–485.

Leibbrandt, V. D., Johnston, L. J., Shurson, G. C., Crenshaw, J. D., Libal, G. W. and Arthur, R. D. (2001). Effect of nipple drinker water flow rate and season on performance of lactating swine. *J. Anim. Sci.* 79(11), 2770-2775.

Li, S., Zheng, J., He, J., Liu, H., Huang, Y., Huang, L., Wang, K., Zhao, X., Feng, B., Che, L., Fang, Z., Li, J., Xu, S., Lin, Y., Jiang, X., Hua, L., Zhuo, Y. and Wu, D. (2023). Dietary fiber during gestation improves lactational feed intake of sows by modulating gut microbiota. *J. Anim. Sci. Biotechnol.* 14(1), 65.

Li, Y. Z., Chénard, L., Lemay, S. P. and Gonyou, H. W. (2005). Water intake and wastage at nipple drinkers by growing-finishing pigs. *J. Anim. Sci.* 83(6), 1413-1422.

Liu, Y., Espinosa, C. D., Abelilla, J. J., Casas, G. A., Lagos, L. V., Lee, S. A., Kwon, W. B., Mathai, J. K., Navarro, D. M. D. L., Jaworski, N. W. and Stein, H. H. (2018a). Non-antibiotic feed additives in diets for pigs: A review. *Anim. Nutr.* 4(2), 113–125.

Liu, Y., Kil, D. Y., Perez-Mendoza, V. G., Song, M. and Pettigrew, J. E. (2018b). Supplementation of different fat sources affects growth performance and carcass composition of finishing pigs. *J. Anim. Sci. Biotechnol.* 9, 56.

Magnabosco, D., Bernardi, M. L., Wentz, I., Cunha, E. C. P. and Bortolozzo, F. P. (2016). Low birth weight affects lifetime productive performance and longevity of female swine. *Livest. Sci.* 184, 119–125.

Magowan, E., O'Connell, N. E. and McCann, E. E. (2007). *The Effect of Drinker Design on the Performance, Behaviour and Water Usage of Growing Pigs*. Agri-Food and Bioscience Institute, Hillsborough. Report prepared for UFU and PPDC Committees.

Moturi, J., Hosseindoust, A., Ha, S. H., Tajudeen, H., Mun, J. Y. and Kim, J. S. (2022). Effects of age at first breeding and dietary energy levels during the rearing period of replacement gilts on reproductive performance. *Anim. Prod. Sci.* 62(16), 1581-1589.

Nakano, T., Brennan, J. J. and Aherne, F. X. (1987). Leg weakness and osteochondritis in swine: A review. *Can. J. Anim. Sci.* 67(4), 883-901.

National Research Council. (1998). *Nutritional Requirements of Swine*. 10th edition. National Academies Press, Washington, DC.

National Research Council. (2012). *Nutritional Requirements of Swine*. 11th edition. National Academies Press, Washington, DC.

Nedovic, V., Kalusevic, A., Manojlovic, V., Levic, S. and Bugarski, B. (2011). An overview of encapsulation technologies for food applications. *Procedia Food Sci.* 1, 1806-1815.

Nitrayova, S., Heger, J., Patras, P. and Brestensky, M. (2006). Effect of body weight and pig individuality on apparent ileal digestibility of amino acids and total nitrogen. *Slovac J. Anim. Sci.* 39, 65-68.

Noblet, J. (2006). Recent advances in energy evaluation of feeds for pigs. In P. C. Garnsworthy and J. Wiseman (Eds.), *Recent Advances in Animal Nutrition*. Nottingham University Press, Nottingham, pp. 1-26.

Noblet, J. and Shi, X. S. (1993). Comparative digestibility of energy and nutrients in growing pigs fed ad libitum and adult sows fed at maintenance. *Livest. Prod. Sci.* 34(1-2), 137-152.

Noel, S. J., Jorgensen, H. J. H. and Bach Knudsen, K. E. (2022). The use of near-infrared spectroscopy (NIRS) to determine the energy value of individual feedstuffs and mixed diets for pigs. *Anim. Feed Sci. Technol.* 283, 115156.

Nowland, T. L., Kirkwood, R. N., Plush, K. J., Barton, M. D. and Torok, V. A. (2021a). Exposure to maternal faeces in lactation influences piglet enteric microbiota, growth and survival pre-weaning. *J. Anim. Sci.* 99(7), skab170.

Nowland, T. L., Kirkwood, R. N., Torok, V. A., Plush, K. J. and Barton, M. D. (2021b). Characterisation of early microbial colonisers within the spiral colon of pre- and post-natal piglets. *Life* 11(4), 312.

Nowland, T. L., Kirkwood, R. N. and Pluske, J. R. (2022). Review: Can early-life establishment of the piglet intestinal microbiota influence production outcomes? *Animal* 16, 100368.

O'Connell, N. E., Beattie, V. E. and Watt, D. (2005). Influence of regrouping strategy on performance, behavior and carcass parameters in pigs. *Livest. Prod. Sci.* 97(2-3), 107-115.

Odakura, A. M., Caldara, F. R., Burbarelli, M. F. C., Almeida Paz, I. C. L., Garcia, R. G., Oliveira Dos Santos, V. M., de Brito Mandu, D. F., Braz, J. M. and Lourenço da Silva, M. I. (2023). Dietary supplementation of eubiotic fiber based on lignocellulose on performance and welfare of gestating and lactating sows. *Animals* 13(4), 695.

Oliviero, C., Kokkonen, T., Heinonen, M., Sankari, S. and Peltoniemi, O. (2009). Feeding sows with high fibre diet around farrowing and early lactation: Impact on intestinal activity, energy balance related parameters and litter performance. *Res. Vet. Sci.* 86(2), 314-319.

Paternostre, L., Baeten, V., Ampe, B., Millet, S. and De Boever, J. (2021). The usefulness of NIRS calibrations based on feed and feces spectra to predict nutrient content, digestibility and net energy of pig feeds. *Anim. Feed Sci. Technol.* 281, 115091.

Patience, J. F., Engele, K., Beaulieu, A. D., Gonyou, H. W. and Zijlstra, R. T. (2004). Variation: Costs and consequences. *Adv. Pork Prod.* 15, 257-266.

Pedersen, T. F., van Vliet, S., Bruun, T. S. and Theil, P. K. (2020). Feeding sows during the transition period–is a gestation diet, a simple transition diet, or a lactation diet the best choice? *Transl. Anim. Sci.* 4(1), 34–48.

Peltoniemi, O. A. T., Kothe, S., Kauffold, J., Valros, A. and Oliviero, C. (2014). Farrowing complications increase stillborn rate and neonatal mortality in the pig. *Reprod. Domest. Anim.* 49, 39–101.

Peng, J. J., Somes, S. A. and Rozeboom, D. W. (2007). Effect of system feeding and watering on performance of lactating sows. *J. Anim. Sci.* 85(3), 853-860.

Pérez-Ciria, L., Carcò, G., Miana-Mena, F. J., Mitjana, O., Falceto, M. V. and Latorre, M. A. (2021). Immunocastration in gilts: A preliminary study of the effect of the second dose administration time on growth, reproductive tract development, and carcass and meat quality. *Animals* 11(2), 510.

Phillips, P. A. and Phillips, M. H. (1999). Effect of dispenser on water intake of pigs at weaning. *Trans. ASAE* 42, 1471–1474.

Plush, K., Glencorse, D., Alexopoulos, J., Tritton, S., Kirkwood, R. and D'Souza, D. (2019). Effect of dextrose supplementation in the pre-ovulatory sow diet to reduce seasonal influences on litter birth weight variation. *Animals* 9(12), 1009.

Pluske, J. R., Hansen, C. F., Payne, H. G., Mullan, B. P., Kim, J. C. and Hampson, D. J. (2007). Gut health in the pig. In J. E. Paterson and J. A. Barker (Eds.), *Manipulating Pig Production XI*. Australasian Pig Science Association, Werribee, pp. 147–158.

Poulsen Nautrup, B., Van Vlaenderen, I. and Mah, C. K. (2020). The effect of immunization against gonadotropin-releasing factor in market gilts: Meta-analyses of parameters relevant for pig producers, pork packers and retailers/consumers. *Res. Vet. Sci.* 131, 159-172.

Prunier, A., de Braganca, M. M. and Le Dividich, J. (1997). Influence of hgh ambient temperature on performance of reproductive sows. *Livest. Prod. Sci.* 52(2), 123–133.

Qi, Y., Zheng, T., Yang, S., Zhang, Q., Li, B., Zeng, X., Zhong, Y., Chen, F., Guan, W. and Zhang, S. (2023). Maternal sodium acetate supplementation promotes lactation performance of sows and their offspring growth performance. *Anim. Nutr.* 14, 213-224.

Quesnel, H., Brossard, L., Valancogne, A. and Quiniou, N. (2008). Influence of some sow characteristics on within-litter variation of piglet birth weight. *Animal* 2(12), 1842–1849.

Quiniou, N. and Noblet, J. (1999). Influence of high ambient temperatures on performance of multiparous lactating sows. *J. Anim. Sci.* 77(8), 2124–2134.

Quiniou, N., Dagorn, J. and Gaudré, D. (2002). Variation of piglets' birth weight and consequences on subsequent performance. *Livest. Prod. Sci.* 78(1), 63–70.

Quiniou, N., Renaudeau, D., Dubois, S. and Noblet, J. (2000). Influence of high ambient temperatures on food intake and feeding behaviour of multiparous lactating sows. *Anim. Sci.* 70(3), 471–479.

Rehfeldt, C., Kuhn, G., Nürnberg, G., Kanitz, E., Schneider, F., Beyer, M., Nürnberg, K. and Ender, K. (2001). Effects if exogenous somatotropin during early gestation on maternal performance, fetal growth, and compositional traits in pigs. *J. Anim. Sci.* 79(7), 1789–1799.

Renaudeau, D. and Noblet, J. (2001). Effects of exposure to high ambient temperature and dietary protein level on sow milk production and performance of piglets. *J. Anim. Sci.* 79(6), 1540–1548.

Revell, D. K., Williams, I. H., Mullan, B. P., Ranford, J. L. and Smits, R. J. (1998a). Body composition at farrowing and nutrition during lactation affect the performance of primiparous sows: I. Voluntary feed intake, weight loss, and plasma metabolites. *J. Anim. Sci.* 76(7), 1729-1737.

Revell, D. K., Williams, I. H., Mullan, B. P., Ranford, J. L. and Smits, R. J. (1998b). Body composition at farrowing and nutrition during lactation affect the performance of primiparous sows: II. Milk composition, milk yield, and pig growth. *J. Anim. Sci.* 76(7), 1738-1743.

Rezaei, R., Gabriel, A. S. and Wu, G. (2022). Dietary supplementation with monosodium glutamate enhances milk production by lactating sows and the growth of suckling piglets. *Amino Acids* 54(7), 1055–1068.

Rutherfurd, S. M., Moughan, P. J. and Morel, P. C. H. (1997). Assessment of the true ileal digestibility of reactive lysine as a predictor of lysine uptake from the small intestine of the growing pig. *J. Agric. Food Chem.* 45(11), 4378–4383.

Schader, C., Muller, A., El-Hage Scialabba, N., Hecht, J., Isensee, A., Erb, K. H., Smith, P., Makkar, H. P. S., Klocke, P., Leiber, F., Schwegler, P., Stolze, M. and Niggli, U. (2015). Impacts of feeding less food-competing feedstuffs to livestock on global food system sustainability. *J. R. Soc. Interface* 12(113), 20150891.

Schrögel, P. and Wätjen, W. (2019). Insects for food and feed-safety aspects related to mycotoxins and metals. *Foods* 8(8), 288.

Shen, J., Chen, Y., Wang, Z., Zhou, A., He, M., Mao, L., Zou, H., Peng, Q., Xue, B., Wang, L., Zhang, X., Wu, S. and Lv, Y. (2014). Coated zinc oxide improves intestinal immunity function and regulates microbiota composition in weaned piglets. *BJN* 111(12), 2123–2134.

Shurson, G. C., Hung, Y. T., Jang, J. C. and Urriola, P. E. (2021). Measures matter – determining the true nutri-physiological value of feed ingredients for swine. *Animals* 11(5), 1259.

Stenhouse, C., Hurst, E., Mellanby, R. J. and Ashworth, C. J. (2022). Associations between maternal vitamin D status and porcine litter characteristics throughout gestation. *J. Anim. Sci. Biotechnol.* 13(1), 106.

Soladoye, P. O., Shand, P. J., Aalhus, J. L., Gariépy, C. and Juárez, M. (2014). Review: Pork velly quality, bacon properties and recent consumer trends. *Can. J. Anim. Sci.* 95(3), 325–340.

Sugiyama, T., Kusuhara, S., Chung, T. K., Yonekura, H., Azem, E. and Hayakama, T. (2013). Effects of 25-hydroxy-cholecalciferol on the development of osteochondrosis in swine. *Anim. Sci. J.* 84(4), 341–349.

Theil, P. K., Farmer, C. and Feyera, T. (2022a). Review: Physiology and nutrition of late gestating and transition sows. *J. Anim. Sci.* 100(6), 1–15.

Theil, P. K., Krogh, U., Bruun, T. S. and Feyera, T. (2022b). Feeding the modern sow to sustain high productivity. *Mol. Reprod. Dev.* 90(7), 517–532.

Thulin, A. J. and Brumm, M. C. (1991). Water: The forgotten nutrient. In E. R. Miller, D. E. Ulrey, and A. J. Lewis (Eds.), *Swine Nutrition*. Butterworth-Heinemann, Stoneham, MA, pp. 315–324.

Tolosa, A. F., DeRouchey, J. M., Tokach, M. D., Goodband, R. D., Woodworth, J. C., Gebhardt, J. T., Ritter, M. J. and Pilcher, C. M. (2021). A meta-analysis to understand the relationship between pig body weight and variation from birth to market. *Animals* 11(7), 2088.

Torrey, S. and Widowski, T. M. (2006). A note on piglets' preferences for drinker types at two weaning ages. *App. Anim. Beh. Sci.* 100(3–4), 333–341.

Tydlitát, D., Vinkler, A. and Czanderlová, L. (2008). Influence of crude protein intake on the duration of delivery and litter size in sows. *Acta Vet. Brno* 77(1), 25–30.

Vallet, J. L., Miles, J. R., Rempel, L. A., Nonneman, D. J. and Lents, C. A. (2015). Relationships between day one piglet serum immunoglobulin immunocrit and subsequent growth, puberty attainment, litter size, and lactation performance. *J. Anim. Sci.* 93(6), 2722–2729.

van Barneveld, R. J. (1999). Physical and chemical characteristics of grains related to variability in energy and amino acid availability in pigs: A review. *Aust. J. Agr. Res.* 50, 667–687.

van Barneveld, R. J. (2010). Mineral supplementation in modern pork production systems for optimum health and efficiency – the need for renewed focus. In *Proceedings of the FeetFirst Sow Lameness Symposium II*. Minneapolis, MN, pp. 66–81.

van Barneveld, R. J. and Ru, Y. J. (2003). Measurement of total and reactive lysine content of canola meal using near infrared spectroscopy. In J. E. Paterson (Ed.), *Manipulating Pig Production IX*. Australasian Pig Science Association, Werribee, p. 157.

van Barneveld, R., Nuttall, J., Flinn, P. C. and Osborne, B. (1998). NIR reflectance measurement of the digestible energy content of cereals for growing pigs. *J. Near Infrared Spec.* 7, 1–7.

van Barneveld, R. J., Nuttall, J. D., Flinn, P. C. and Osborne, B. G. (1999). Near infrared reflectance measurement of the digestible energy content of cereals for growing pigs. *J. Near Infrared Spec.* 7(1), 1–7.

van Barneveld, R. J., Hewitt, R. J. E. and D'Souza, D. N. (2017). Advances in understanding pig nutritional requirements and metabolism. In J. Wiseman (Ed.), *Achieving Sustainable Production of Pig Meat (Vol. 2): Animal Breeding and Nutrition*. Burleigh Dodds Science Publishing Limited, Cambridge.

van Barneveld, R. J., Graham, H. and Diffey, S. (2018). Predicting the nutritional quality of feed ingredients for pigs using near infra-red spectroscopy (NIRS) and chemical analysis. *Anim. Prod. Sci.* 58(4), 709–718.

van Barneveld, R. J., Hewitt, R. J. E. and D'Souza, D. N. (2023). Net protein contribution from an intensive Australian Pork supply chain. *Anim. Prod. Sci.* 63(18), 1837-1850. https://doi.org/10.1071/AN23057.

van den Brand, H., van Enckevort, L. C., van der Hoeven, E. M. and Kemp, B. (2009). Effects of dextrose plus lactose in the sows diet on subsequent reproductive performance and within litter birth weight variation. *Reprod. Domest. Anim.* 44(6), 884–888.

Vande Pol, K. D., Grohmann, N. S., Weber, T. E., Ritter, M. J. and Ellis, M. (2022a). Effect of drinker type on water disappearance of nursery pigs. *Trans. Ani. Sci.* 6(1), 1–7.

Vande Pol, K. D., Grohmann, N. S., Weber, T. E., Ritter, M. J. and Ellis, M. (2022b). Influence of high cyclic ambient temperature and water drinker design on growth performance and water disappearance of growing-finishing pigs. *Trans. Ani. Sci.* 6(3), 1–13.

Veldkamp, T. and Vernooij, A. G. (2021). Use of insect products in pig diets. *J. Insects Food Feed* 7, 1–14.

Wang, K., Liu, J., Zhao, L., Li, Y. and Li, C. (2023). Effects of dietary valine-to-lysine ratio on placenta growth, the placental nutrient transporters, and reproductive performance in sows. *Anim. Prod. Sci.* 63(8), 742–750.

Wang, M., Wanying, Z., Zhinghing, W., Jijun, L., Zhaohui, C. and Na, L. (2017). Comparison experiment of total water consumption and water leakage of different types of drinker for nursery pig. *Tran. Chin. Soc. Ag. Eng.* 33, 242–247.

Wang, W., Yu, Z., Yin, X., Wang, Z., Xu, S., Shi, C., Zang, J., Liu, H. and Wang, F. (2023). Effect of dietary starch-to-fat ratio on lipid metabolism, inflammation, and microbiota of multiparous sow and newborn piglets. *Agriculture* 13(5), 1069.

Weldon, W. C., Lewis, A. J., Louis, G. F., Kovar, J. L., Giesemann, M. A. and Miller, P. S. (1994). Postpartum hypophagia in primiparous sows: I. Effects of gestation feeding level on feed intake, feeding behaviour, and plasma metabolite concentrations during lactation. *J. Anim. Sci.* 72(2), 387–394.

Westendorf, M. L. (2000). Food waste as swine feed. In M. L. Westendorf (Ed.), *Food Waste to Animal Feed*. Iowa State University Press, Ames, IA, pp. 69–89.

Wijesiriwardana, U. A., Plush, K. J., Tritton, S., Pluske, J. R., Cottrell, J. J. and Dunshea, F. R. (2021). Feeding sows lucerne, or diets with similar energy and nutritional profiles to lucerne, improves the pre-weaning performance of piglets. *Agriculture* 11(11), 1146.

Williams, N. H., Patterson, J. and Foxcroft, G. (2005). Non-negotiables of gilt development. *Adv. Pork Prod.* 16, 281–289.

Willis, S. and Collman, G. (2007). Providing chilled drinking water to lactating sows using a modified milk vat. APL project 2006/2154. Australian Pork Limited, Deakin. https://australianpork.infoservices.com.au/downloads/2006-2154 (accessed 6 July 2023).

Wolter, B. F., Ellis, M., Corrigan, B. P. and DeDecker, J. M. (2002). The effect of birth weight and feeding of supplemental milk replacer to piglets during lactation on preweaning and postweaning growth performance and carcass characteristics. *J. Anim. Sci.* 80(2), 301–308.

Wu, G., Ott, T. L., Kanbe, D. A. and Bazer, F. W. (1999). Animal acid composition of the fetal pig. *J. Nutr.* 129(5), 1031–1038.

Wu, G., Bazer, F. W., Davis, T. A., Jaegar, L. A., Johnson, G. A., Kim, S. W., Knabe, D. A., Meininger, C. J., Spencer, T. E. and Yin, Y.-L. (2007). Important roles for the arginine family of amino acids in swine nutrition and production. *Livest. Sci.* 112(1–2), 8–22.

Wu, G., Bazer, F. W., Burghardt, R. C., Johnson, G. A., Kim, S. W., Li, X. L., Satterfield, M. C. and Spencer, T. E. (2010). Impacts of amino acid on pregnancy outcome in pigs: Mechanisms and implications for swine production. *J. Anim. Sci.* 88(13), E195–E204.

Wu, G., Bazer, F. W., Johnson, G. A., Knabe, D. A., Burghardt, R. C., Spencer, T. E., Li, X. L. and Wang, J. J. (2011). Important roles for L-glutamine in swine nutrition and production. *J. Anim. Sci.* 89(7), 2017–2030.

Ytrehus, B., Grindflek, E., Teige, J., Stubsjøen, E., Grøndalen, T., Carlson, C. S. and Ekman, S. (2004). The effect of parentage on the prevalence, severity and location of lesions of osteochondrosis in swine. *J. Vet. Med. A Physiol. Pathol. Clin. Med.* 51(4), 188–195.

Zhan, X., Fletcher, L., Dingle, S., Baracuhy, E., Wang, B., Huber, L. A. and Li, J. (2021). Choline supplementation influences ovarian follicular development. *Front. Biosci. (Landmark Ed.)* 26(12), 1525–1536.

Zhao, L., Li, Y., Li, Z., Wu, S., Huang, K., Chen, J. and Li, C. (2019). Effect of the valine-to-lysine ratio on the performance of sows and piglets in a hot, humid environment. *J. Therm. Biol.* 81, 89–97.

Chapter 2

Gastrointestinal development in pigs: implications for nutrition and performance

John O'Doherty, Dillon Kiernan and Torres Sweeney, University College Dublin, Ireland

1 Introduction

The digestive system of pigs shares fundamental similarities with that of other non-ruminant mammals. Comprising the stomach, divided into four regions (oesophageal, cardiac, fundic, and pyloric), it plays a vital role in food breakdown through the release of hydrochloric acid (HCl) and digestive enzymes. This acidic environment serves both as a defence against ingested pathogens and as a facilitator of enzyme activation, notably pepsin, for protein digestion. The small intestine, serving as the core site for digestion and absorption, is where a diverse array of digestive enzymes, derived from the stomach, pancreas, and the small intestine itself, work to convert food into readily absorbable molecules - such as amino acids, monosaccharides, and fatty acids. These molecules are subsequently efficiently absorbed through transporters found on the surface of enterocytes (Daniel, 2004). However, enzymatic digestion ceases upon entering the large intestine, where microbial enzymes take over. The colon hosts a predominantly Firmicutes and Bacteroidetes bacterial population (Isaacson et al., 2012; Gao and Song, 2018), with microbial fermentation producing short-chain fatty acids (SCFAs), including acetate, propionate, and butyrate. These

http://dx.doi.org/10.19103/AS.2024.0140.02

SCFAs significantly contribute to the pig's energy needs and colonic mucosal health (Canani et al., 2011; Guilloteau et al., 2010).

The gastrointestinal tract (GIT) is dynamic, undergoing numerous changes throughout a pig's life, driven by prenatal, neonatal, and post-weaning factors. Notably, the most profound changes occur in the first 24 h post-birth and during the immediate post-weaning period. Weaning, typically around 4 weeks of age, poses challenges due to the pig's immature gastric capacity (Cranwell, 1985; Kiernan et al., 2023a). However, strategies to bolster stomach development in pre-weaned pigs, such as increasing acid and enzyme secretion, remain underexplored with limited research in this scientific area. Weaning is a critical phase in pig production, marked by separation from the sow, dietary changes, and environmental and social adjustments. These changes disrupt intestinal health, leading to morphological and physiological alterations, including inflammation, villous atrophy, crypt hyperplasia, and reduced brush border activity (Lallès et al., 2004; O'Doherty et al., 2021). Consequently, the pig's GIT undergoes rapid changes in size, protein turnover, microbiota composition, and digestive and immune functions (de Lange et al., 2010), collectively reducing the digestive and absorptive capacity of small intestinal enterocytes. This can result in growth setbacks or weight loss post weaning (O'Doherty et al., 2021).

Understanding the developmental nuances of the pig's intestinal tract during prenatal, neonatal, and post-weaning phases is crucial and will be discussed in this chapter. It empowers swine nutritionists to formulate effective dietary strategies to regulate intestinal function, ultimately enhancing growth, health, and feed efficiency. This chapter delves into GIT development, its influences on intestinal health, and the nutritional and management approaches to support GIT maturation and optimize pig performance during these critical life stages.

Furthermore, the small intestine plays a pivotal role in nutrient absorption, guided by specialized transporters like SGLT1, GLUT5, PEPT1, CD36, and FABP2 (Wood and Trayhurn, 2003; Daniel, 2004; Hajri and Abumrad, 2002; Glatz and van der Vusse, 1996). The regulation of these transporters is influenced by factors such as developmental stages, weaning, and dietary changes (Ashida et al., 2002; Howard et al., 2004; Rodriguez et al., 2004; Vigors et al., 2014; Heim et al., 2014a), and holds significant implications for absorption efficiency. Optimizing these transporters and improving digestibility not only enhances our understanding of digestive physiology but also offers promise for innovative approaches to swine nutrition.

2 Prenatal development of the foetal pig's gastrointestinal tract

The gestational period of a pig is crucial for the development of its GIT, a system essential for its subsequent survival and growth. From days 45 to 110, this intricate system undergoes significant morphological and functional changes that not only prepare the foetal pig for birth but also ensure its adaptability to postnatal life. In this section, we delve into the processes of stomach and intestine development, shedding light on the myriad factors influencing their growth, from hormonal cues to the nutritive properties of the amniotic fluid. The profound impact of these prenatal changes underscores the importance of understanding and potentially optimizing them to ensure robust postnatal health and productivity in pigs.

2.1 Prenatal stomach development

In the prenatal period, several pivotal changes unfold within the pig's stomach to prepare it for birth. The concentration of chymosin surges five-fold, while pH levels fall from above 7 to a range of 2–3 from day 95 of gestation until birth. This shift in stomach pH is crucial for adapting the piglet to abrupt exposure to environmental microorganisms and the transition from parenteral to enteral nutrition at birth (Sangild et al., 2002). The decrease in pH is instrumental for enzyme activation and serves as a defence mechanism against pathogenic microbes. Prochymosin is detectable as early as day 80 of gestation and plays a pivotal role in milk coagulation in the stomach. Conversely, pepsin A levels remain relatively low during the prenatal phase (Sangild et al., 2000). Cortisol, a prenatal hormone, regulates these changes, stimulating the release of acid, gastrin, and gastric intrinsic factor in foetal pigs (Sangild et al., 1994). This hormonal cascade begins around day 100 of gestation, intensifying with parturition (Silver and Fowden, 1989). Notably, ghrelin-secreting X/A-like cells are abundant in the fundic gland region of the prenatal pig's stomach, while gastrin is present in both the GIT and plasma of the foetal pig from day 93 of gestation (Xu and Cranwell, 1991; Willemen et al., 2013).

2.2 Prenatal small intestine

The development of the small intestine during the prenatal phase is highly regulated by intrinsic programming and greatly influenced by environmental factors such as luminal stimuli and hormonal cues. The ingestion of amniotic fluid by the foetus, which begins before birth, plays a vital role in shaping this development. Amniotic fluid serves as the initial 'diet' for the developing foetus, impacting GIT development with both physical and trophic effects. Remarkably, amniotic fluid contributes up to 20% of foetal energy and protein supply (Trahair and Sangild, 2000). This fluid is rich in proteins and growth factors, influencing GIT development (Sangild et al., 2013). Notably, the disruption of this process, such as when foetal fluid ingestion is obstructed late in gestation, results in reduced body weight, especially in intestinal weight, aminopeptidase A activity, and glucose absorption (Sangild et al., 2002). Maternal diet can also impact the metabolic profile of amniotic fluid (Wan et al., 2017). One pivotal growth factor found in amniotic fluid is insulin-like growth factor-1 (IGF-1), a key regulator of conceptus growth during pregnancy (Jeong et al., 2014). The levels of IGF-1 correlate with the sow's feed intake during pregnancy (Musser et al., 2004).

Enzyme activity in the foetal pig's intestine is dynamic. Lactase is highly active, particularly at term, with lactase activity gradually increasing throughout gestation. Maltase and sucrase activities are relatively low, but maltase activity experiences an approximate 30% surge in the final week of gestation (Sangild et al., 2002). Aminopeptidase N activity significantly increases from day 80 of gestation, and the small intestine starts absorbing D-glucose at day 49 of gestation, with a remarkable four-fold increase in the rate of glucose uptake in the final 21 days of gestation (Sangild et al., 2002). The increase in intestinal absorption capacity, coupled with rapid intestinal growth, leads to a seven-fold increase in total glucose uptake capacity per kilogram of body weight (Sangild et al., 2002).

3 Postnatal gastrointestinal tract growth

The postnatal development of the pig's GIT is a dynamic process, particularly in the first 2 weeks of life, with the most significant changes occurring in the initial 24 h after birth. These changes are driven by growth factors and hormones present in colostrum. However, the GIT faces various challenges and physiological stresses, especially at birth and weaning. To meet these demands and adapt to environmental changes, the GIT, primarily the small intestine, undergoes accelerated tissue growth and functional maturation, starting from mid-gestation and continuing until 6–8 weeks after birth (Rome et al., 2002).

3.1 Postnatal stomach development

The number of stomach parietal cells, crucial for H+ secretion, steadily increases with age from birth to weaning, with a significant surge observed between days 7 and 14 post weaning (Xu et al., 1992). However, the piglet's limited parietal cell abundance and immaturity lead to a restricted capacity to maintain low gastric pH. As a result, lactic acid and acetate, produced from lactose breakdown by stomach bacteria, play a vital role in maintaining gastric acidity in suckling piglets.

Both parietal cells and enteroendocrine cell numbers increase with age during the suckling phase (Trevisi et al., 2021). Stomach tissue weight in newborn piglets significantly increases by the first and third day after birth, surpassing the increments in body weight (Xu et al., 1992). The gastric acid secretory capacity per unit of gastric tissue mass doubles during the first 3 days after birth, leading to a more than three-fold increase in total gastric acid secretory capacity (Xu and Cranwell, 1990). Additionally, the concentration of gastrin in the gastric antral tissue triples, while circulating gastrin concentration increases by about 1.5-fold in newborn pigs during the first 7 days after birth (Xu and Cranwell, 1991). The fundic gland region of the piglet's stomach is rich in ghrelin-secreting X/A-like cells, with X/A-like cell density doubling in this region within the first 28 days after birth. By day 28, the density of X/A-like cells in the cardiac and pyloric gland regions matches that of the fundic gland region on day 1 (Vitari et al., 2012). Interestingly, pIgR, TLR2, and TLR4 expression increase linearly in the stomach during the suckling phase, indicating their role in stomach gut-associated lymphoid tissue (GALT) maturation (Trevisi et al., 2013).

Proteolytic enzyme capacity in the piglet is initially low due to limited acid secretory capacity and low pepsin A secretion. Prochymosin, the precursor to chymosin, is the dominant enzyme at birth, characterized by strong milk-clotting ability but weak proteolytic activity. This low proteolytic environment is essential for preserving the integrity of immunoglobulins, hormones, and growth factors in colostrum and milk as they pass through the stomach (Xu et al., 2000). Subsequently, chymosin concentration declines steadily, becoming undetectable by 2 months of age. Simultaneously, pepsinogen A concentration rises. The gastric secretory capacity for protease enzymes increases nine-fold during the first 7 postnatal days, with pepsin A becoming the predominant protease in the stomach from 4 to 5 weeks of age. Gastric acid secretory capacity intensifies with age and stomach weight. Cholecystokinin (CCK) and gastrin plasma concentrations rise progressively during the suckling phase (Marion et al., 2003).

3.2 Postnatal small intestine

At birth, the piglet's small intestine undergoes rapid transformations as it transitions from placental to enteral nutrition. In the first 24 h of suckling, the intestinal weight, length, total surface area, and estimated tissue thickness of the piglet increase significantly, with most of this growth attributed to the rapid expansion of the mucosal component throughout the intestine (Xu et al., 1992). A marked increase in gut thickness results in a 100% increase in absorptive area at the microscopic level and approximately a 27% increase in nominal surface area. Villus height and width also experience substantial increases within the first 3 days after birth (Xu et al., 1992). Crypt depth shows an initial surge after birth and continues to increase steadily (Elefson et al., 2021).

A noteworthy feature of the piglet's small intestine is its high permeability at birth, allowing for the uptake of macromolecules, including intact immunoglobulins, and their transport into the bloodstream. This process, known as 'intestinal closure', typically occurs within 24–36 h, although this timeline may vary. Intestinal closure begins within 6–12 h after colostrum consumption. Notably, factors such as fasting after birth can prolong the closure process (Weström et al., 1984). The ability to transport large immunoglobulins and other macromolecules is a specialized function of enterocytes, which internalize luminal content through endocytosis, either selectively (involving the fragment crystallizable receptor FcRn binding to IgG) or non-selectively as enterocytes engulf various luminal macromolecules (Stirling et al., 2005).

Maltase and sucrase activities are relatively low at birth but steadily increase during the suckling period. Lactase activity levels decrease throughout suckling but remain the most active enzyme (Cao et al., 2014; Marion et al., 2005). Peptidases, including aminopeptidase A, aminopeptidase N, and dipeptidyl peptidase IV, are highly active in the new-born piglet, with aminopeptidase A and aminopeptidase N showing uniform activity throughout the suckling phase (Marion et al., 2005).

3.3 Development of intestinal microbiota after birth

Historically, it was widely believed that the GIT of prenatal pigs remained a sterile environment. However, recent research has challenged this notion, suggesting that amniotic fluid might make a minor contribution to the colonization of the intestine before birth (Leblois et al., 2017; Jiménez et al., 2008). Following birth, piglets primarily acquire microbes from sources such as the sow's milk, her nipples, and the surrounding environment. However, as lactation progresses, piglets develop a GIT microbiota that closely mirrors that of their mothers rather than reflecting the characteristics of their rearing environment (Chen et al., 2018). The sow's colostrum and milk play a crucial role in shaping the GIT microbiota of piglets. This influence is attributed to the microbial content, nutrient composition, and the presence of prebiotic oligosaccharides in the sow's milk (Chen et al., 2018; Salcedo et al., 2016). Consequently, the microbiota of suckling piglets is prominently characterized by genera such as Bacteroidaceae, Clostridiaceae, Lachnospiraceae, Lactobacillaceae, and Enterobacteriaceae (Frese et al., 2015). The first few days post-birth are marked by a high diversity in the piglet's intestinal microbiota, reflecting a blend of maternal and environmental bacterial influences. As they grow, the establishment of their adult faecal microbiota can be categorized into three phases. The initial phase, covering the first week, sees a dominance of aerobes and facultative anaerobes, constituting approximately 0.8 of the total microbiota (Swords et al., 1993). This colonization occurs rapidly, with a significant bacterial count observed in the distal colon within just 12 h of birth (Swords et al., 1993).

During the second phase, which spans from the end of the first week to weaning, there is a noticeable shift towards anaerobic species like Lactobacillus and Streptococcus (Konstantinov et al., 2006). Factors like colostrum immunoglobulins play a role in this transition by preventing specific antigens from accessing the intestine (Brandtzaeg, 2002). By the 48-h mark post-birth, strict anaerobes start to dominate aerotolerant bacteria, constituting up to 90% of the total bacteria (Swords et al., 1993). In the third phase, which extends from weaning until piglets fully adapt to dry food, species like Lactobacillus and Streptococcus become dominant, primarily due to their ability to break down milk-derived substrates (Mathew et al., 1998). This period also witnessed an increase in the diversity of anaerobic bacteria (Inoue et al., 2005).

Importantly, maternal dietary choices can influence this microbial development. For instance, the inclusion of fermented liquid feed in the sow's diet has been observed to reduce harmful bacteria in her faeces 7 days post-parturition (Demecková et al., 2002). Furthermore, incorporating probiotics into a sow's diet can indirectly benefit her offspring. These beneficial bacteria are acquired by piglets through the sow's milk and faeces, significantly influencing their microbial composition (Kiernan et al., 2023b). Over time, a piglet's microbiota aligns more with its mother's than its immediate environment. This suggests a potential entero-mammary connection, indicating that the health of a mother's gut can dictate the microbial quality of her milk. This alignment of the mother's and offspring's microbiota can lead to improved health and performance in the piglets. Therefore, optimizing the sow's microbiota during her gestation and lactation phases becomes pivotal for the well-being of both her and her offspring. While the current data is promising, continued research is essential to corroborate these findings and explore further benefits.

3.4 Development of the post-weaning stomach

The pig's ability to maintain a low gastric pH through HCl secretion is hindered by the immaturity of parietal cells in the fundic gland region at weaning. The number of H+/K+-ATPase positive cells, crucial for H+ secretion, increases steadily with age until weaning. A significant increase is observed 7–14 days post weaning (Trevisi et al., 2021). This coincides with ghrelin expression in the stomach, which is age-dependent. Ghrelin mRNA levels increase steadily from birth until 7–10 days after weaning when a significant surge in ghrelin is observed (Du et al., 2007). *In vitro*, ghrelin significantly amplifies mRNA expression and activity of H+/K+-ATPase, suggesting ghrelin's potential to stimulate gastric acid secretion through an alternative mechanism to vagal and enteric neuron stimulation (Du et al., 2007).

The initial fasting or low feed intake associated with the post-weaning phase results in a reduction of gastrointestinal hormones. Ghrelin, CCK, somatostatin, and gastrin are notably depressed in the immediate post-weaning period (Marion et al., 2003). The stomach retention time increases in the initial days post weaning, likely due to the dietary shift from liquid to solid food and the stress of weaning. Two weeks post weaning, the retention time considerably decreases compared to the first 3 days post weaning. By 3 weeks post weaning, stomach emptying becomes rapid (De Snoeck et al., 2011). The dietary change associated with weaning dramatically alters the gastrointestinal bacterial composition. The total bacterial content of the stomach increases, while the composition shifts. Before weaning, the non-glandular region is colonized by large numbers of Lactobacillus spp. However, after weaning, there is a gradual decrease in these numbers (Guevarra et al., 2019).

3.5 Postweaning small intestine

The villi and crypts lining the epithelium of the small intestine are crucial for digestion and absorption processes. Their structure and function after weaning are influenced by various factors (Pluske et al., 1997; O'Doherty et al., 2021). Diet composition has only marginal effects on small intestinal morphology in weaned pigs, with feed intake level being the most significant determinant of mucosal function and integrity (Vente-Spreeuwenberg et al., 2003). Low feed intake in piglets leads to reduced villous height and increased crypt depth due to a lack of luminal stimulation (Pluske et al., 1997). Villous height minimizes after 2–5 days post weaning, associated with cell loss and reduced absorption (Hedemann et al., 2003). Intestinal villous height begins recovery approximately 4 days after feeding is restarted, taking more than 10 days to fully recover (Boudry et al., 2004). Crypt hyperplasia, indicative of increased cell production, occurs in the immediate post-weaning period (Heo et al., 2013).

Post weaning, the activity of many brush-border and pancreatic digestive enzymes, such as sucrase, lactase, aminopeptidase A and N, and lipase, is reduced (Lallès et al., 2004). Maintaining GIT integrity requires a continuous supply of nutrients. However, due to reduced feed intake post weaning, this supply diminishes. The change in diet composition, with sow's milk being highly digestible and the main energy sources being fat and lactose, to a post-weaning diet that is solid, has higher dry matter content and primarily relies on starch as the main energy source, reduces digestibility (Kim et al., 2012). Insufficient enzyme production in the piglet, adapted to digest nutrients from sow's milk, contributes to this reduced digestibility (Heo et al., 2013). Adequate production of lipase, amylase, and other necessary enzymes needed for post-weaning diet digestion is not achieved until approximately 3--4 weeks post weaning (Torrallardona, 2010). The reduced enzyme production affects brush border enzymes' activity, linked to villous atrophy and crypt hyperplasia, and the immature GIT's inability to produce sufficient enzymes for efficient digestion of unfamiliar nutrients in the post-weaning diet (Bauer et al., 2011). This decreased enzyme production results in reduced digestibility and affects the composition of pancreatic juice, rich in digestive enzymes, secreted into the proximal small intestine (Lallès et al., 2004).

Many studies have reported that weaning stress impairs intestinal barrier function, increases intestinal permeability, and triggers an inflammatory response in the GIT (Boudry et al., 2004; Bauer et al., 2011; Wijtten et al., 2011). Intestinal barrier function relies on the mucosal layer and tight junctions. The mucosal layer, the primary barrier between the internal and external environment, protects the epithelial layer against microbes and antigens (Farhadi et al., 2003; Kim et al., 2012). Tight junctions, formed by transmembrane protein complexes like claudins and occludins and cytosolic proteins, create a physical barrier (Hu et al., 2013). The cascade of events affecting intestinal barrier function involves the mucosal layer preventing pathogenic adhesion (Kaper et al., 2004; Smith et al., 2010). The intestinal epithelial cells excrete defensive compounds into the mucosal layer, including mucins, defensins, antibodies, and more (Linden et al., 2008). However, if pathogenic bacteria breach the mucous layer, they can penetrate the tight junctions, releasing toxins and triggering inflammation and post-weaning diarrhoea (Pluske et al., 2002). Increased production of pro-inflammatory cytokines such as TNF-α, IFN-γ, and IL-6 is associated with increased intestinal permeability, while anti-inflammatory cytokines like IL-10 decrease intestinal permeability (Hu et al., 2013). The increased cell turnover in the intestine results in fewer mature enterocytes, where tight junctions are not fully developed, leading to an increase in intestinal permeability (Wijtten et al., 2011). Therefore, it is crucial to improve small intestinal morphology by enhancing feed intake in the immediate post-weaning period to reduce intestinal permeability, pathogenic susceptibility, and intestinal inflammation and this will be discussed later in the chapter.

3.6 Weaning-induced disruptions in large intestinal microbiota and health in young pigs

The weaning phase, typically occurring between 3 and 4 weeks of age in commercial pig production, introduces a turbulent period for young pigs. The stress of weaning can disrupt the balance of the GIT microbiota (Li et al., 2018). The microbiota, previously adapted to a milk-based diet, must quickly adapt to a diet centred around plant materials. This swift change, coupled with the inherent stress of weaning, creates opportunities for pathogenic bacteria to establish themselves, potentially leading to conditions like diarrhoea, which in severe cases can result in pig mortality (Luppi et al., 2016). Interestingly, certain bacterial genera found in suckling pigs, including Lactobacillaceae, Ruminococcaceae, and Prevotellaceae, have been associated with a reduced incidence of post-weaning diarrhoea (Dou et al., 2017). This discovery highlights the potential benefits of enhancing the presence of these beneficial bacteria to mitigate post-weaning health challenges. Considering the significant impact of the sow's microbiota on piglets (Chen et al., 2018), strategies aimed at positively influencing the sow's microbiota can have far-reaching effects, benefiting not only the GIT health of piglets but also the overall health and performance of the sow (Dowley et al., 2022; Leonard et al., 2011; Heim et al., 2014b; Rattigan et al., 2023; Kiernan et al., 2023b).

Furthermore, the transition phase following weaning often results in shifts in the composition of the GIT microbiota. These changes can be attributed to weaning stress, alterations in diet, and fluctuations in feed intake. Research has shed light on these shifts, revealing a temporary decrease in beneficial bacteria, notably Lactobacillus spp., during the immediate post-weaning period (Guevarra et al., 2018; Konstantinov et al., 2006; Pieper et al., 2008). Additionally, the dietary shift from milk to plant-derived polysaccharides brings about compositional changes in the faecal microbiota, including a decrease in Bacteroides spp. and an increase in Prevotella spp. (Frese et al., 2015). Such dysbiosis can render pigs more susceptible to conditions like post-weaning diarrhoea (Gresse et al., 2017). It is evident that nutritional strategies hold the key to modulating these microbial dynamics. However, precise dietary interventions in weaned pigs can be challenging, as they are intricately linked to external variables, including environmental stressors (Lallés et al., 2007; O' Doherty et al., 2021).

4 Dietary modulation for enhanced stomach development and functionality in young pigs

The significance of optimizing the development and functionality of the porcine stomach, especially in post-weaned piglets, is often underestimated. The stomach plays a pivotal role in initiating digestion through enzymatic secretions and acts as an initial barrier against ingested pathogens owing to its acidic environment. Efforts are being directed towards enhancing stomach function, particularly in post-weaned pigs, with the ultimate goal of reducing stomach pH. This reduction mitigates pathogen survival and ensures an ideal pH environment for enzyme activation, thereby facilitating superior nutrient utilization from the ingested feed.

One established strategy for achieving this is the incorporation of weak organic acids into piglet diets, which effectively lowers dietary pH (Canibe et al., 2005; Suiryanrayna and Ramana, 2015; Wei et al., 2021; Nowak et al., 2021). Research has also indicated that providing piglets access to creep feed during the suckling phase enhances the maximum acid output of their stomachs, highlighting the potential of dietary interventions to augment the acid secretion capacity of pre-weaned piglets (Cranwell, 1985).

Additionally, the role of the calcium-sensing receptor (CaSR) as a critical regulator of acid secretion in response to luminal pH is explored. The CaSR can also be activated by select dietary components such as calcium, magnesium, and specific amino acids, including tryptophan and phenylalanine (Busque et al., 2005). Tryptophan emerges as a key player, with research indicating that it leads to increased CaSR expression in pig stomach tissue samples, subsequently stimulating gastrin, somatostatin, and H+, K+-ATPase activity. Tryptophan also influences ghrelin expression, resulting in increased feed intake when orally infused. Augmenting the inclusion rate of tryptophan in pig diets has been associated with amplified feed intake and greater effects on plasma ghrelin expression than food volume. Maternal supplementation of sow diets with tryptophan during gestation and lactation positively impacts milk yield, average daily feed intake, average daily gain, milk calcium, and 5-HT concentration, raising questions about the potential to increase Ca^{2+} levels in milk and stimulate CaSR in the offspring's stomach (Zhang et al., 2007; Miao et al., 2019).

A novel research avenue is proposed, here involving maternal dietary intervention to enhance acid production in the pre-weaned pig stomach. Given the guaranteed milk consumption by piglets before weaning, this approach presents advantages over supplementing via creep feed in the farrowing house. Furthermore, the impact of creep feed with varying acid-binding capacities on parietal cell development is also suggested as a research endeavour.

This chapter underscores the critical role of increasing feed intake during the post-weaning period for optimal gastrointestinal health in pigs. Creep feeding, which introduces highly digestible and palatable solid feed to piglets in the farrowing house, is emphasized as a strategic preparation for the transition to solid post-weaning diets. A strong correlation exists between pre-weaning creep feed intake and subsequent post-weaning feed intake and growth.

By focusing on stomach function optimization, these strategies hold promise for improving the health and performance of post-weaned pigs, ultimately benefiting the swine industry as a whole. Continued research and implementation of these approaches are essential to further advance the understanding of piglet stomach development and its implications for pig production. Table 1 presents a comprehensive list of compounds, including amino acids, chemicals, and other bioactive elements, aimed at enhancing the functionality of the pig's stomach. These compounds play a crucial role in promoting optimal stomach health, aiding in digestion, and ensuring the overall well-being of the pig.

4.1 Dietary strategies for enhancing growth and health in newly weaned piglets

Newly weaned piglets face nutritional challenges during the transition from sow's milk to solid feed. This section explores various dietary strategies that

Table 1 Compounds, including amino acids, organic acids, and other bioactive elements, aimed at enhancing the functionality of the pig's stomach

Compound	Pig type	Effect on stomach	Other effects	Reference
Calcium formate	Post-weaned pig	Reduction in parietal cell number and expression of *ATP4A* gene and an increase in *SST* cells in the fundic gland region. Reduced stomach weight.	Not recorded.	(Bosi et al., 2006)
Cysteamine	Suckling pig and *in vitro*	*In vitro*: increased ghrelin mRNA expression. Suckling pig: increased ghrelin and gastrin and decreased somatostatin mRNA expression.	Increased ADG creep feed consumption and bodyweight gain. Decreased diarrhoea rates.	(Du et al., 2012)
Formic acid	Post-weaned pig	No effect on number of parietal cells. No effect on stomach weight.	Increased ADG, tended to increase ADFI and tended to decrease G:F. Formic acid inclusion had only minor effect on jejunum microbiota. High formic acid inclusion increased the microbiota diversity. Formic acid supplementation decreased the abundance of *Gemella*, *Lactobacillus,* and *Parvimimoas*.	(Luise et al., 2017)
Potassium diformate	Post-weaned pig	Reduced stomach weight, HCl concentration, and pepsin activity. Increased lactic concentration, somatostatin, and activity of H^+/K^+-ATPase but did not change gastrin activity.	Improved bodyweight after 35 days.	(Xia et al., 2015)
Sodium butyrate	Suckling/ post-weaned pig	Increased the number of parietal cell per gland in the fundic gland region. Increased number of SST-positive cells and tended to increase gastrin when supplemented post weaning but not during suckling phase in the fundic gland region. Increased gastric mucosa thickness when fed post weaning. No difference in gastric pH amongst treatments.	Supplementation before weaning improved final body weight D12 PW. Improved daily gain when supplemented before or after weaning or in combination.	(Mazzoni et al., 2008)

Thymol (introduced directly into the stomach)	Post-weaned/grower pig	Stimulated gastric proliferation activity and altered digestive activity. SST, SSTRs, peptide transporter 1, and calpain 9 most strongly upregulated genes.	Not recorded.	(Colombo et al., 2014a)
Tryptophan (at 0.12%, 0.19%, 0.26% inclusion)	Post-weaned pig	0.26% and 0.19% increased plasma ghrelin and expression of ghrelin mRNA in fundic gland region compared to 0.12%.	Improved weight gain, food intake, and feed conversion. Plasma lectin decreased in pigs fed 0.19% versus 0.12%.	(Zhang et al., 2007a)
Tryptophan or phenylalanine	*In vitro* perfusion using finisher pig stomachs	Increased gastrin, SST, and H^+/K^+-ATPase activity and increased CaSR expression. Effect was dependent on Ca^{2+}.	Not recorded.	(Xian et al., 2018)
ZnO	Post-weaned pig/*in vitro*	*In vitro*: increased ghrelin production from gastric mucosal cells. Post-weaned pig: increased plasma ghrelin, IFG-1, and CCK.	Increased food intake and growth.	(Yin et al., 2009)

ZnO, zinc oxide; mRNA, messenger RNA HCl, hydrochloric acid; SST, somatostatin; SSTR, somatostatin receptor; CaSR, calcium-sensing receptor; IFG-, insulin-like growth factor 1; CCK, cholecystokinin; ADG, average daily gain; ADFI, average daily feed intake; G:F, gain to feed; PW, post weaning.

influence feed intake, growth performance, and gastrointestinal health in these young animals. We will discuss the benefits of dairy products, the role of fermentable carbohydrates, the impact of leguminous plant proteins, the use of exogenous enzymes, and the challenges posed by high protein content in their diets.

4.1.1 Benefits of dairy products

Dairy products are known for their palatability and ease of digestibility. Lactose, a sugar present in milk, plays a pivotal role in the diet of young weaned piglets. At weaning, the pig's capacity to digest lactose diminishes (Pluske et al., 2003). Consequently, a portion of the lactose consumed by pigs reaches their large intestine, where it undergoes fermentation by microbes. This fermentation process yields lactic acid and SCFAs, which confer benefits to the pig's gastrointestinal health and metabolism (Pierce et al., 2006; Liu et al., 2018).

The concept of employing lactose as a prebiotic in pig nutrition is gaining momentum. This is particularly significant as the industry endeavours to reduce antibiotic usage and lower levels of zinc oxide in pig diets. Recent studies demonstrate that lactose can stimulate the growth of beneficial gut bacteria, even in individuals with lactose digestion challenges (MacFarlane et al., 2008; Pierce et al., 2006). Some studies suggest that these benefits extend beyond altering the gut bacteria composition and encompass the augmentation of various gut bacterial metabolites, including butyrate (Pierce et al., 2006). Butyrate is of particular importance as it serves as an energy source for colonic epithelial cells and possesses anti-inflammatory properties, among other advantages (Venema, 2012). Furthermore, the acidic environment created by this fermentation process in the gut modulates the gut bacteria composition by lowering the pH, thereby inhibiting the proliferation of harmful bacteria (Fukuda et al., 2011; Pierce et al., 2006).

However, determining the precise dietary lactose requirements for piglets is challenging. Although piglets initially receive a diet of maternal milk and then transition to lower lactose feed post weaning, excessive lactose intake can lead to issues such as diarrhoea.

In summary, lactose stands as an indispensable component of the diet for newly weaned pigs. Its most pronounced impact on their health and growth manifests during the first 2 weeks after weaning when their inherent lactose-digesting capacity is at its peak. Beyond the 14-day mark, variations in lactose levels might not yield as noticeable effects on growth and health. Zhao et al. (2021) propose specific lactose levels for different phases of piglet growth: 200 g/kg for the initial 7 days, 150 g/kg for the subsequent 7 days, and none after the first 14 days. It is crucial to acknowledge that these recommended levels can be challenging to implement in practice due to cost considerations. Consequently, many pig diets do not adhere to these lactose recommendations, opting for alternative ingredients, additives, and nutrients to substitute for lactose (Zhao et al., 2021).

4.1.2 Fermentable carbohydrates

Incorporating fermentable carbohydrates into the diet, including sugar beet pulp, non-digestible oligosaccharides (NDO), and resistant starch, has shown potential in enhancing gut health in piglets. Such carbohydrates augment the retention time of digesta within the GIT, thereby enhancing enzymatic hydrolysis of nutrients. An appropriate balance of insoluble and soluble dietary fibre can influence the microbial profile, reducing Enterobacteriaceae and thus contributing to gastrointestinal well-being (Trachsel et al., 2019; Lallès et al., 2007a; Molist et al., 2014).

Beta-glucans, polysaccharides prevalent in substrates like yeast, algae, and specific cereals, possess anti-inflammatory and antioxidant properties, contributing to various physiological benefits (Luo et al., 2019; O'Doherty et al., 2021). The efficacy of these beta-glucans as bioactive compounds varies depending on parameters such as solubility, molecular weight, and structural characteristics (Venardou et al., 2023).

Non-digestible oligosaccharides are oligosaccharides with 2–20 monomers and characterized by β-links between monosaccharide units, serve as substrates for beneficial microorganisms such as Bifidobacteria and Lactobacilli. This fosters a conducive microbial environment and promotes associated health advantages (Xing et al., 2020). Additionally, NDOs have mechanisms that mitigate the colonization of pathogenic bacteria and modulate the host's immune responses (Capitán-Cañadas et al., 2014). Certain NDO types have anti-adhesive properties, disrupting pathogen binding to the intestinal epithelial surface (Newburg et al., 2005).

Resistant starch, a form of starch resistant to enzymatic digestion in the stomach and small intestine, becomes a substrate for bacterial fermentation in the large intestine (Tan et al., 2021). Notably, RS2, one of the four types of resistant starch, has shown efficacy in promoting the proliferation of faecal Lactobacilli and Bifidobacteria in pigs, although its optimal inclusion rate is contingent upon variables such as the pig's physiological weight (Metzler-Zebeli et al., 2019).

4.1.3 Exogenous enzyme supplementation

The post-weaning phase in pig farming is crucial, historically addressed using in-feed antibiotics, pharmacological doses of zinc oxide, and copper in starter diets. Yet, with the growing concerns about antimicrobial resistance and environmental implications, the industry is transitioning towards alternative strategies.

One viable approach is the integration of exogenous enzymes into piglet diets. Post weaning, piglets face a myriad of non-milk-based dietary components and lack endogenous enzymes for optimal digestion. By introducing carbohydrases such as xylanase, β-glucanase, β-mannanase, and α-galactosidase, the digestibility of substrates within the non-starch polysaccharide (NSP) fraction of the diet is improved, addressing the challenges posed by plant-based diets rich in NSPs (Masey O'Neill et al., 2014; Högberg and Lindberg, 2006).

While digestion in the GIT primarily depends on endogenous enzymes from various secretions, weaning results in reduced secretion of essential enzymes. Some, like lipase and chymotrypsin, remain below optimal levels even 2 weeks post weaning (Lindemann et al., 1986). Supplementing with exogenous enzymes can help mitigate these disruptions, enhancing the digestion of post-weaning dietary components.

While numerous studies support the benefits of enzyme supplementation, including phytase, carbohydrases, and proteases, in elevating digestibility and promoting growth (Yáñez et al., 2013; Nortey et al., 2015; Torres-Pitarch et al., 2017), results regarding consistent gut maturation and overall health in weaned piglets remain variable (Kim et al., 2004; Zijlstra et al., 2004; L'Anson et al., 2013; Torres-Pitarch et al., 2017).

4.1.4 Challenges of high protein content

The early post-weaning phase in pig farming is characterized by the need for specific dietary adjustments to ensure optimal growth and gut health. A promising method to address this is to modify the crude protein (CP) levels in the diet. The significance of such a change hinges on the fact that undigested proteins can eventually find their way to the large intestine. Here, they become substrates for proteolytic bacteria, which can produce harmful metabolites such as ammonia, amines, indoles, and phenols. These metabolites have been directly linked to post-weaning diarrhoea in previous studies (de Lange et al., 2010; O'Doherty et al., 2021).

The National Research Council (NRC) suggests that the diet of weaned pigs should have CP concentrations ranging between 20% and 23% (NRC, 2012). However, this recommendation may not account for the limited enzymatic activity in piglets aged 3–4 weeks, resulting in the inability to fully digest and assimilate such protein levels. As a consequence, undigested proteins make their way into the large intestine, creating a conducive environment for the proliferation of pathogenic bacteria (O'Doherty et al., 2017; Rattigan et al., 2020).

Evidence from various studies emphasises the benefits of reducing CP levels. Such adjustments have been observed to decrease the instances of diarrhoea in controlled settings (Wellock et al., 2007; Kim et al., 2011; Wu et al., 2015; Rattigan et al., 2020). Nevertheless, it's essential to note that the benefits in terms of gut health must be balanced against potential impacts on growth performance. For instance, diets with a modest reduction to 18.9% CP, when bolstered with crystalline amino acids (AA), do not seem to affect growth negatively. However, a more drastic reduction to 17% CP may lead to suboptimal growth outcomes (Opapeju et al., 2008; Yue and Qiao, 2008). In conclusion, adjusting the CP levels post-weaning appears to be a viable approach to reducing post-weaning diarrhoea in piglets. However, it is crucial to strike a balance, ensuring the overall health and performance of the animal are not compromised.

4.1.5 Negative effects of leguminous plant proteins

The inclusion of leguminous plant proteins, such as soya bean meal and peas, can have adverse effects on post-weaning piglet growth and health. These proteins may trigger delayed hypersensitivity reactions and antigenic responses, leading to morphological alterations like villous atrophy and crypt hyperplasia, resulting in diminished growth rates. Nonetheless, high dietary lactose levels can mitigate some of these detrimental effects, enabling increased incorporation of leguminous plant proteins without compromising performance or health (Pierce et al., 2004).

Optimizing the diet of newly weaned piglets is pivotal for their growth and health. A profound understanding of these dietary strategies and their effects can facilitate the development of nutritionally balanced diets that support a successful transition from sow's milk to solid feed.

5 Digestion and absorption of nutrients

The transport of nutrients in the small intestine is a complex process involving various transporters and pathways. Monosaccharides are transported by sodium-glucose-linked transporters (SGLT) and glucose transporters (GLUT) (Wood and Trayhurn, 2003). Small peptides (di- and tripeptides) are transported by peptide transporters (PEPT) (Daniel, 2004), and unesterified long-chain fatty acids are transported by cluster of differentiation 36 (CD36) (Hajri and Abumrad, 2002) and fatty acid-binding proteins (FABP) (Glatz and Van der Vusse, 1996). Glucose and galactose are transported from the lumen of the intestine into enterocytes by SGLT1, whereas fructose is transported by GLUT5. Glucose, galactose, and fructose, once accumulated in the cells, are released from the cytosol across the basolateral membrane into the systemic system by GLUT2 (Helliwell et al., 2000; Kellett, 2001). The protein levels of SGLT1 and GLUT2 are crucial to the absorption and uptake of glucose in the small intestine, and their protein levels are upregulated by the expression of the SGLT1 and GLUT2 genes (Rodriguez et al., 2004).

The cellular transport of small peptides from the intestinal lumen into enterocytes is primarily mediated by PEPT1, which is located on the brush border membrane of intestinal enterocytes (Daniel, 2004). The intracellular transported small peptides are then rapidly hydrolyzed to free amino acids by intracellular peptidases. The intracellular free amino acids are then transported via amino acid transport systems across the basolateral membrane into the bloodstream (Ganapathy et al., 2001; Fan et al., 2004).

The uptake of unesterified long-chain fatty acids occurs through saturable, carrier-mediated modes. The most prominent and best-characterized carriers are CD36 (Hajri and Abumrad, 2002) and FABP2 (Glatz and van der Vusse, 1996). Free fatty acids and monoglycerides recombine into triacylglycerol within the enterocytes. Then, they enter the lymphatic fluid as droplets called chylomicrons, which are then taken up by the lacteals in the intestine (Hajri and Abumrad, 2002).

5.1 Manipulating intestinal nutrient transporters through nutritional interventions

The foetal small intestine of many mammals is known to actively transport glucose (Ferraris and Diamond, 1997). In pigs, intestinal transport rates of glucose, galactose, and fructose (per mg intestine) are highest at birth, with a steep decline after the onset of suckling (Puchal and Buddington, 1992). This transient decline is most probably connected to the dilution of mature enterocytes in the small intestine due to the rapid cell proliferation in the crypt triggered by suckling of colostrum or milk (Freeman, 1995).

The ontogenetic regulation of intestinal PEPT1 gene expression might be induced post-partum by suckling and later weaning, in an adaptive response to changes in the diet, from high-protein milk to an adult diet containing more carbohydrates than protein. Peptide transport is present in late gestation and dramatically increases to peak around birth, followed by a general decline to the end of weaning and then an increase to an intermediate level at adulthood (Ashida et al., 2002). Food deprivation observed after weaning causes a decreased concentration of nutrients, including amino acids and small peptides, in the gut lumen, and it generally increases the functional activity of intestinal peptide and amino acid transporters primarily through the increased gene expression of these transporters (Howard et al., 2004).

Digestibility is a factor that expresses the potential of an ingredient to be digested, absorbed, and processed by the animal's digestive system, whether in terms of nutrients or energy. Nutrient digestibility can be increased by improving the villus height in the small intestinal epithelium and by increasing the number of nutrient transporters (Rodriguez et al., 2004). Specific feed ingredients like laminarin (beta-glucan) have been shown to increase nutrient transporters (GLUT1, GLUT2, and SGLT1) in the small intestine of weaned pigs (Heim et al., 2014). The upregulation of these glucose transporters with laminarin supplementation may be attributed to its chemical structure (β-(1→3)-linked glucans with β-(1→6)-linked side chains) and/or molecular weight (approximately 5 kDa). This increase in nutrient transporters corresponds with an increase in the ileal digestibility of gross energy (Heim et al., 2014). Similar increases in nutrient transporters have been demonstrated with the use of the exogenous enzyme phytase (Vigors et al., 2014).

Digestion of feed is also thought to be an important mechanism behind the differences in residual feed intake (RFI) and feed efficiency in pigs (Herd et al., 2004). In combination with increased total tract digestibility, low RFI pigs (more feed efficient) had increased ileal digestibility of gross energy, suggesting increased nutrient uptake in the small intestine (Vigors et al., 2016). The more feed-efficient pigs had increased gene expression of the enzymes sucrase-isomaltase and the nutrient transporters SGLT1, GLUT2, and FABP2 in the small intestine. These results suggest an increased nutrient breakdown and subsequent transport in more efficient pigs. The higher gene expression of these nutrient transporters implies an increased expression of the corresponding protein that would increase the transport of glucose and fatty acids across the intestinal wall. The negative correlations between RFI and nutrient transporters could be important in the future selection of pigs using genomic selection.

6 Conclusion

The GIT of pigs undergoes substantial developmental changes from prenatal to post-weaning stages, significantly impacting health and overall performance. Prenatally, GIT development is primed for birth, with distinct enzymatic adaptations and a dominant presence of lactase. The postnatal period sees GIT maturation facilitated by colostral growth factors, impacting both stomach cellular composition and small intestinal morphology. Post-weaning represents a pivotal phase in a piglet's development, characterized by profound shifts in gastric pH regulation, a decline in parietal cell abundance, and changes in the intestinal microbiota, notably the reduction of beneficial lactobacilli. It becomes paramount to utilize targeted dietary interventions during this window. Such strategies not only counteract the physiological challenges but also optimize GIT function and stabilize the microbiome, fostering overall swine health. Further insight into small intestine nutrient transport has elucidated the roles of specific transporters, including SGLT1 and PEPT1, and emphasized the potential modulatory influence of dietary constituents.

To conclude, the chapter emphasizes the pig's GIT development's pivotal role and underscores the necessity of strategic dietary interventions, particularly post weaning, in advancing swine health and promoting sustainable production practices.

7 Where to look for further information

The following articles and reviews formulate a further narrative on enhancing swine gastrointestinal health from birth through weaning. O'Doherty et al. (2024) explore how glucans, vitamin D, and selenium supplements can fortify piglets during weaning. Kiernan et al. (2023a) assess the effects of probiotics or synbiotics in the maternal diet on sow and piglet health. Dowley et al. (2022) demonstrate the advantages of specific maternal supplements for piglet gut health. Kiernan et al. (2023b) investigate stomach gene expression in post-weaned piglets, revealing key physiological changes. Together, these articles highlight the significance of targeted nutritional strategies in the pivotal early stages of swine development.

O'Doherty, J., Dowley, A., Conway,E. and Sweeney, T. (2024). Nutritional strategies to mitigate post-weaning challenges in pigs: A focus on glucans, vitamin D, and selenium. *Animals*, 14(1), 13. https://doi.org/10.3390/ani14010013.

Dowley, A., O'Doherty, J. V., Mukhopadhya, A., Conway, E., Vigors, S., Maher, S., ... & Sweeney, T. (2022). Maternal Supplementation With a Casein Hydrolysate and Yeast Beta-glucan From Late Gestation Through Lactation Improves Gastrointestinal Health of Piglets at Weaning. Science Reports, 12, 17407.

Kiernan, D. P., O'Doherty, J. V. and Sweeney, T. (2023). The effect of maternal probiotic or synbiotic supplementation on sow and offspring gastrointestinal microbiota, health, and performance. *Animals*, 13, 2996. https://doi.org/10.3390/ani13192996.

Kiernan, D. P., O'Doherty, J. V.,Connolly, K. R., Ryan, M. and Sweeney, T. (2023). Exploring the differential expression of a set of key genes involved in the regulation and functioning of the stomach in the post-weaned pig. *Veterinary Sciences*, 10(7), 473. https://doi.org/10.3390/vetsci10070473

8 References

Ashida, K., Katsura, T., Motohashi, H., Saito, H. and Inui, K. (2002). Thyroid hormone regulates the activity and expression of the peptide transporter PepT1 in Caco-2 cells. *American Journal of Physiology - Gastrointestinal and Liver Physiology* 282(4), G617–G623.

Bauer, E., Metzler-Zebeli, B. U., Verstegen, M. W. and Mosenthin, R. (2011). Intestinal gene expression in pigs: Effects of reduced feed intake during weaning and potential impact of dietary components. *Nutrition Research Reviews* 24(2), 155–175.

Boudry, G., Péron, V., Le Huërou-Luron, I., Lallès, J. P. and Sève, B. (2004). Weaning induces both transient and long-lasting modifications of absorptive, secretory, and Barrier properties of piglet intestine. *The Journal of Nutrition* 134(9), 2256–2262.

Bosi, P., Mazzoni, M., De Filippi, S., Trevisi, P., Casini, L., Petrosino, G. and Lalatta-Costerbosa, G. (2006). A continuous dietary supply of free calcium formate negatively affects the parietal cell population and gastric RNA expression for H+/K+-ATPase in weaning pigs. *The Journal of Nutrition* 136(5), 1229–1235.

Brandtzaeg, P. E. (2002). Current understanding of gastrointestinal immunoregulation and its relation to food allergy. *Annals of the New York Academy of Sciences* 964, 13–45.

Busque, S. M., Kerstetter, J. E., Geibel, J. P. and Insogna, K. (2005). L-type amino acids stimulate gastric acid secretion by activation of the calcium-sensing receptor in parietal cells. *American Journal of Physiology-Gastrointestinal and Liver Physiology* 289(4), G664–G669.

Canani, R. B., Costanzo, M. D., Leone, L., Pedata, M., Meli, R. and Calignano, A. (2011). Potential beneficial effects of butyrate in intestinal and extraintestinal diseases. *World Journal of Gastroenterology* 17(12), 1519–1528.

Cao, M., Che, L., Wang, J., Yang, M., Su, G., Fang, Z., Lin, Y., Xu, S. and Wu, D. (2014). Effects of maternal over- and undernutrition on intestinal morphology, enzyme activity, and gene expression of nutrient transporters in newborn and weaned pigs. *Nutrition* 30(11–12), 1442–1447.

Canibe, N., Højberg, O., Højsgaard, S. and Jensen, B. B. (2005). Feed physical form and formic acid addition to the feed affect the gastrointestinal ecology and growth performance of growing pigs. *Journal of Animal Science* 83(6), 1287–1302.

Capitán-Cañadas, F., Ortega-González, M., Guadix, E., Zarzuelo, A., Suárez, M. D., de Medina, F. S. and Martínez-Augustin, O. (2014). Prebiotic oligosaccharides directly modulate proinflammatory cytokine production in monocytes via activation of TLR4. *Molecular Nutrition and Food Research* 58(5), 1098–1110.

Chen, X., Xu, J., Ren, E., Su, Y. and Zhu, W. (2018). Co-occurrence of early gut colonization in neonatal piglets with microbiota in the maternal and surrounding delivery environments. *Anaerobe* 49, 30–40.

Colombo, M., Priori, D., Gandolfi, G., Boatto, G., Nieddu, M., Bosi, P. and Trevisi, P. (2014). Effect of free thymol on differential gene expression in gastric mucosa of the young pig. *Animal* 8(5), 786–791.

Cranwell, P. D. (1985). The development of acid and pepsin (EC 3.4.23.1) secretory capacity in the pig; the effects of age and weaning. 1. Studies in anaesthetized pigs. *British Journal of Nutrition* 54(1), 305–320.

Daniel, H. (2004). Molecular and integrative physiology of intestinal peptide transport. *Annual Review of Physiology* 66, 361–384.

de Lange, C. F. M., Pluske, J., Gong, J. and Nyachoti, C. M. (2010). Strategic use of feed ingredients and feed additives to stimulate gut health and development in young pigs. *Livestock Science* 134(1–3), 124–134.

De Snoeck, S., van der Wolf, P. J., Swart, W., Heiiman, E. and Ebbinge, B. (2011). The effect of the application of mono-lauric acid with glycerol mono-laurate in weaned piglets, on the use of antimicrobials in sow herds[MS1] . In *Proceedings of the 9th*

International Conference on Epidemiology and Control of Foodborne Pathogens and Antimicrobial Resistance in Pigs and Pork, pp. 346–348.

Demecková, V., Kelly, D., Coutts, A. G. P., Brooks, P. H. and Campbell, A. (2002). The effect of fermented liquid feeding on the faecal microbiology and colostrum quality of farrowing sows. *International Journal of Food Microbiology* 79(1-2), 85–97.

Dou, S., Gadonna-Widehem, P., Rome, V., Hamoudi, D., Rhazi, L., Lakhal, L., Larcher, T., Bahi-Jaber, N., Pinon-Quintana, A., Guyonvarch, A., Huërou-Luron, I. L. and Abdennebi-Najar, L. (2017). Characterisation of early-life fecal microbiota in susceptible and healthy pigs to post-weaning diarrhea. *PLoS ONE* 12(1), e0169851.

Dowley, A., O'Doherty, J. V., Mukhopadhya, A., Conway, E., Vigors, S., Maher, S., Ryan, M. T. and Sweeney, T. (2022). Maternal supplementation with a casein hydrolysate and yeast beta-glucan from late gestation through lactation improves gastrointestinal health of piglets at weaning. *Scientific Reports* 12(1), 17407.

Du, G. M., Shi, Z. M., Wei, X. H., Liu, M. J., Zhang, L. and Zhao, R. Q. (2007). Expression of gastric ghrelin and H+-K+-Atpase mrna in weanling piglets and effect of ghrelin on H+-K+-Atpase expression and activity in gastric mucosal cells in vitro. *Research in Veterinary Science* 82(1), 99–104.

Du, G., Shi, Z., Xia, D., Wei, X., Zhang, L., Parvizi, N. and Zhao, R. (2012). Cysteamine improves growth performance and gastric ghrelin expression in preweaning piglets. *Domestic Animal Endocrinology* 42(4), 203–209.

Elefson, S. K., Lu, N., Chevalier, T., Dierking, S., Wang, D., Monegue, H. J., Matthews, J. C., Jang, Y. D., Chen, J., Rentfrow, G. K., Adedokun, S. A. and Lindemann, M. D. (2021). Assessment of visceral organ growth in pigs from birth through 150 kg. *Journal of Animal Science* 99(9), skab249.

Fan, M. Z., Matthews, J. C., Etienne, N. M. P., Stoll, B., Lackeyram, D. and Burrin, D. G. (2004). Expression of apical membrane l-glutamate transporters in neonatal porcine epithelial cells along the small intestinal crypt-villus axis. *American Journal of Physiology-Gastrointestinal and Liver Physiology* 287(2), G385–G398.

Farhadi, A., Banan, A., Fields, J. and Keshavarzian, A. (2003). Intestinal Barrier: An interface between health and disease. *Journal of Gastroenterology and Hepatology* 18(5), 479–497.

Ferraris, R. P. and Diamond, J. (1997). Regulation of intestinal sugar transport. *Physiological Reviews* 77(1), 257–302.

Freeman, T. C. (1995). Parallel patterns of cell-specific gene expression during enterocyte differentiation and maturation in the small intestine of the rabbit. *Differentiation* 59(3), 179–192.

Frese, S. A., Parker, K., Calvert, C. C. and Mills, D. A. (2015). Diet shapes the gut microbiome of pigs during nursing and weaning. *Microbiome* 3, 28.

Fukuda, S., Toh, H., Hase, K., Oshima, K., Nakanishi, Y., Yoshimura, K., Tobe, T., Clarke, J. M., Topping, D. L., Suzuki, T., Taylor, T. D., Itoh, K., Kikuchi, J., Morita, H., Hattori, M. and Ohno, H. (2011). Bifidobacteria can protect from enteropathogenic infection through production of acetate. *Nature* 469(7331), 543–547.

Ganapathy, V., Ganapathy, M. E. and Leibach, F. H. (2001). Intestinal transport of peptides and amino acids. In *Current Topics in Membranes* (Ed. K. E. D. M. Barrett), pp. 379-412. Academic Press, New York, NY.

Gao, X. and Song, J. (2018). Bacterial community shifts in the gastrointestinal tract of pigs fed diets with different additives. *Microorganisms* 6(4), 98.

Glatz, J. F. and Van Der Vusse, G. J. (1996). Cellular fatty acid-binding proteins: Their function and physiological significance. *Progress in Lipid Research* 35(3), 243-282.

Gresse, R., Chaucheyras-Durand, F., Fleury, M. A., Van de Wiele, T., Forano, E. and Blanquet-Diot, S. (2017). Gut microbiota dysbiosis in postweaning piglets: Understanding the keys to health. *Trends in Microbiology* 25(10), 851–873.

Guevarra, R. B., Hong, S. H., Cho, J. H., Kim, B. R., Shin, J., Lee, J. H., Kang, B. N., Kim, Y. H., Wattanaphansak, S., Isaacson, R. E., Song, M. and Kim, H. B. (2018). The dynamics of the piglet gut microbiome during the weaning transition in association with health and nutrition. *Journal of Animal Science and Biotechnology* 9, 54.

Guevarra, R. B., Lee, J. H., Lee, S. H., Seok, M. J., Kim, D. W., Kang, B. N., Johnson, T. J., Isaacson, R. E. and Kim, H. B. (2019). Piglet gut microbial shifts early in life: Causes and effects. *Journal of Animal Science and Biotechnology* 10(1), 1. https://doi.org/10.1186/s40104-018-0308-3.

Guilloteau, P., Martin, L., Eeckhaut, V., Ducatelle, R., Zabielski, R. and Van Immerseel, F. (2010). From the gut to the peripheral tissues: The multiple effects of butyrate. *Nutrition Research Reviews* 23(2), 366–384.

Hajri, T. and Abumrad, N. A. (2002). Fatty acid transport across membranes: Relevance to nutrition and metabolic pathology. *Annual Review of Nutrition*, 22, 383–415. https://doi.org/10.1146/annurev.nutr.22.020402.130846

Hedemann, M. S., Højsgaard, S. and Jensen, B. B. (2003). Small intestinal morphology and activity of intestinal peptidases in piglets around weaning. *Journal of Animal Physiology and Animal Nutrition*, 87(1), 32–41. https://doi.org/10.1046/j.1439-0396.2003.00405.x

Heim, G., Walsh, A. M., Sweeney, T., Doyle, D. N., O'Shea, C. J., Ryan, M. T. and O'Doherty, J. V. (2014a). Effect of seaweed-derived laminarin and fucoidan and zinc oxide on gut morphology, nutrient transporters, nutrient digestibility, growth performance and selected microbial populations in weaned pigs. *British Journal of Nutrition* 111(9), 1577–1585.

Heim, G., Sweeney, T., O'Shea, C. J., Doyle, D. N. and O'Doherty, J. V. (2014b). Effect of maternal supplementation with seaweed extracts on growth performance and aspects of gastrointestinal health of newly weaned piglets after challenge with enterotoxigenic Escherichia coli K88. *British Journal of Nutrition* 112(12), 1955–1965.

Helliwell, P. A., Richardson, M., Affleck, J. and Kellett, G. L. (2000). Regulation of Glut5, Glut2 and intestinal brush-border fructose absorption by the extracellular signal-regulated kinase, P38 mitogen-activated kinase and phosphatidylinositol 3-kinase intracellular signalling pathways: Implications for adaptation to diabetes. *Biochemical Journal* 350(1), 163–169.

Heo, J. M., Opapeju, F. O., Pluske, J. R., Kim, J. C., Hampson, D. J. and Nyachoti, C. M. (2013). Gastrointestinal health and function in weaned pigs: A review of feeding strategies to control post-weaning diarrhoea without using in-feed antimicrobial compounds. *Journal of Animal Physiology and Animal Nutrition* 97(2), 207–237.

Herd, R. M., Oddy, V. H. and Richardson, E. C. (2004). Biological basis for variation in residual feed intake in beef cattle. 1. Review of potential mechanisms. *Australian Journal of Experimental Agriculture* 44(5), 423–430.

Högberg, A. and Lindberg, J. E. (2006). The effect of level and type of cereal non-starch polysaccharides on the performance, nutrient utilization, and gut environment of pigs around weaning. *Animal Feed Science and Technology* 127(3–4), 200–219.

Howard, A., Goodlad, R. A., Walters, J. R. F., Ford, D. and Hirst, B. H. (2004). Increased expression of specific intestinal amino acid and peptide transporter mrna in rats fed by TPN is reversed by Glp-2. *Journal of Nutrition* 134(11), 2957-2964.

Hu, C. H., Xiao, K., Luan, Z. S. and Song, J. (2013). Early weaning increases intestinal permeability, alters expression of cytokine and tight junction proteins, and activates mitogen-activated protein kinases in pigs. *Journal of Animal Science* 91(3), 1094–1101.

Inoue, R., Tsukahara, T., Nakanishi, N. and Ushida, K. (2005). Development of the intestinal microbiota in the piglet. *The Journal of General and Applied Microbiology* 51(4), 257-265.

Isaacson, R., Kim, H. B. and Weisel, C. P. (2012). Antibiotics in feed and their role in animal agriculture. In Eds Gould, I. M. and van der Meer, J. W.M. *Antibiotic Policies: Fighting Resistance*, pp. 1-18. Springer, NewYork, USA.

Jeong, W., Song, G., Bazer, F. W. and Kim, J. (2014). Insulin-like growth factor I induces proliferation and migration of porcine trophectoderm cells through multiple cell signaling pathways, including protooncogenic protein kinase 1 and mitogen-activated protein kinase. *Molecular and Cellular Endocrinology* 384(1-2), 175-184.

Jiménez, E., Marín, M. L., Martín, R., Odriozola, J. M., Olivares, M., Xaus, J., Fernández, L. and Rodríguez, J. M. (2008). Is meconium from healthy newborns actually sterile? *Research in Microbiology* 159(3), 187-193.

Kaper, J. B., Nataro, J. P. and Mobley, H. L. (2004). Pathogenic Escherichia coli. *Nature Reviews. Microbiology* 2(2), 123-140.

Kellett, G. L. (2001). The facilitated component of intestinal glucose absorption. *Journal of Physiology* 531(3), 585-595.

Kiernan, D. P., O'Doherty, J. V., Connolly, K. R., Ryan, M. and Sweeney, T. (2023a). Exploring the differential expression of a set of key genes involved in the regulation and functioning of the stomach in the post-weaned pig. *Veterinary Sciences* 10(7), 473.

Kiernan, D. P., O'Doherty, J. V. and Sweeney, T. (2023b). The effect of maternal probiotic or synbiotic supplementation on sow and offspring gastrointestinal microbiota, health, and performance. *Animals* 13(19), 2996.

Kim, J. C., Hansen, C. F., Mullan, B. P. and Pluske, J. R. (2012). Nutrition and pathology of Weaner pigs: Nutritional strategies to support barrier function in the gastrointestinal tract. *Animal Feed Science and Technology* 173(1-2), 3-16. https://doi.org/10.1016/j.anifeedsci.2011.12.022.

Konstantinov, S. R., Awati, A. A., Williams, B. A., Miller, B. G., Jones, P., Stokes, C. R., Akkermans, A. D., Smidt, H. and de Vos, W. M. (2006). Post-natal development of the porcine microbiota composition and activities. *Environmental Microbiology* 8(7), 1191-1199.

Lallès, J. P., Boudry, G., Favier, C., Le Floc'h, N., Luron, I., Montagne, L., Oswald, I. P., Pié, S., Piel, C. and Sève, B. (2004). Gut function and dysfunction in young pigs: Physiology. *Animal Research* 53, 301-316.

Lallès, J. P., Bosi, P., Smidt, H. and Stokes, C. R. (2007). Nutritional management of gut health in pigs around weaning. *Proceedings of the Nutrition Society* 66(2), 260-268.

L'Anson, K. A., Choct, M. and Brooks, P. H. (2013). Effect of feed processing and enzyme supplementation on diet digestibility and performance of male weaner pigs fed wheat-based diets in dry or liquid form. *Animal Production Science* 53(6), 531.

Leblois, J., Massart, S., Li, B., Wavreille, J., Bindelle, J. and Everaert, N. (2017). Modulation of piglets' microbiota: Differential effects by a high wheat bran maternal diet during gestation and lactation. *Scientific Reports* 7(1), 7426.

Leonard, S. G., Sweeney, T., Bahar, B., Lynch, B. P. and O'Doherty, J. V. (2011). Effects of dietary seaweed extract supplementation in sows and post-weaned pigs on performance, intestinal morphology, intestinal microflora and immune status. *British Journal of Nutrition* 106(5), 688–699.

Li, Y., Guo, Y., Wen, Z., Jiang, X., Ma, X. and Han, X. (2018). Weaning stress perturbs gut microbiome and its metabolic profile in piglets. *Scientific Reports* 8(1), 18068.

Lindemann, M. D., Cornelius, S. G., El Kandelgy, S. M., Moser, R. L. and Pettigrew, J. E. (1986). Effect of age, weaning and diet on digestive enzyme levels in the piglet. *Journal of Animal Science* 62(5), 1298–1307.

Linden, S. K., Sutton, P., Karlsson, N. G., Korolik, V. and McGuckin, M. A. (2008). Mucins in the mucosal barrier to infection. *Mucosal Immunology* 1(3), 183–197.

Liu, H., Wang, J., He, T., Becker, S., Zhang, G., Li, D. and Ma, X. (2018). Butyrate: A double-edged sword for health? *Advances in Nutrition* 9(1), 21–29.

Luise, D., Motta, V., Salvarani, C., Chiappelli, M., Fusco, L., Bertocchi, M., Mazzoni, M., Maiorano, G., Costa, L. N., Van Milgen, J., Bosi, P. and Trevisi, P. (2017). Long-term administration of formic acid to weaners: Influence on intestinal microbiota, immunity parameters and growth performance. *Animal Feed Science and Technology* 232, 160–168.

Luo, J., Chen, D., Mao, X., He, J., Yu, B., Cheng, L. and Zeng, D. (2019). Purified β-glucans of different molecular weights enhance growth performance of LPS-challenged piglets via improved gut barrier function and microbiota. *Animals* 9(9), 602.

Luppi, A., Gibellini, M., Gin, T., Vangroenweghe, F., Vandenbroucke, V., Bauerfeind, R., Bonilauri, P., Labarque, G. and Hidalgo, Á. (2016). Prevalence of virulence factors in enterotoxigenic Escherichia coli isolated from pigs with post-weaning diarrhea in Europe. *Porcine Health Management* 2, 20.

MacFarlane, G. T., Steed, H. and MacFarlane, S. (2008). Bacterial metabolism and health-related effects of galacto-oligosaccharides and other prebiotics. *Journal of Applied Microbiology* 104(2), 305–344.

Marion, J., Romé, V. R., Savary, G. R., Thomas, F. O., Le Dividich, J. and Le Huërou-Luron, I. (2003). Weaning and feed intake alter pancreatic enzyme activities and corresponding mRNA levels in 7-D-old piglets. *The Journal of Nutrition* 133(2), 362–368.

Marion, J., Petersen, Y. M., Romé, V., Thomas, F., Sangild, P. T., Le Dividich, J. and Le Huërou-Luron, I. (2005). Early weaning stimulates intestinal brush border enzyme activities in piglets, mainly at the posttranscriptional level. *Journal of Pediatric Gastroenterology and Nutrition* 41(4), 401–410.

Masey O'Neill, H. V., Smith, J. A. and Bedford, M. R. (2014). Multicarbohydrase enzymes for non-ruminants. *Asian-Australasian Journal of Animal Sciences* 27(2), 290–301.

Mathew, A. G., Chattin, S. E., Robbins, C. M. and Golden, D. A. (1998). Effects of a direct-fed yeast culture on enteric microbial populations, fermentation acids, and performance of weanling pigs. *Journal of Animal Science* 76(8), 2138–2145.

Mazzoni, M., Le Gall, M., De Filippi, S., Minieri, L., Trevisi, P., Wolinski, J., Lalatta-Costerbosa, G., Lallès, J. P., Guilloteau, P. and Bosi, P. (2008). Supplemental sodium butyrate stimulates different gastric cells in weaned pigs. *The Journal of Nutrition* 138(8), 1426–1431.

Metzler-Zebeli, B. U., Canibe, N., Montagne, L., Freire, J., Bosi, P., Prates, J. A. M., Tanghe, S. and Trevisi, P. (2019). Resistant starch reduces large intestinal pH and promotes fecal lactobacilli and bifidobacteria in pigs. *Animal* 13(1), 64–73.

Miao, J., Adewole, D., Liu, S., Xi, P., Yang, C. and Yin, Y. (2019). Tryptophan supplementation increases reproduction performance, milk yield, and milk composition in lactating sows and growth performance of their piglets. *Journal of Agricultural and Food Chemistry* 67(18), 5096–5104.

Molist, F., Van Oostrum, M., Pérez, J. F., Mateos, G. G., Nyachoti, C. M. and Van Der Aar, P. J. (2014). Relevance of functional properties of dietary fibre in diets for weanling pigs. *Animal Feed Science and Technology* 189, 1–10.

Musser, R. E., Davis, D. L., Dritz, S. S., Tokach, M. D., Nelssen, J. L., Minton, J. E. and Goodband, R. D. (2004). Conceptus and maternal responses to increased feed intake during early gestation in pigs. *Journal of Animal Science* 82(11), 3154–3161.

Newburg, D. S., Ruiz-Palacios, G. M. and Morrow, A. L. (2005). Human milk glycans protect infants against enteric pathogens. *Annual Review of Nutrition* 25, 37–58.

Nowak, P., Zaworska-Zakrzewska, A., Frankiewicz, A. and Kasprowicz-Potocka, M. (2021). The effects and mechanisms of acids on the health of piglets and weaners – a review. *Annals of Animal Science* 21(2), 433–455.

Nortey, T. N., Owusu-Asiedu, O. and Zijlstra, R. T. (2015). Effects of xylanase and phytase on digestion site of low-density diets fed to weaned pigs. *Livestock Research and for Rural Development* 27, 133.

NRC. (2012). *Nutrient Requirements of Swine: 11th Revised Edition*. The National Academies Press, Washington, DC.

O'Doherty, J. V., Bouwhuis, M. A. and Sweeney, T. (2017). Novel marine polysaccharides and maternal nutrition to stimulate gut health and performance in post-weaned pigs. *Animal Production Science* 57(12), 2376–2385.

O'Doherty, J. V., Venardou, B., Rattigan, R. and Sweeney, T. (2021). Feeding marine polysaccharides to alleviate the negative effects associated with weaning in pigs. *Animals* 11(9), 2644.

Opapeju, F. O., Rademacher, M., Blank, G. and Nyachoti, C. M. (2008). Effect of low-protein amino acid-supplemented diets on the growth performance, gut morphology, organ weights and digesta characteristics of weaned pigs. *Animal*, 2(10), 1457-1464. https://doi.org/10.1017/S175173110800270X

Pieper, R., Janczyk, P., Zeyner, A., Smidt, H., Guiard, V. and Souffrant, W. B. (2008). Ecophysiology of the developing total bacterial and lactobacillus communities in the terminal small intestine of weaning piglets. *Microbial Ecology* 56(3), 474–483.

Pierce, K. M., Callan, J. J., McCarthy, P. and O'Doherty, J. V. O. (2004). Effects of high dietary concentration of lactose and increased soybean meal inclusion in starter diets for piglets. *Animal Science* 79(3), 445–452.

Pierce, K. M., Callan, J. J., Brophy, P. O., McCarthy, P., Sweeney, T., Fitzpatrick, E. C., Byrne, C., Ni Cheallaigh, S. and O'Doherty, J. V. (2006). The effect of lactose and inulin on intestinal morphology, microbiology, and volatile fatty acids of the weanling pig. *Animal Science* 82, 311–318.

Pluske, J. R., Hampson, D. J. and Williams, I. H. (1997). Factors influencing the structure and function of the small intestine in the weaned pig: A review. *Livestock Production Science* 51(1–3), 215–236.

Pluske, J. R., Pethick, D. W., Hopwood, D. E. and Hampson, D. J. (2002). Nutritional influences on some major enteric bacterial diseases of pig. *Nutrition Research Reviews* 15(2), 333–371.

Pluske, J. R., Black, B., Pethick, D. W., Mullan, B. P. and Hampson, D. J. (2003). Effects of different sources and levels of dietary fiber in diets on performance, digesta characteristics and antibiotic treatment of pigs after weaning. *Animal Feed Science and Technology* 107(1-4), 129-142.

Puchal, A. A. and Buddington, R. K. (1992). Postnatal development of monosaccharide transport in pig intestine. *American Journal of Physiology* 262, G895-G902.

Rattigan, R., Sweeney, T., Maher, S., Ryan, M. T., Thornton, K. and O'Doherty, J. V. (2020). Effects of reducing dietary crude protein concentration and supplementation with either laminarin or zinc oxide on the growth performance and intestinal health of newly weaned pigs. *Animal Feed Science and Technology* 270, 114693.

Rattigan, R., Lawlor, P. G., Cormican, P., Crespo-Piazuelo, D., Cullen, J., Phelan, J. P., Ranjitkar, S., Crispie, F. and Gardiner, G. E. (2023). Maternal and/or post-weaning supplementation with Bacillus altitudinis spores modulates the microbial composition of colostrum, digesta and faeces in pigs. *Scientific Reports* 13(1), 8900.

Rodriguez, S. M., Guimaraes, K. C., Matthews, J. C., McLeod, K. R., Baldwin, R. L. and Harmon, D. L. (2004). Influence of abomasal carbohydrates on small intestinal sodium-dependent glucose cotransporter activity and abundance in steers. *Journal of Animal Science* 82(10), 3015-3023.

Rome, S., Barbot, L., Windsor, E., Kapel, N., Tricottet, V., Huneau, J. F., Reynes, M., Gobert, J. G. and Tome, D. (2002). The regionalization of PepT1, Nbat and Eaac1 transporters in the small intestine of rats are unchanged from birth to adulthood. *Journal of Nutrition* 132(5), 1009-1011.

Salcedo, J., Frese, S. A., Mills, D. A. and Barile, D. (2016). Characterization of porcine milk oligosaccharides during early lactation and their relation to the fecal microbiome. *Journal of Dairy Science* 99(10), 7733-7743.

Sangild, T., Silver, M., Fowden, A. L., Turvey, A. and Foltmann, B. (1994). Adrenocortical stimulation of stomach development in the prenatal pig. *Biology of the Neonate* 65(6), 378-389.

Sangild, P. T., Fowden, A. L. and Trahair, J. F. (2000). How does the foetal gastrointestinal tract develop in preparation for enteral nutrition after birth? *Livestock Production Science* 66(2), 141-150.

Sangild, P. T., Schmidt, M., Elnif, J., Bjornvad, C. R., Westrom, B. R. and Buddington, R. K. (2002). Prenatal development of gastrointestinal function in the pig and the effects of fetal esophageal obstruction. *Pediatric Research* 52(3), 416-424.

Sangild, P. T., Thymann, T., Schmidt, M., Stoll, B., Burrin, D. G. and Buddington, R. K. (2013). Invited review: The preterm pig as a model in pediatric gastroenterology. *Journal of Animal Science* 91(10), 4713-4729.

Silver, M. and Fowden, A. L. (1989). Pituitary-adrenocortical activity in the fetal pig in the last third of gestation. *Quarterly Journal of Experimental Physiology* 74(2), 197-206.

Smith, F., Clark, J. E., Overman, B. L., Tozel, C. C., Huang, J. H., Rivier, J. E., Blisklager, A. T. and Moeser, A. J. (2010). Early weaning stress impairs development of mucosal Barrier function in the porcine intestine. *American Journal of Physiology-Gastrointestinal and Liver Physiology* 298(3), G352-G363.

Stirling, C. M. A., Charleston, B., Takamatsu, H., Claypool, S., Lencer, W., Blumberg, R. S. and Wileman, T. E. (2005). Characterization of the porcine neonatal Fc receptor-potential use for trans-epithelial protein delivery. *Immunology* 114(4), 542-553.

Suiryanrayna, M. V. A. N. and Ramana, J. V. (2015). A review of the effects of dietary organic acids fed to swine. *Journal of Animal Science and Biotechnology* 6, 45.

Swords, W. E., Wu, C. C., Champlin, F. R. and Buddington, R. K. (1993). Postnatal changes in selected bacterial groups of the pig colonic microflora. *Biology of the Neonate* 63(3), 191–200.

Tan, F. P. Y., Beltranena, E. and Zijlstra, R. T. (2021). Resistant starch: Implications of dietary inclusion on gut health and growth in pigs: A review. *Journal of Animal Science and Biotechnology* 12(1), 124.

Torrallardona, D. (2010). Spray dried animal plasma as an alternative to antibiotics in weanling pigs: A review. *Asian-Australasian Journal of Animal Sciences* 23(1), 131–148.

Torres-Pitarch, A., Hermans, D., Manzanilla, E. G., Bindelle, J., Everaert, N., Beckers, Y., Torrallardona, D., Bruggeman, G., Gardiner, G. E. and Lawlor, P. G. (2017). Effect of feed enzymes on digestibility and growth in weaned pigs: A systematic review and meta-analysis. *Animal Feed Science and Technology* 233, 145–159.

Trachsel, J., Briggs, C., Gabler, N. K., Allen, H. K. and Loving, C. L. (2019). Dietary resistant potato starch alters intestinal microbial communities and their metabolites, and markers of immune regulation and Barrier function in swine. *Frontiers in Immunology* 10, 1381.

Trahair, J. F. and Sangild, P. T. (2000). Fetal organ growth in response to oesophageal infusion of amniotic fluid, colostrum, milk or gastrin-releasing peptide: A study in fetal sheep. *Reproduction, Fertility, and Development* 12(1–2), 87–95.

Trevisi, P., Gandolfi, G., Priori, D., Messori, S., Colombo, M., Mazzoni, M., Lallès, J. P. and Bosi, P. (2013). Age-related expression of the polymeric immunoglobulin receptor (pIgR) in the gastric mucosa of young pigs. *PLoS One* 8(11), e81473.

Trevisi, P., Luise, D., Correa, F., Messori, S., Mazzoni, M., Lallès, J. P. and Bosi, P. (2021). Maternal antibiotic treatment affects offspring gastric sensing for umami taste and ghrelin regulation in the pig. *Journal of Animal Science and Biotechnology* 12(1), 31.

Venardou, B., O'Doherty, J. V., Garcia-Vaquero, M., Kiely, C., Rajauria, G., McDonnell, M. J., Ryan, M. T. and Sweeney, T. (2023). In vitro evaluation of brown seaweed Laminaria spp. as a source of antibacterial and prebiotic extracts that could modulate the gastrointestinal microbiota of weaned pigs. *Animals* 13(5), 823.

Venema, K. (2012). Intestinal fermentation of lactose and prebiotic lactose derivatives, including human milk oligosaccharides. *International Dairy Journal* 22(2), 123–140.

Vente-Spreeuwenberg, M. A., Verdonk, J. M., Verstegen, M. W. and Beynen, A. C. (2003). Villus height and gut development in weaned piglets receiving diets containing either glucose, lactose, or starch. *British Journal of Nutrition* 90(5), 907–913.

Vigors, S., Sweeney, T., O'Shea, C. J., Browne, J. A. and O'Doherty, J. V. (2014). Improvements in growth performance, bone mineral status and nutrient digestibility in pigs following the dietary inclusion of phytase are accompanied by modifications in intestinal nutrient transporter gene expression. *British Journal of Nutrition* 112(5), 688–697.

Vigors, S., Sweeney, T., O'Shea, C. J., Kelly, A. K. and O'Doherty, J. V. (2016). Pigs that are divergent in feed efficiency differ in intestinal enzyme and nutrient transporter gene expression, nutrient digestibility, and microbial activity. *Animal* 10(11), 1848–1855.

Vitari, F., Di Giancamillo, A., Deponti, D., Carollo, V. and Domeneghini, C. (2012). Distribution of ghrelin-producing cells in the gastrointestinal tract of pigs at different ages. *Veterinary Research Communications* 36(1), 71–80.

Wan, J., Jiang, F., Zhang, J., Xu, Q., Chen, D., Yu, B., Mao, X., Yu, J., Luo, Y. and He, J. (2017). Amniotic fluid metabolomics and biochemistry analysis provides novel insights into the diet-regulated foetal growth in A pig model. *Scientific Reports* 7, 44782.

Wei, X., Bottoms, K. A., Stein, H. H., Blavi, L., Bradley, C. L., Bergstrom, J., Knapp, J., Story, R., Maxwell, C., Tsai, T. and Zhao, J. (2021). Dietary organic acids modulate gut microbiota and improve growth performance of nursery pigs. *Microorganisms* 9(1), 110.

Wellock, I. J., Fortomaris, P. D., Houdijk, J. G. M. and Kyriazakis, I. (2007). Effect of weaning age, protein nutrition, and enterotoxigenic Escherichia coli challenge on the health of newly weaned piglets. *Livestock Science* 108(1-3), 102-105.

Weström, B. R., Svendsen, J., Ohlsson, B. G., Tagesson, C. and Karlsson, B. W. (1984). Intestinal transmission of macromolecules (bsa and Fitc-labelled dextrans) in the neonatal pig. *Biology of the Neonate* 46(1), 20-26.

Wijtten, P. J., Van Der Meulen, J. and Verstegen, M. W. (2011). Intestinal Barrier function and absorption in pigs after weaning: A review. *British Journal of Nutrition* 105(7), 967-981.

Willemen, S. A., De Vos, M., Huygelen, V., Fransen, E., Tambuyzer, B. R., Casteleyn, C., Van Cruchten, S. and Van Ginneken, C. (2013). Ghrelin in the gastrointestinal tract and blood circulation of perinatal low and normal weight piglets. *Animal* 7(12), 1978-1984.

Wu, Y., Jiang, Z., Zheng, C., Wang, L., Zhu, C., Yang, X., Wen, X. and Ma, X. (2015). Effects of protein sources and levels in antibiotic-free diets on diarrhea, intestinal morphology, and expression of tight junctions in weaned piglets. *Animal Nutrition*, 1(3), 170-176. https://doi.org/10.1016/j.aninu.2015.08.013.

Wood, I. S. and Trayhurn, P. (2003). Glucose transporters (GlUT and SGLT): Expanded families of sugar transport proteins. *British Journal of Nutrition* 89(1), 3-9.

Xia, S., Yao, W., Zou, B., Lu, Y., Lu, N., Lei, H. and Xia, D. (2015). Effects of potassium diformate on the gastric function of weaning[MS1] piglets. *Animal Production Science*, 56 (7), 1161-1166.

Xian, Y., Zhao, X., Wang, C., Kang, C., Ding, L., Zhu, W. and Hang, S. (2018). Phenylalanine and tryptophan stimulate gastrin and somatostatin secretion and H^+-K^+-ATPase activity in pigs through calcium-sensing receptor. *General and Comparative Endocrinology* 267, 1-8.

Xing, Y., Li, K., Xu, Y., Wu, Y., Shi, L., Guo, S., Yan, S., Jin, X. and Shi, B. (2020). Effects of galacto-oligosaccharide on growth performance, faecal microbiota, immune response and antioxidant capability in weaned piglets. *Journal of Applied Animal Research* 48(1), 63-69.

Xu, R. J., Wang, F. and Zhang, S. H. (2000). Postnatal adaptation of the gastrointestinal tract in neonatal pigs: A possible role of milk-borne growth factors. *Livestock Production Science*, 66, 95-107.

Xu, R.-J., Tungthanathanich, P., Birtles, M. J., Mellor, D. J., Reynolds, G. W. and Simpson, H. V. (1992). Growth and morphological changes in the stomach of newborn pigs during the first three days after birth. *Journal of Developmental Physiology*, 17(1), 7-14.

Xu, R.-J. and Cranwell, P. D. (1991). Gastrin in fetal and neonatal pigs. *Comparative Biochemistry and Physiology Part B: Comparative Biochemistry*, 98(4), 615-621. https://doi.org/10.1016/0305-0491(91)90264-E.

Yáñez, J. L., Landero, J. L., Owusu-Asiedu, A., Cervantes, M. and Zijlstra, R. T. (2013). Growth performance, diet nutrient digestibility, and bone mineralization in weaned pigs fed pelleted diets containing thermostable phytase. *Journal of Animal Science* 91(2), 745-754.

Yin, J., Li, X., Li, D., Yue, T., Fang, Q., Ni, J., Zhou, X. and Wu, G. (2009). Dietary supplementation with zinc oxide stimulates ghrelin secretion from the stomach of young pigs. *The Journal of Nutritional Biochemistry* 20(10), 783–790.

Yue, L. Y. and Qiao, S. Y. (2008). Effects of low-protein diets supplemented with crystalline amino acids on performance and intestinal development in piglets over the first 2 weeks after weaning. *Livestock Science*, 115(2–3), 144–152.

Zhang, H., Yin, J., Li, D., Zhou, X. and Li, X. (2007). Tryptophan enhances ghrelin expression and secretion associated with increased food intake and weight gain in weanling pigs. *Domestic Animal Endocrinology* 33(1), 47–61.

Zhao, J., Zhang, Z., Zhang, S., Page, G. and Jaworski, N. W. (2021). The role of lactose in weanling pig nutrition: A literature and meta-analysis review. *Journal of Animal Science and Biotechnology* 12(1), 10.

Zijlstra, R. T., Li, S., Owusu-Asiedu, A., Simmins, P. H. and Patience, J. F. (2004). Effect of carbohydrase supplementation of wheat- and canola-meal-based diets on growth performance and nutrient digestibility in group-housed weaned pigs. *Canadian Journal of Animal Science* 84(4), 689–695.

Part 2

Feed formulation

Chapter 3

Developing nutritional guidelines for pigs

Jung Yeol Sung, Opeadura Timileyin Osunbami and Olayiwola Adeola, Purdue University, USA

1 Introduction

1.1 Pig feeding and nutrition from birth to market weight

Pig nutrition could be defined as the science of feed preparation through careful selection of suitable ingredients and its feeding to meet the needs of pigs at different phases of growth or life stages. This involves selecting the right type and amount of feed, monitoring feed intake as well as adjusting the diet as needed. Providing a balanced diet that meets the pig's nutritional requirements at each stage of life is essential for optimal performance and profitability. Pig feeding and nutrition is a crucial aspect of pig farming, as it directly impacts growth, health, and productivity from birth to market weight. Proper nutrition ensures that pigs reach their full genetic potential, produce high-quality meat, and remain healthy throughout their lifecycle. A holistic nutritional guideline focuses on all aspects of pig production from birth to market weight, cutting across matching feeder design to age, feeder space, reduction of feed wastage, and batch homogeneity at market weight (López-Vergé et al., 2018). Aside from providing a balanced diet, nutritional guidelines must also consider feed intake and how to curtail feed wastage. Nutritional guidelines should be regularly reviewed and updated as new scientific evidence and industry trends emerge (Alonso et al., 2010). Setting nutritional guidelines involve considering various vital factors that interplay to ensure the guidelines are effective, sustainable,

http://dx.doi.org/10.19103/AS.2024.0140.05

and appropriate for the targeted stage of growth. Some of these key factors are described in the following sections.

1.1.1 Breed and genetic traits affecting nutrient and energy requirements

Different pig breeds have unique genetic traits affecting nutrient requirements (Emmans and Kyriazakis, 2000). Human selection, rather than natural selection for unique traits in pigs, must be accompanied by appropriate and accurate prediction of how changes in genotype could alter the nutritional needs of the animal. For example, breeds with a higher potential for lean growth require higher amino acids and energy to support their growth. Also, selection for a higher growth rate or efficiency would consequently increase feed intake, heat production, and oxygen consumption.

1.1.2 Production goals

Nutritional guidelines should be set with specific production goals in mind, such as weight gain, lean meat production, and reproductive performance. Any nutritional guideline to be integrated into the production system of pigs must be economically viable with improving feed efficiency and environmentally safe by reducing nutrient excretion and subsequent pollution (Lu et al., 2017). Improving feed efficiency is of high priority for the pork industry because feed cost is an integral production component (Patience et al., 2015).

1.1.3 Growth stage and weight

The nutritional requirements of pigs change throughout their growth stages, from nursing piglets to finishing pigs. Piglets are usually weaned from their mother's milk at 21–28 days of age. At this stage, they require a diet that is high in protein, energy, and easily digestible nutrients. A starter feed containing at least 18% protein is recommended, with additional supplements such as vitamins and minerals to support healthy growth and development. As pigs grow and mature, their nutritional requirements change. From 8 to 10 weeks of age until market weight, pigs require a diet that is lower in protein but higher in energy. A balanced mix of carbohydrates, fats, and fibers is recommended to maintain body condition and support lean tissue growth. In the weeks leading up to market weight, it is important to adjust the diet to ensure the production of high-quality meat that is both tender and flavorful. The concentration of carbohydrates such as muscle glycogen is highly correlated to fried and roasted flavor in pork, and this can be achieved by reducing the protein intake and increasing energy intake, typically by feeding a high-energy diet consisting of grains and fats (Alonso et al., 2010; Aaslyng and Meinert, 2017). This helps to increase fat deposition and improve the overall quality of the meat. The composition of the fatty acids in the diet mirrors the fatty acid composition of pork (Alonso et al., 2010).

1.1.4 Nutrient interactions

The interplay between different nutrients is another crucial factor when setting nutritional guidelines for pigs. For example, an excess of a single essential amino acid can negatively impact the utilization of another amino acid, leading to deficiencies, toxicity, or imbalances (Floc'h et al., 2012). A comprehensive understanding of nutrient interactions may present an opportunity to refine previous nutritional recommendations based on the physiological needs of the animal. Nutrient recommendations should, therefore, be guided by the knowledge of the involvement of nutrients in the modulation of various metabolic pathways. For example, micronutrients serve as cofactors, whereas amino acids are essential precursors for neurotransmitters or immune system modulators (Wang et al., 2020). Furthermore, there is evidence of the complex interactions between carbohydrate sources and dietary protein levels, as seen with the microbial fermentation and production of short-chain fatty acids along the gastrointestinal system and their effects on gut health (Pieper et al., 2014).

1.1.5 Feed ingredients and quality

The type and quality of feed ingredients used can impact nutrient requirements. For example, the amino acid profile of the protein source can influence lysine requirements. Additionally, feed quality and digestibility can affect nutrient absorption and utilization. It is well established that in feedstuffs of lower quality, more undigested dietary proteins will flow into the hindgut and likely increase the amount of fermentable protein, hence promoting fermentation and selective growth of proteolytic bacteria (Pieper et al., 2016).

1.1.6 Feed cost and availability

Feed cost can significantly impact the nutritional guidelines of pigs, as it can affect the quality and quantity of available feed (Patience et al., 2015). Generally, pigs require a balanced diet that provides adequate protein, carbohydrates, fats, vitamins, and minerals to support growth, development, and overall health. When feed costs are high, farmers may have to reduce the amount or quality of feed provided to the animals to save money. This can result in a feed that is inadequate in one or more nutrients, leading to nutritional deficiencies and health problems. On the other hand, when feed costs are low, producers may be able to provide higher quality and quantity of feed, which can lead to better growth rates, reproductive performance, and overall health (Agyekum and Nyachoti, 2017).

To ensure that pigs receive adequate nutrition regardless of feed costs, it is important to develop a feeding program that considers the specific nutritional requirements, current cost, and availability of feed ingredients. This may necessitate adjusting the types and amounts of feed ingredients used, as well as incorporating alternative feed sources or other supplements as needed. Woyengo et al. (2014) indicated that using alternative feed ingredients could only reduce cost if their supplementation for replacing conventional feedstuffs returned a lower price per unit of net energy or digestible lysine. This could be achieved by lowering the content of the anti-nutritional factors commonly associated with alternative feedstuffs and their combined supplementation with exogenous enzymes, such as phytase and fiber-degrading enzymes.

2 Formulating diets to maximize feed efficiency and performance but minimize nutrient excretion

Feed efficiency can be defined as the cumulative effectiveness with which pigs utilize nutrients from various dietary sources for maintenance, lean gain, and lipid accretion. Key steps must be considered when formulating diets to optimize feed efficiency with the simultaneous reduction in nutrient excretion. Such steps are characterized in the following sections.

2.1 Basis of diet formulation

Pig diets are formulated based on the principle that pigs cannot utilize 100% of the energy or nutrients provided. Formulating diets based on total energy or nutrient basis may result in inadequate energy or nutrient provision, leading to suboptimal performance. Conversely, excessive dietary energy and nutrient concentrations can increase energy and nutrient excretion, increasing the feed costs. Due to energy being the most expensive component of pig diets, it is essential to determine the available energy of feeds accurately (Sung and Kim, 2021). The gross energy of feeds is estimated using a bomb calorimeter, while digestible energy, metabolizable energy, and net energy are calculated by subtracting energy losses from gross energy (Osunbami and Adeola, 2022). Net energy is the most commonly used energy system in the swine industry since it accounts for post-absorption energy losses, including heat increment resulting from metabolic utilization of feeds. The net energy system's primary principle is to formulate diets with low fiber and protein concentrations and high starch and fat concentrations because fat and starch are more energetically efficient than fiber and protein (Noblet et al., 1994). This approach helps to minimize feed costs while optimizing performance. In summary, formulating pig diets involves considering the pig's limited ability to utilize energy and nutrients, accurately determining the available energy of feeds, and utilizing the net energy system to optimize performance and minimize feed costs.

Amino acids in pig diets are formulated based on a standardized ileal digestible (SID) basis. This takes into account the fact that digestion and absorption of amino acids mainly occur in the small intestine and that microbial fermentation in the large intestine changes the amino acid profile (Kong and Adeola, 2014). Because SID amino acid is calculated by correcting apparent ileal digestible (AID) amino acid for basal endogenous losses of amino acid, SID amino acid has been reported to be more additive compared to AID amino acid (Sung et al., 2022a). In contrast to amino acid, phosphorus is formulated based on a standardized total tract digestible (STTD) basis because digestion and absorption of phosphorus mainly occur in the small intestine, but phosphorus absorption in the large intestine has been recently reported (Zhai et al., 2022). In addition, the ease of measuring total tract digestibility compared to ileal digestibility further justifies the use of the STTD basis. In NRC (2012), calcium requirements of pigs are expressed as total calcium basis. Recently, the STTD calcium requirements have been clarified because of the availability of updated information on calcium digestibility in feed ingredients (González-Vega et al., 2016). For this reason, it is recommended to formulate diets based on STTD calcium basis for precise feed formulation (Shurson et al., 2022).

Another pig-diet-formulating principle is low-protein diet supplemented with crystalline amino acids and exogenous phytase. Crystalline amino acids can be produced through microbial (L-form of amino acid) or chemical synthesis (DL-form of amino acid). The availability of crystalline amino acids at a reasonable price enables precision pig nutrition. In the past, protein sources such as soybean meal were included in diets to meet requirements of limiting amino acid although requirements of other amino acids are met. However, the introduction of crystalline amino acids can partially replace protein sources such as soybean meal without negatively affecting growth performance, which is beneficial in different aspects. First, producers can save feed costs because protein sources are relatively expensive. Second, reducing dietary protein is known to maintain healthy status of the gut, particularly for nursery pigs. If peptides and proteins are not digested and absorbed by the end of the small intestine, they flow into the large intestine and are hydrolyzed into amino acids by microbial proteases and peptidases (Sung et al., 2023). The resulting amino acid can be decarboxylated, which releases amine and carbon oxide or deaminated into ammonia and keto acids (Fig. 1). Subsequently, the fate of the keto acids would be incorporated into microbial proteins or fermented into fatty acids, indole, phenol, etc. by the microbial population in the large intestine. Some products such as amine, ammonia, indole, and phenol are reported to be potentially detrimental at their high concentration (Oliphant and Allen-Vercoe, 2019). Furthermore, protein fermentation in the large intestine is one of the

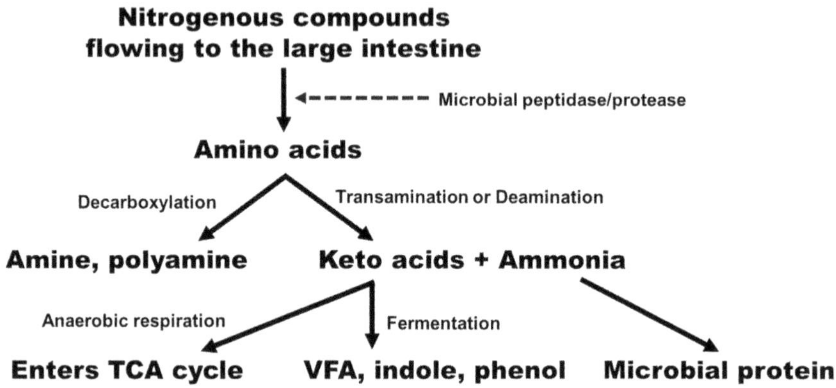

Figure 1 The fate of indigestible nitrogen in the large intestine. Source: Modified from: Lammers-Jannink (2022).

factors causing diarrhea, particularly in early post-weaning (Heo et al., 2008). Third, greenhouse gas emissions and carbon footprint can be decreased by low-protein diets supplemented with crystalline amino acids. this is because synthesizing feed-grade amino acids decreases land use, carbon footprint, and greenhouse gas emissions compared to cultivating soybeans and oil extraction, which is required for soybean meal production (Pomar et al., 2021; Shurson et al., 2022). Fourth, reducing protein can decrease nitrogen excretion, which is beneficial to the environment (Wang et al., 2018). The reason that low-protein diets are beneficial to gut health and can reduce nitrogen excretion is because low-protein diets generally decrease the amount of dietary ileal indigestible protein. However, reducing dietary protein does not always guarantee a decrease in ileal indigestible protein. For this reason, it might be better to formulate diets toward reducing ileal indigestible protein concentrations (Sung et al., 2023).

Since the introduction of exogenous phytase in 1991, it has been supplemented in most commercial pig diets because the digestion capacity of phosphorus in pigs is low due to the insufficient activity of endogenous phytase (Mroz et al., 1994). The supplementation of phytase can decrease the inclusion rate of inorganic phosphates, which are regarded as limited resources. Furthermore, partially replacing inorganic phosphates with exogenous phytase reduces phosphorus excretion by improving phosphorus digestibility in plant-derived feed ingredients such as corn and soybean meal.

2.2 Diets for nursery pigs

The main principle for formulating diets for nursery pigs is to maximize digestible nutrients while minimizing anti-nutrients and indigestible nutrients. Given that the digestive system, including gut morphology, enzyme secretion, and nutrient transporters of nursery pigs, is immature, especially just after weaning, it is essential to include feedstuffs with highly digestible protein and carbohydrate in their diets. Protein sources for nursery pigs can be classified into animal- and plant-derived ingredients (Menegat et al., 2019). Animal proteins such as fish meal, meat and bone meal, poultry meal, and spray-dried plasma proteins are commonly used in nursery pig diets due to their high digestible protein concentration and balanced amino acid profiles (Almeida et al., 2013; Kerr et al., 2019; Sung et al., 2022b). Typically, raw animal protein sources undergo steam heating during the rendering process to inactivate harmful pathogens and reduce moisture content. However, the demand for animal byproducts has decreased due to concerns about virus transmission through animal proteins (Shurson and Urriola, 2022). In contrast to animal proteins, plant protein sources usually contain anti-nutrients, with soybean meal being the most used protein supplement. Soybean meal contains oligosaccharides (raffinose, stachyose, and verbascose), trypsin inhibitor, tannins, allergenic proteins (glycinin and β-conglycinin), and phytate (Li et al., 1990; Woyengo et al., 2017). As nursery pigs are relatively less tolerant to these anti-nutrients, conventional soybean meal is partially replaced with alternative feed ingredients such as fermented or enzyme-treated soybean meal, soy protein concentrate, or isolate, where anti-nutrients are partially removed by processing (Cervantes-Pahm and Stein, 2010).

Starch in commonly used feed ingredients, such as corn and wheat, is typically highly digestible for pigs. However, nursery pigs may be unable to fully digest starch due to digestive system immaturity. To address this issue, lactose, whey permeate, or dried whey are added to nursery pig diets to provide lactose. This is because the activity of the lactase enzyme is high in earlier post-weaning stages, considering that pigs are fed sow's milk until weaning. A recent study by Jang et al. (2021) found that the growth of pigs weaned at 21 and 25 days of age and weighing 7–11 kg was maximized when their diets contained 17.0% and 14.4% whey permeate, respectively. However, because lactose sources can be relatively expensive, their inclusion rate should be determined based on ingredient availability and least-cost formulation.

2.3 Diets for growing-finishing pigs

Most pig diets are composed of feeds given to growing-finishing pigs. As pigs gain weight, their absolute daily nutrient requirements increase, but the nutrient concentration of their diet decreases as they consume more feed each day. It is essential to use a phase-feeding approach that closely matches the changing nutrient requirements of the pigs throughout their growth cycle (Kansas State University Nutrition Guideline A) to prevent nutrient waste or deficiency. This approach optimizes feed utilization and ensures that the pigs receive the right balance of nutrients at each stage of their development. Traditionally, 5-6 phase feeding from 25 kg to market weight was considered ideal. However, recent research suggests that 2–4 feeding phases may provide similar economic returns (Menegat et al., 2020). This is because some pigs may be underfed due to various factors, requiring compensatory growth. Conducting 5-6 feeding phases may limit the provision of amino acid concentration, preventing underfed pigs from achieving compensatory growth.

The potential for lean gain in pigs may differ between barrows and gilts, and therefore, split-sex feeding can be an effective strategy to optimize growth performance while minimizing feed costs. After 9–10 weeks of age, pigs can be housed separately according to their sex. Depending on the market target, the nutrient composition of pig diets can be adjusted, and pigs can be moved out based on constant weight or time (Kansas State University Nutrition Guideline A). Under constant day marketing strategies, improving the gain through feeding high-quality feeds can potentially improve economic output although the cost of feed per weight gain is increased. However, this is not the case under constant weight marketing strategies. The nutrient composition of feeds also depends on whether the target is lean gain or body weight gain. Although the primary goal of producers is to maximize lean gain, during the summer when the demand for pork is high and prices are elevated, it may be more advantageous to focus on maximizing total body weight rather than lean gain.

2.4 Diets for boars

Normal growing-finishing pigs are typically slaughtered around 180 days after birth. However, boars and gilts have a longer life span and reach a higher body weight than market weight for growing-finishing pigs. In order to support their heavier weight, the skeletal structure of boars and gilts needs to be robust enough, which requires an adequate supply of calcium and phosphorus. Therefore, boars and gilts generally receive a greater amount of calcium and phosphorus compared to growing-finishing pigs. This need for a higher amount of calcium and phosphorus is supported by previous studies that have shown that the requirements for maximizing bone ash are greater than those for optimizing growth performance (González-Vega et al., 2016; Vier et al., 2019). However, information on boar diets is limited, as they make up a small proportion of the pig industry. The primary goal of feeding programs for boars is to maximize semen quality. When determining the daily feed allowance, consideration should be given to the body weight and condition of mature boars. Once they reach maturity, a diet containing 0.62% SID lysine is sufficient to optimize reproductive performance. During the growing phase, additional lysine is required (PIC Nutrition and Feeding Guideline, 2021). Studies have shown that supplementing with 0.3 ppm of organic selenium can maintain sperm motility during storage (Speight et al., 2012). Furthermore, feeding 90 g of omega-3 fatty acids per day for 16 weeks has been demonstrated to increase total sperm production (Estienne et al., 2008). Although previous research has reported improved semen quality and volume in boars through L-carnitine supplementation, the effect was inconsistent across studies (Kozink et al., 2004; Wähner et al., 2004; Jacyno et al., 2007).

2.5 Diets for gilts and sows

A well-designed nutritional program for replacement gilts is crucial as their reproductive performance and longevity as sows greatly depend on them (Australian Pork Limited, 2017). It is important to provide these gilts with enough dietary protein to facilitate the development of their reproductive tract. The optimal supply of minerals is essential because the longevity of sows depends in part on their bone health. This is because poor bone health can lead to lameness and other health problems that can reduce productivity and longevity in the breeding herd. As mentioned previously in the section on boars, it is important to provide gilts with higher levels of calcium and phosphorus than what is typically provided to growing pigs. This is necessary to support optimal bone mineralization in gilts, which is essential for their long-term health and reproductive performance. If gilts are too underweight or over-conditioned despite receiving optimal nutrition, they may not be suitable for breeding. This is because gilts with poor growth may experience abnormal feed intake during lactation, delayed puberty, and immature reproductive tract development, which can impact their fertility and ability to produce healthy offspring (Vallet et al., 2016). Careful monitoring of gilt growth and condition and informed breeding decisions based on their overall health and development is thus important.

Typically, the recommended target body weight for gilts to be considered for breeding is between 135 and 155 kg (Kansas State University Nutrition Guideline B). This weight range is considered optimal for ensuring that gilts have reached an appropriate level of maturity and are capable of supporting a healthy pregnancy and producing healthy offspring. Gestating sows have specific energy and nutrient requirements, which are necessary for the development of their mammary glands and conceptus, as well as the maintenance and growth of the conceptus (NRC, 2012). These requirements increase significantly during days 70–90 of gestation due to the rapid development of the mammary glands and conceptus (McPherson et al., 2004; Kim et al., 2013). As a result, the NRC (2012) provides energy and nutrient recommendations separately before and after 90 days of gestation. Sow parity is another factor that affects their nutrient requirements. As the parity increases, the nutrient requirements decrease because maternal growth is not as active in sows with high parity as it is in those with low parity. Typically, gestating sows are fed a restricted amount of feed (2-3 kg per day) based on their body condition. This is to prevent issues such as difficulty in delivery, disruptions to the optimal reproductive cycle, lower subsequent reproductive performance, and excessive fat deposition. However, feed restriction may lead to welfare concerns because the amount of feed given to gestating sows is typically lower than their voluntary feed intake. To address these issues, high-fiber feed ingredients such as sugar beet pulp are included in the diets of gestating sows, which may help increase satiety and prevent potential complications (Meunier-Salaün et al., 2001).

'Bump feeding' is a feeding practice in which some producers increase the feed intake of gestating sows from day 90 of gestation until farrowing (Goncalves et al., 2016). The purpose of this practice is to meet the increased demand for energy and nutrients that occurs in late-gestating sows, due to the growth of the conceptus and mammary gland. However, the effect of bump feeding on feed intake and weight loss of sows during lactation period, litter birth weight, litter size, and milk production is not consistent (de Oliveira Araújo et al., 2020). In contrast to bump feeding, sows are typically given unrestricted access to feed during the transition period, which typically lasts 5-7 days before farrowing. This feeding strategy is aimed at providing sufficient energy to support the development of the mammary gland and conceptus, as well as the production of colostrum (Theil et al., 2022). Depending on the farm's specific feeding program, either a lactation diet or a specific transition diet may be provided during this period.

After giving birth, the energy and nutrient demands of lactating sows increase significantly to support milk production. With current breeding practices, hyper-prolific sows give birth to a large number of piglets, which translates to a higher demand for milk production. Insufficient feed intake or inadequate dietary energy and nutrient supply during this period can result in negative energy balance and subsequent body mobilization. Excessive body weight loss due to this mobilization can lead to poor reproductive performance in the future, and therefore, prevention of this condition is crucial. For this reason, additional 2-4% fats or oils, particularly for oils with high linoleic acid or α-linolenic acid, are needed to be included in lactating sow diets to increase dietary energy concentration, which potentially improves litter growth performance and milk fat output (Kansas State University Nutrition Guideline C; Rosero et al., 2016). Strathe et al. (2017) suggested that standardized ileal digestible protein concentration of lactating sow diets should be greater than 13.5% to improve litter body weight gain by increasing milk yield and milk protein.

3 Key challenges in formulating nutritional guidelines and possible solutions

3.1 Heat stress

Heat stress (HS) has become an increasingly prevalent challenge in pig management because of global warming and the need to accentuate production in warm climatic regions to meet the increasing world's pork demand (Renaudeau et al., 2012). With rising temperatures and increased frequency of heat waves, pigs experience severe HS, which can negatively impact their health and productivity (Fig. 2). Pigs are particularly susceptible to HS due to their limited ability to dissipate heat. Due to the lack of sweat glands, pigs rely on panting and evaporation through their skin to regulate body temperature (Bracke, 2011). When the ambient temperature exceeds their thermoneutral zone, pigs may reduce their feed intake. Therefore, the primary index of HS can first be observed with the reduction in feed intake. Huynh et al. (2005) reported a daily reduction in voluntary feed intake that ranges from 81 to 106 g/pig for every 1°C increase above the upper critical temperature. The economic effect of HS in lactating sows differs from those associated with gestating sows, of which a decrease in milk yield and low birth weight of piglets are the respective commonly associated effects (Black et al., 1993; Liu et al., 2022). Similarly, HS in boars could produce low semen quality (Liu et al., 2022). It has been observed that HS does not only cause physiological and metabolic imbalances in live animals but also potentially impacts the characteristics of their carcass and meat quality (Zhang et al., 2020).

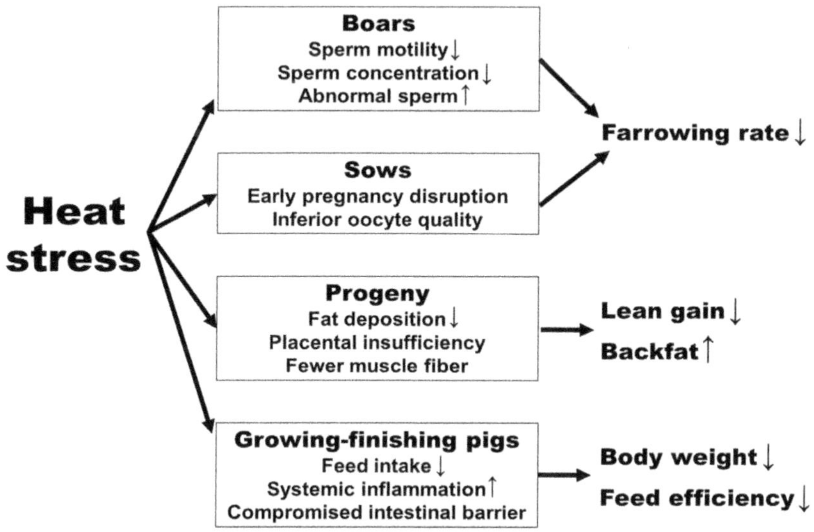

Figure 2 Effects of heat stress on pigs. Source: Modified from: Liu et al. (2022).

This could be due to its effect on postmortem muscle glycolysis, altering the rate and extent of this process and ultimately affecting muscle pH (Downing et al., 2017).

As global temperatures continue to rise, the risk of HS in pigs is expected to increase. Although multiple strategies are needed to manage HS, there is growing recognition of the need to adapt pig management practices to mitigate the effects of HS by adjusting dietary guidelines. Under the condition of HS, production performance can only be improved through modifications of diet composition to either promote a higher intake or compensate for low feed consumption (Renaudeau et al., 2012). Nutritional recommendations may need to be altered to provide more easily digestible nutrients, increase the energy concentration of the feed, and ensure adequate hydration (Zhang et al., 2020). Research has shown that HS shifts energy metabolism toward increased use of carbohydrates while reducing lipid oxidation, and that effective functioning of insulin plays a vital role in adjusting to HS, hence nutritional interventions aiming to alleviate feed consumption under HS should promote the use of glucose and insulin sensitivity (Rhoads et al., 2013; Cottrell et al., 2015).

Additionally, the inclusion of certain feed additives, such as electrolytes, betaine, and antioxidants, may be beneficial in mitigating the negative effects of HS via nutrition (Cottrell et al., 2015; Downing et al., 2017). Other examples of feed additives targeted at improving HS in intensive pig production systems are thiazolidinediones and chromium, facilitating insulin sensitivity to glucose (Rhoads et al., 2013). For example, Peña et al. (2019) reported that using an antioxidant blend reduced the DNA-damaged sperm from 16 to 10 and 7% when fed to boars for 42 or 84 days, respectively. Also, L-arginine was reported to exhibit antioxidant ability with an impact on improving semen quality and libido in boars under HS. Chen et al. (2018) indicated that the supplementation of L-arginine improved semen quality with the recommendation of 0.8–1.0% as the optimal dosage to elicit this effect.

Overall, understanding the impact of HS on pig nutrition and adjusting dietary guidelines accordingly is critical for promoting optimal growth, health, and welfare in pig production. The increasing prevalence of HS in pig management is a clear example of how global warming is impacting animal agriculture and highlights the need for sustainable and adaptive management practices to ensure the continued health and welfare of livestock.

3.2 Presence of toxigenic fungi and mycotoxins in feed ingredients

Mycotoxins are toxic compounds produced as secondary metabolites by fungi, and can contaminate various feed ingredients, such as corn, wheat, barley, and soybean meal, used in pig nutrition. Mycotoxins such as aflatoxins, fumonisins, ochratoxin A, trichothecenes, and zearalenone are reported to be the most prevalent in plant-based animal feed (Tolosa et al., 2021). They are heat stable and could persist even after heat treatments of raw material above 150°C (Marin et al., 2013). The presence of mycotoxins in pig feed could have negative effects on pig health, growth, and productivity. The severity of the negative effect associated with mycotoxin could vary based on the specific type. For instance, aflatoxins and fumonisins reduced pig weight gain by 22% and 10%, respectively (Andretta et al., 2012). These mycotoxins can interfere with nutrient absorption, metabolism, and utilization, leading to reduced feed intake, impaired immune function, and decreased nutrient digestibility and efficiency (Pierron et al., 2016). The negative effect of mycotoxins on weight gain were reported by Andretta et al. (2012) to be greater in male compared to female pigs and younger pigs are also more susceptible to the effect of mycotoxin, indicating an interaction of mycotoxin with age and sex. For example, weight gain was reduced by 19% in male pigs, whereas the reduction was 15% in the female counterpart.

To mitigate the adverse effects of mycotoxins on pig nutrition, nutritional guidelines need to be adjusted. Specifically, the nutrient requirements of pigs may need to be increased to compensate for the mycotoxin-induced reduction in nutrient absorption and utilization. Moreover, nutritional strategies that enhance immune function, reduce inflammation and oxidative stress, and support liver and kidney function may be necessary to counteract the adverse effects of mycotoxins in pigs. Several approaches, which could include selecting high-quality feed ingredients, monitoring and testing feed for mycotoxin contamination, using feed additives and binders to detoxify or reduce the absorption of mycotoxins, implementing good storage and feed management practices to prevent fungal growth, and mycotoxin production can be taken to manage mycotoxin contamination in pig feed (Wambacq et al., 2016; Debevere et al., 2019; Tso et al., 2019). Andretta et al. (2012) reported that adjusting the nutritional content of diets can influence the growth of pigs challenged with mycotoxins. A 1 g increase in methionine per kg of animal metabolic weight corresponded to an increase of 1.2–1.8 g in the daily weight gain of pigs challenged with mycotoxin. This occurrence suggests that methionine could exert an antitoxic property for counteracting the effect of mycotoxins on growth retardation and this could be linked to the antioxidant capability of methionine. Methionine, a sulfur-containing amino acid, is evolutionarily known to act as an endogenous antioxidant in protein and increasing methionine content in proteins through mistranslation could obliterate reactive oxygen species (Aledo, 2019). In a similar manner, the exogenous supplementation of methionine is catalytically structured to react against oxidizing species as an antioxidant agent. Methionine is a precursor for cysteine, and its antioxidant role is associated with glutathione, an essential metabolite whose availability is dependent on cysteine level (Andretta et al.,

2012). Other examples of antioxidant feed additives that could mitigate the oxidative stress caused by mycotoxins in feed include vitamin E or selenium (Liu et al., 2016; Zhu et al., 2022). Antioxidants such as vitamin E and selenium play crucial roles in counteracting reactive oxygen species and protecting cells from oxidative damage. Vitamin E, a lipid-soluble antioxidant, acts primarily in the cell membrane by neutralizing lipid peroxyl radicals, thereby resulting in reduced lipid peroxidation and protection of liver function in animals exposed to mycotoxins (Frankič et al., 2008; Sodhi et al., 2008). Selenium, an essential trace element, is a component of glutathione peroxidase, an antioxidant enzyme which catalyzes the breakdown of peroxides and protects cells from oxidative stress (Liu et al., 2018). Selenium supplementation has demonstrated protective effects against mycotoxin-induced hepatotoxicity and immunosuppression (Zhang et al., 2022). Moreso, it is important to carefully appropriate dosage and duration of antioxidant supplementation based on the specific mycotoxin challenge and age of pigs in future nutritional guidelines.

In summary, mycotoxin contamination in feed ingredients can affect nutritional guidelines in pigs by reducing nutrient absorption, protein synthesis, and deposition, thereby compromising the health and performance of pigs (Kipper et al., 2020). Nutritional strategies that address the adverse effects of mycotoxins on pig nutrition and health should be implemented to mitigate the negative impact of mycotoxin contamination on pig production. Further research should seek to elucidate the optimal combinations and interactions of different antioxidants in the context of mycotoxin mitigation.

3.3 Post-weaning diarrhea

Post-weaning diarrhea in pigs is commonly caused by pathogenic strains of *Escherichia coli*, with enterotoxigenic *E. coli* expressing F4 or F18 fimbriae being the primary culprits (Kim et al., 2022). These pathogenic strains adhere to epithelial cells in the small intestine and produce enterotoxins, which can be heat labile or heat stable. Although the mechanisms of enterotoxins differ, they both bind to receptors on the epithelial cells, resulting in alterations in mineral channels and basolateral transporters. This ultimately leads to a loss of cellular water and electrolytes, increasing gut permeability and decreasing the expression of tight junction proteins. The gut becomes more susceptible to pathogenic invasion, leading to systemic inflammation, including an increase in circulating white blood cells, neutrophils, and lymphocytes, as well as elevated pro-inflammatory cytokines (TNF-α, IL-6, etc.) and acute phase proteins.

Clinical signs of post-weaning diarrhea include watery or liquid feces, reduced appetite, dehydration, and poor growth performance. Antibiotics have been frequently added to animal feeds to treat post-weaning diarrhea. The exact mode of action of antibiotics is not entirely clear, but they may inhibit the proliferation of pathogenic bacteria such as *E. coli* in the gut by directly inhibiting protein synthesis, disrupting bacterial cell function, or modulating the microbiota to prevent *E. coli* survival. Supplementation of antibiotics into animal feeds has been reported to alleviate post-weaning diarrhea and improve growth performance and immune status of nursery pigs (Nyachoti et al., 2012; He et al., 2020).

Due to the concerns about the development of antimicrobial resistance, many countries have ceased using antibiotics in animal feeds as growth promoters (Van Boeckel et al., 2015). To replace antibiotics in feeds, some studies have suggested that the use of high levels of zinc oxide (above 2000 ppm) or copper sulfate (above 200 ppm) can mitigate the negative effects of pathogen challenges and diarrhea while also enhancing growth performance in weaned pigs. However, high dietary concentrations of zinc and copper are not entirely bioavailable in young pigs, resulting in undigested zinc and copper being excreted, potentially leading to environmental pollution. Furthermore, the administration of high doses of zinc oxide may contribute to the development of antimicrobial resistance among bacteria (Ciesinski et al., 2018). As a consequence, researchers have conducted extensive studies to identify alternatives to antibiotics and zinc or copper. Various nutritional interventions can be classified according to their mode of action.

Probiotics and prebiotics are regarded as potential alternatives to antibiotics and zinc oxide, as they can promote the development of a healthy gut microbiota. Probiotics are live microorganisms that provide health benefits to host animals when consumed in adequate amounts (Tran et al., 2018). They can be categorized into *Bacillus*, lactic acid-producing bacteria, or yeast. Successful colonization of the small or large intestine can enhance the growth of beneficial bacteria while suppressing the growth of harmful pathogens like *E. coli*, provided they can survive the acidic environment of the stomach. Furthermore, probiotics may stimulate production of organic acids in the intestine (Canibe et al., 2022). Prebiotics are non-digestible substrates selectively utilized by the host microbiota, providing a health benefit to the host. Fructo-oligosaccharide, galacto-oligosaccharide, mannan-oligosaccharide, and inulin are examples of prebiotics (Kim et al., 2022). Instead of directly providing beneficial bacteria to the gut, prebiotics supplemented to diets are utilized by beneficial bacteria, promoting their growth. Additionally, probiotics and prebiotics can improve immune systems either directly or indirectly. They can reduce pro-inflammatory cytokines, increase tight junction protein expression, or improve gut morphology (Broadway et al., 2015; Liu et al., 2020).

Plants or plant extracts are also being explored as an alternative to antibiotics and zinc oxide in pig diets. They can be used alone or in combination with other compounds (Canibe et al., 2022). Examples of plants or plant extracts include essential oils like carvacrol and thymol, other botanical compounds, and tannins. Previous studies (Liu et al., 2013; Tian and Piao, 2019) have shown that supplementing these products can reduce the incidence of diarrhea and inflammation in weaned pigs. Although their mode of action is not yet fully understood, these compounds may exert antimicrobial, immune-boosting, or antioxidant effects. Furthermore, the supplementation of organic acid, peptides, dietary fiber, or algae can alleviate post-weaning diarrhea (Nyachoti et al., 2012; Jiang et al., 2015; Liu et al., 2018; Xu et al., 2018). Major feed additives that may alleviate diarrhea are summarized in Table 1.

However, the effectiveness of these alternatives to antibiotics and zinc oxide in alleviating diarrhea or improving growth performance and immune system is not consistent. The inconsistency in the effectiveness can be attributed to several factors, including variations in their activity, differences in the dose and duration of supplementation, variations in the status of the gut microbiota and immune system among individual pigs, and environmental factors.

3.4 Slowing down growth of pigs

The primary goal of nutrition in pig production is typically to maximize the growth performance of pigs. However, there may be situations where producers wish to slow down the growth of finishing pigs. During the COVID-19 pandemic, some packing plants were temporarily shut down, which resulted in the inability to harvest pigs that had reached the market weight. To address this situation, producers had to make the finishing diets to be less efficient, to slow down pig growth and prevent reaching market weight before processing capacity became available (Helm et al., 2021; Rao et al., 2021). Not slowing down the growth of pigs may result in economic losses when processing plants reopen. This is because pigs that exceed the market weight may be subject to discounts or inability to be processed due to limitations in plant infrastructure.

From a nutritional standpoint, three potential methods could slow down the growth of finishing pigs. The first method is to increase dietary fiber concentration. Because dietary fiber is not well digested by mammalian enzymes and hinders digestion process of other nutrients, available energy and nutrients would be reduced, which decreases body weight gain of finishing pigs. In the study of Helm et al. (2021), 70 kg pigs were fed diets with graded concentrations of neutral detergent fiber in diets (from 8.7% to 25.0%) by increasing the inclusion rate of soybean hulls for 28 days. Increasing dietary fiber concentration did not significantly decrease body weight gain but gain-to-feed ratio and loin muscle area decreased, which may negatively

Table 1 Nutritional intervention of alleviating post-weaning diarrhea

Nutritional intervention	Type	Dose	Initial body weight, kg	Improvement	Reference
Probiotics	*Lactobacillus rhamnosus*	1.00 × 10^{12} CFU/day	5.2	Reduced diarrhea score and fecal coliform count; increased fecal bifidobacteria and lactobacilli count	Li et al. (2012)
Probiotics	*Bacillus subtilis*	2.56 × 10^{9} CFU/kg	6.2	Reduced diarrhea frequency, fecal beta-hemolytic coliform count, white blood cell, and mRNA expression of IL-6 and PTGS2 in ileal mucosa	He et al. (2020)
Probiotics	Mixture of *Bacillus amyloliquefaciens* and *Bacillus subtilis*	1.50 × 10^{5} CFU/kg	6.4	Increased ileal gene mRNA expression of ZO-1 and CD14	Becker et al. (2020)
Prebiotics	Mannan-oligosaccharide	0.2%	6.7	Increased gain-to-feed ratio. Improved fecal score	Castillo et al. (2008)
Prebiotics	Chito-oligosaccharide	100 mg/kg	4.7	Reduced diarrhea frequency. Increased fecal Lactobacilli count and ileum villus height	Liu et al. (2008)
Prebiotics	Algae-derived beta-glucan	108 mg/kg	7.7	Reduced gut permeability, diarrhea score, serum TNF-α, and mRNA expression of IL-6	Kim et al. (2019)
Plant extract	Capsicum oleoresin Garlic botanical Turmeric oleoresin	10 mg/kg	6.3	Reduced diarrhea score, diarrhea frequency, serum TNF-α, and count of white blood cell or neutrophil	Liu et al. (2013)
Essential oil	Blend of cinnamaldehyde and thymol	100 mg/kg	8.1	Reduced diarrhea frequency. Increased energy and protein digestibility, villus height to crypt depth ratio in the jejunum, and serum superoxide dismutase	Tian and Piao (2019)
Lysozyme	Lysozyme	0.10%	4.9	Decreased serum TNF-α and count of coliform and *E. coli* K88+ in ileal mucosa and colon digesta. Increased ileal villus height	Nyachoti et al. (2012)
Yeast	Live yeast	0.30%	4.5	Increased gain-to-feed ratio, jejunal villus height to crypt depth ratio, serum IgA, and CD4+ to CD8+ ratio	Jiang et al. (2015)

affect profitability. Despite the decrease in loin muscle area, the nonsignificant effect of neutral detergent fiber on body weight gain in finishing pigs can be partially attributed to the increased size of the gut and organs needed to adapt to fibrous diets.

The second method is to restrict indispensable amino acid intake. Considering that indispensable amino acids are required for protein accretion, restriction of indispensable amino acid can decrease body weight of pigs. To restrict indispensable amino acid intake, Helm et al. (2021) decreased the inclusion rate of soybean meal from 16% to 0%, while Rao et al. (2021) decreased dietary standardized ileal digestible lysine concentration from 0.70% (11.5% soybean meal) to 0.18% (no soybean meal). Helm et al. (2021) and Rao et al. (2021) reported that pigs fed diets without soybean meal for 28 days weighed 11.5 kg or 22.9 kg less, respectively, compared to pigs fed the standard diets. Reducing the amount of soybean meal in diets can be advantageous in terms of feed cost savings, as corn is typically less expensive than soybean meal. While reducing the inclusion rate of soybean meal can lead to cost savings for producers, it may not be an economically viable strategy under normal market conditions due to lower gain-to-feed ratio. However, the purpose of this strategy is to match the body weight of pigs feasible for processing in anomalous situations such as COVID-19 (Rao et al., 2021).

The third method is to reduce dietary electrolyte balance by adding chloride sources such as calcium chloride or ammonium chloride. Dietary electrolyte balance is the balance of sodium, potassium, and chloride. In the study by Helm et al. (2021), diets were supplemented with 0%, 2%, or 4% calcium chloride, resulting in dietary electrolyte balance values of 95, −161, and −404, respectively. Pigs fed the diet containing 4% of calcium chloride for 28 days weighed 25.8 kg less compared to pigs fed the diet without calcium chloride. The decrease in body weight primarily is attributed to lower feed intake (average daily feed intake: 2.8–1.4 kg/d) because negative dietary electrolyte balance may have resulted in acidosis and lower appetite. However, the study by Helm et al. (2021) demonstrated that increasing dietary calcium chloride concentration led to a linear decrease in gain-to-feed ratio. Additionally, if calcium chloride is used as a supplement in diets, it is important to note that an additional source of phosphorus needs to be included to maintain the same ratio of calcium to standardized total tract digestible phosphorus.

3.5 Potential detrimental effect of high dietary leucine concentration

Branched-chain amino acids include leucine, isoleucine, and valine. Amino acids absorbed in the small intestine are primarily used for protein synthesis.

However, any excess or unused amino acids are broken down through catabolism to produce energy. Because branched-chain amino acids have similar chemical structures, they are broken down through a common metabolic pathway (Harris et al., 2005). The catabolic pathway of branched-chain amino acids involves two steps: transamination and decarboxylation. The branched-chain amino acid transaminase in skeletal muscle converts branched-chain amino acids into their corresponding alpha-keto acids: leucine is converted to alpha-ketoisocaproate (KIC), isoleucine to alpha-keto-methylvalerate (KMV), and valine to alpha-ketoisovalerate (KIV). The resulting alpha-keto acids are transported to the liver, where they undergo irreversible decarboxylation by the branched-chain alpha-keto acid dehydrogenase (BCKDH) enzyme complex. This results in the formation of acyl-CoA derivatives of the branched-chain amino acids. Valine is a glucogenic amino acid, meaning that its acyl-CoA derivative can be used for gluconeogenesis. In contrast, leucine is ketogenic and its acyl-CoA derivative is converted into a ketone bodies precursor. Isoleucine can be converted into either glucose or acyl-CoA derivatives. Overall, the catabolism of branched-chain amino acids provides a source of energy and intermediates for other metabolic processes in the body.

If pigs are fed diets with excessive leucine concentration, their growth can be retarded for two reasons (Kwon et al., 2019). Firstly, high levels of dietary leucine increase the production of KIC, which in turn activates the activity of BCKDH. Consequently, the conversion of KMV and KIV into acyl-CoA is also activated by BCKDH, leading to a decrease in the amount of isoleucine and valine available for protein synthesis. The second involves isoleucine's effect on the synthesis of serotonin, an appetite-regulating hormone. Leucine competes with tryptophan, a large neutral amino acid, for transport across the blood–brain barrier via the common uptake transporter (LAT1). When leucine concentration in the diet is high, there may be decreased transport of tryptophan due to competition, leading to a decrease in serotonin synthesis.

In the USA, corn, soybean meal, and corn DDGS are the main ingredients used in pig diets. However, compared to other ingredients, corn products have higher concentrations of leucine and a higher ratio of leucine to other amino acids (Kwon et al., 2019). This leads to more leucine than required and an imbalanced ratio of leucine to other amino acids in typical U.S. pig diets. Because reducing leucine concentration in pig diets is difficult, supplementing with amino acids that are hindered by leucine, such as isoleucine, valine, and tryptophan, may help alleviate the negative effects of excessive leucine. Kwon et al. (2020) reported that increasing the ratio of valine to lysine in diets from 60% to 80% resulted in a linear decrease in plasma urea nitrogen and KIC concentrations in growing pigs. This suggests that the efficiency of amino acid utilization was improved while the catabolism of branched-chain amino acids was reduced. Increasing the ratio of tryptophan to lysine from 18% to 28% linearly increased feed intake and body weight gain of growing pigs (Kwon et al., 2022). However, the effect of tryptophan was more significant in diets containing 200% or 300% of the recommended tryptophan to lysine ratio compared to a diet with 100% of the recommended ratio. Furthermore, increasing the ratio of tryptophan to lysine linearly increased hypothalamic serotonin concentration regardless of dietary leucine concentration. These amino acid additions, of course, must be assessed on production cost effectiveness.

4 Conclusion and future trends

Dynamic feeding standards for pigs ingesting mycotoxin-contaminated feed is an area that requires further research to address the significant gaps in knowledge. While some studies have examined the impact of mycotoxin exposure on animal growth, there is still a need for more comprehensive investigations to establish precise feeding protocols that support pig health and productivity under these circumstances. Future research should focus on developing methods for accurately measuring the dietary requirements of pigs affected by mycotoxins. This information will be valuable in developing more effective feed formulations that address the specific nutritional needs of these animals. One potential approach that warrants further attention is the role of amino acid supplementation in mitigating the negative effects of mycotoxins on pig growth. While the importance of methionine as an essential amino acid in this regard has been established, there is a need to explore the potential benefits of other sulfur-containing amino acids such as cysteine. Furthermore, the development of more rapid diagnostic procedures for feed contaminated with mycotoxins is essential. The current methods for detecting mycotoxin contamination in animal feed are time-consuming and expensive, which limit their widespread use (Li et al., 2021). Rapid diagnostic techniques, such as biosensors, could enable the timely detection of mycotoxin contamination and facilitate the implementation of appropriate management strategies to minimize animal exposure. In addition, future research should focus on developing effective detoxification strategies to mitigate the negative impact of mycotoxins on animal health and performance. Using natural adsorbents, such as activated charcoal or clay and carbon-based nanomaterials (Temba et al., 2016; Horky et al., 2018), has shown promise in reducing mycotoxin levels in animal feed, but their efficacy and potential side effects require further investigation. In conclusion, the determination of dynamic feed standards for pigs challenged with mycotoxins requires a multi-disciplinary approach to address the significant gaps in knowledge. Future research should explore the role of amino acid supplementation, including cysteine, in mitigating the negative impact of mycotoxins on animal growth and develop rapid diagnostic procedures for detecting mycotoxin contamination in animal feed. Furthermore, developing effective detoxification strategies using natural adsorbents warrants further investigation.

Linear programming is a widely used method in diet formulation to determine the least-cost diet that meets the minimum nutritional requirements of pigs. However, simply focusing on cost minimization may not be sufficient to achieve various objectives related to pig production. For example, in addition to cost minimization, diet formulation may need to consider minimizing environmental impact, reducing heat production to alleviate heat stress, or meeting specific growth performance targets. A multi-objective optimization approach may require more advanced optimization algorithms beyond conventional linear programming methods, such as deep learning algorithms (Pomar et al., 2021).

The introduction of low-protein diets supplemented with crystalline amino acids and supplemented with exogenous phytase has resulted in a reduction of environmental footprint of pork production (Shurson et al., 2022). Nevertheless, it is imperative to continue exploring ways to further minimize the environmental impact of pig industry. Precision feeding has the potential to address this issue. The traditional method of phase feeding is used to control variations in nutrient requirements over time (Pomar et al., 2021). However, the use of multiple diets may increase feed costs, and the current phase feeding system may not account for variations in lean gain potential between pigs housed in the same barn. As a result, the use of traditional phase feeding methods may result in some pigs being underfed, while others may be overfed, leading to variations in growth performance and increased nutrient excretion. Precision feeding has emerged as a potential solution to address these issues. Precision feeding differs from traditional feeding methods by focusing on the groups or individual pig's nutrient requirements and growth performance on a daily basis, rather than on the population as a whole (Pomar and Remus, 2019).

With the advancements in technology, the utilization of smart sensors and big data analysis has made it possible to automatically record the feed intake and growth of individual or groups of pigs. This recorded data can be used to facilitate real-time modeling procedures to achieve precision nutrition. Andretta et al. (2014) implemented precision feeding by providing tailored diets on a daily basis to individual pigs or groups of pigs during the growing-finishing period. This was achieved by mixing low- and high-nutrient concentration diets based on recorded information and a modeling procedure for estimating nutrient requirements. Compared to conventional three-phase feeding, the precision feeding reduced nitrogen intake and excretion by 7% and 12%, respectively. The precision feeding system has shown potential for reducing the environmental footprint of the pig industry without compromising growth performance. However, further validation is needed to assess its effectiveness across all phases of production. Future research should focus on validating the precision feeding system for different stages of pig production, as well as determining its impact on the overall profitability of the operation. This will require the development of practical and cost-effective methods for collecting and analyzing real-time data on individual pig nutrient and energy requirements and growth performance.

5 Where to look for further information

The following articles provide a good overview of the subject:

- Kansas State University Nutrition Guideline A. Key concepts of a successful grow-finish nutritional program. Available at https://www.asi.k-state.edu/extension/swine/swinenutritionguide/growfinish/GF1_KeyConcepts.pdf.
- Kim, K., Song, M., Liu, Y. and Ji, P. 2022. Enterotoxigenic *Escherichia coli* infection of weaned pigs: Intestinal challenges and nutritional intervention to enhance disease resistance. *Front. Immunol.* 13:885253.
- Patience, J. F., Rossoni-Serão, M. C. and Gutiérrez, N. A. 2015. A review of feed efficiency in swine: biology and application. *J. Anim. Sci. Biotechnol.* 6:33.
- Shurson, G. C., Pelton, R. E. O. .Yang, Z., Urriola, P. E. and Schmitt, J. 2022. Environmental impacts of eco-nutrition swine feeding programs in spatially explicit geographic regions of the United States. *J. Anim. Sci.* 100:skac356.

6 References

Aaslyng, M. D. and Meinert, L. 2017. Meat flavour in pork and beef – from animal to meal. *Meat Sci.* 132:112–117.

Agyekum, A. K. and Nyachoti, C. M. 2017. Nutritional and metabolic consequences of feeding high-fiber diets to swine: a review. *Engineering* 3(5):716–725.

Aledo, J. C. 2019. Methionine in proteins: the Cinderella of the proteinogenic amino acids. *Protein Sci.* 28(10):1785–1796.

Almeida, F. N., Htoo, J. K., Thomson, J. and Stein, H. H. 2013. Comparative amino acid digestibility in US blood products fed to weanling pigs. *Anim. Feed Sci. Technol.* 181(1–4):80–86.

Alonso, V., del M. Campo, M., Provincial, L., Roncalés, P. and Beltrán, J. A. 2010. Effect of protein level in commercial diets on pork meat quality. *Meat Sci.* 85(1):7–14.

Andretta, I., Kipper, M., Lehnen, C. R., Hauschild, L., Vale, M. M. and Lovatto, P. A. 2012. Meta-analytical study of productive and nutritional interactions of mycotoxins in growing pigs. *Animal* 6(9):1476–1482.

Andretta, I., Pomar, C., Rivest, J., Pomar, J., Lovatto, P. A. and Radünz Neto, J. 2014. The impact of feeding growing-finishing pigs with daily tailored diets using precision feeding techniques on animal performance, nutrient utilization, and body and carcass composition. *J. Anim. Sci.* 92(9):3925-3936.

Australian Pork Limited 2017. Producers' guide to pig production & nutrition. Available at: https://www.australianpork.com.au/sites/default/files/2021-07/2017-APL-Producers-Guide-to-Pig-Production-and-Nutrition.pdf.

Becker, S. L., Li, Q., Burrough, E. R., Kenne, D., Sahin, O., Gould, S. A. and Patience, J. F. 2020. Effects of an F18 enterotoxigenic Escherichia coli challenge on growth performance,immunological status, and gastrointestinal structure of weaned pigs and the potential protective effect of direct-fed microbial blends. *J. Anim. Sci.* 98(5):skaa113.

Black, J. L., Mullan, B. P., Lorschy, M. L. and Giles, L. R. 1993. Lactation in the sow during heat stress. *Livest. Prod. Sci.* 35(1-2):153-170.

Bracke, M. B. M. 2011. Review of wallowing in pigs: description of the behaviour and its motivational basis. *Appl. Anim. Behav. Sci.* 132(1-2):1-13.

Broadway, P. R., Carroll, J. A. and Burdick Sanchez, N. C. 2015. Live yeast and yeast cell wall supplements enhance immune function and performance in food-producing livestock: a review. *Microorganisms* 3(3):417-427.

Canibe, N., Højberg, O., Kongsted, H., Vodolazska, D., Lauridsen, C., Nielsen, T. S. and Schönherz, A. A. 2022. Review on preventive measures to reduce post-weaning diarrhoea in piglets. *Animals* 12(19):2585.

Castillo, M., Martin-Orue, S. M., Taylor-Pickard, J. A., Perez, J. F. and Gasa, J. 2008. Use of mannanoligosaccharides and zinc chelate as growth promoters and diarrhea preventative in weaning pigs: Effects on microbiota and gut function. *J. Anim. Sci.* 86(1):94-101.

Cervantes-Pahm, S. K. and Stein, H. H. 2010. Ileal digestibility of amino acids in conventional, fermented, and enzyme-treated soybean meal and in soy protein isolate, fish meal, and casein fed to weanling pigs. *J. Anim. Sci.* 88(8):2674-2683.

Chen, J. Q., Li, Y. S., Li, Z. J., Lu, H. X., Zhu, P. Q. and Li, C. M. 2018. Dietary l-arginine supplementation improves semen quality and libido of boars under high ambient temperature. *Animal* 12(8):1611-1620.

Ciesinski, L., Guenther, S., Pieper, R., Kalisch, M., Bednorz, C. and Wieler, L. H. 2018. High dietary zinc feeding promotes persistence of multi-resistant *E. coli* in the swine gut. *PLoS ONE* 13(1):e0191660.

Cottrell, J. J., Liu, F., Hung, A. T., DiGiacomo, K., Chauhan, S. S., Leury, B. J., Furness, J. B., Celi, P., Dunshea, F. R., Cottrell, J. J., Liu, F., Hung, A. T., DiGiacomo, K., Chauhan, S. S., Leury, B. J., Furness, J. B., Celi, P. and Dunshea, F. R. 2015. Nutritional strategies to alleviate heat stress in pigs. *Anim. Prod. Sci.* 55(12):1391-1402.

de Oliveira Araújo, V., de Oliveira, R. A., Silveira, M. de F. A., Silveira, H., da Silva Fonseca, L. K. S., Guimarães, E. B. B., Schinckel, A. P. and Garbossa, C. A. P. 2020. Bump feed for gestating sows is really necessary? *Livest. Sci.* 240:104184.

Debevere, S., De Baere, S., Haesaert, G., Rychlik, M., Fievez, V. and Croubels, S. 2019. Development of an UPLC-MS/MS method for the analysis of mycotoxins in rumen fluid with and without maize silage emphasizes the importance of using matrix-matched calibration. *Toxins* 11(9):519.

Downing, J. A., Kerr, M. J. and Hopkins, D. L. 2017. The effects of pre-transport supplementation with electrolytes and betaine on performance, carcass yield and meat quality of broilers in summer and winter. *Livest. Sci.* 205:16-23.

Emmans, G. C. and Kyriazakis, I. 2000. Issues arising from genetic selection for growth and body composition characteristics in poultry and pigs. *BSAP Occas. Publ.* 27:39-53.

Estienne, M. J., Harper, A. F. and Crawford, R. J. 2008. Dietary supplementation with a source of omega-3 fatty acids increases sperm number and the duration of ejaculation in boars. *Theriogenology* 70(1):70–76.

Floc'h, N. L., Gondret, F., Matte, J. J. and Quesnel, H. 2012. Towards amino acid recommendations for specific physiological and patho-physiological states in pigs. *Proc. Nutr. Soc.* 71(3):425–432.

Frankič, T., Salobir, J. and Rezar, V. 2008. The effect of vitamin E supplementation on reduction of lymphocyte DNA damage induced by T-2 toxin and deoxynivalenol in weaned pigs. *Anim. Feed Sci. Technol.* 141(3-4):274-286.

Goncalves, M. A. D., Dritz, S. S., Tokach, M. D., Piva, J. H., DeRouchey, J. M., Woodworth, J. C. and Goodband, R. D. 2016. Face sheet - impact of increased feed intake during late gestation on reproductive performance of gilts and sows. *J. Swine Health Prod.* 24:264-266.

González-Vega, J. C., Walk, C. L., Murphy, M. R. and Stein, H. H. 2016. Requirement for digestible calcium by 25 to 50 kg pigs at different dietary concentrations of phosphorus as indicated by growth performance, bone ash concentration, and calcium and phosphorus balances. *J. Anim. Sci.* 94(12):5272–5285.

Harris, R. A., Joshi, M., Jeoung, N. H. and Obayashi, M. 2005. Overview of the molecular and biochemical basis of branched-chain amino acid catabolism. *J. Nutr.* 135:1527S–1530S.

He, Y., Kim, K., Kovanda, L., Jinno, C., Song, M., Chase, J., Li, X., Tan, B. and Liu, Y. 2020. Bacillus subtilis: a potential growth promoter in weaned pigs in comparison to carbadox. *J. Anim. Sci.* 98(9):skaa290.

Helm, E. T., Patience, J. F., Romoser, M. R., Johnson, C. D., Ross, J. W. and Gabler, N. K. 2021. Evaluation of increased fiber, decreased amino acids, or decreased electrolyte balance as dietary approaches to slow finishing pig growth rates. *J. Anim. Sci.* 99(7):skab164.

Heo, J. M., Kim, J. C., Hansen, C. F., Mullan, B. P., Hampson, D. J. and Pluske, J. R. 2008. Effects of feeding low protein diets to piglets on plasma urea nitrogen, faecal ammonia nitrogen, the incidence of diarrhoea and performance after weaning. *Arch. Anim. Nutr.* 62(5):343–358.

Horky, P., Skalickova, S., Baholet, D. and Skladanka, J. 2018. Nanoparticles as a solution for eliminating the risk of mycotoxins. *Nanomaterials* 8(9):727.

Huynh, T. T. T., Aarnink, A. J. A., Verstegen, M. W. A., Gerrits, W. J. J., Heetkamp, M. J. W., Kemp, B. and Canh, T. T. 2005. Effects of increasing temperatures on physiological changes in pigs at different relative humidities. *J. Anim. Sci.* 83(6):1385–1396.

Jacyno, E., Kołodziej, A., Kamyczek, M., Kawęcka, M., Dziadek, K. and Pietruszka, A. 2007. Effect of L-Carnititine supplementation on boar semen quality. *Acta Vet. Brno* 76(4):595–600.

Jang, K. B., Duarte, M. E., Purvis, J. M. and Kim, S. W. 2021. Impacts of weaning age on dietary needs of whey permeate for pigs at 7 to 11 kg body weight. *J. Anim. Sci. Biotechnol.* 12(1):111.

Jiang, Z., Wei, S., Wang, Z., Zhu, C., Hu, S., Zheng, C., Chen, Z., Hu, Y., Wang, L., Ma, X. and Yang, X. 2015. Effects of different forms of yeast *Saccharomyces cerevisiae* on growth performance, intestinal development, and systemic immunity in early-weaned piglets. *J. Anim. Sci. Biotechnol.* 6:47.

Kansas State University Nutrition Guideline A. Key concepts of a successful grow-finish nutritional program. Available at: https://www.asi.k-state.edu/extension/swine/swinenutritionguide/growfinish/GF1_KeyConcepts.pdf

Kansas State University Nutrition Guideline B. Gilt Development. Available at: https://www.asi.k-state.edu/extension/swine/swinenutritionguide/sow_nutrition/SN1_GiltDevelopment_Final.pdf

Kansas State University Nutrition Guideline C. Sow Nutrition - Lactation. Available at: https://www.asi.k-state.edu/extension/swine/swinenutritionguide/sow_nutrition/SN4_Lactation_Final.pdf

Kerr, B. J., Urriola, P. E., Jha, R., Thomson, J. E., Curry, S. M. and Shurson, G. C. 2019. Amino acid composition and digestible amino acid content in animal protein by-product meals fed to growing pigs. *J. Anim. Sci.* 97(11):4540–4547.

Kim, K., Song, M., Liu, Y. and Ji, P. 2022. Enterotoxigenic *Escherichia coli* infection of weaned pigs: intestinal challenges and nutritional intervention to enhance disease resistance. *Front. Immunol.* 13:885253.

Kim, S. W., Weaver, A. C., Shen, Y. B. and Zhao, Y. 2013. Improving efficiency of sow productivity: nutrition and health. *J. Anim. Sci. Biotechnol.* 4(1):26.

Kim, K., Ehrlich,A., Perng, V., Chase, J. A., Raybould, H., Li, X., Atwill, E. R., Whelan, R., Sokale, A. and Liu, Y. 2019. Algae-derived β-glucan enhanced gut health and immune responses of weaned pigs experimentally infected with a pathogenic E. coli. *Anim. Feed Sci. Technol.* 248:114–125.

Kipper, M., Andretta, I., Ribeiro, A. M. L., Pires, P. G. da S., Franceschina, C. S., Cardinal, K. M., Moraes, P. de O. and Schroeder, B. 2020. Assessing the implications of mycotoxins on productive efficiency of broilers and growing pigs. *Sci. Agric.* 77(3):e20180236.

Kong, C. and Adeola, O. 2014. Evaluation of amino acid and energy utilization in feedstuff for swine and poultry diets. *Asian-Australas. J. Anim. Sci.* 27(7):917–925.

Kozink, D. M., Estienne, M. J., Harper, A. F. and Knight, J. W. 2004. Effects of dietary L-carnitine supplementation on semen characteristics in boars. *Theriogenology* 61(7–8):1247–1258.

Kwon, W. B., Touchette, K. J., Simongiovanni, A., Syriopoulos, K., Wessels, A. and Stein, H. H. 2019. Excess dietary leucine in diets for growing pigs reduces growth performance, biological value of protein, protein retention, and serotonin synthesis. *J. Anim. Sci.* 97(10):4282–4292.

Kwon, W. B., Soto, J. A. and Stein, H. H. 2020. Effects on nitrogen balance and metabolism of branched-chain amino acids by growing pigs of supplementing isoleucine and valine to diets with adequate or excess concentrations of dietary leucine. *J. Anim. Sci.* 98(11):skaa346.

Kwon, W. B., Soto, J. A. and Stein, H. H. 2022. Effects of dietary leucine and tryptophan on serotonin metabolism and growth performance of growing pigs. *J. Anim. Sci.* 100(1):skab356.

Lammers-Jannink, K. C. M. 2022. Microbial protein metabolism in the monogastric gastrointestinal tract: a review. Understanding gut microbiomes as targets for improving pig gut health. In: Bailey, M. and Stokes, C. (Eds). *Understanding Gut Microbiomes as Targets for Improving Pig Gut Health*. Cambridge (UK): Burleigh Dodds Science Publishing Limited; p. 435–466.

Li, D. F., Nelssen, J. L., Reddy, P. G., Blecha, F., Hancock, J. D., Allee, G. L., Goodband, R. D. and Klemm, R. D. 1990. Transient hypersensitivity to soybean meal in the early-weaned pig. *J. Anim. Sci.* 68(6):1790–1799.

Li, R., Wen, Y., Wang, F. and He, P. 2021. Recent advances in immunoassays and biosensors for mycotoxins detection in feedstuffs and foods. *J. Anim. Sci. Biotechnol.* 12(1):108.

Li, X.-Q., Zhu, Y.-H., Zhang, H.-F., Yue, Y., Cai, Z.-X., Lu, Q.-P., Zhang, L., Weng, X.-G., Zhang, F.-J. and Zhou. D. 2012. Risks associated with high-dose Lactobacillus rhamnosus in an Escherichia coli model of piglet diarrhoea: Intestinal microbiota and immune imbalances. *PLoS One* 7(7):e40666.

Liu, P., Piao,X. S., Kim, S. W., Wang, L., Shen, Y. B., Lee, H. S. and Li, S. Y. 2008. Effects of chito-oligosaccharide supplementation on the growth performance, nutrient digestibility, intestinal morphology, and fecal shedding of Escherichia coli and Lactobacillus in weaning pigs. *J. Anim. Sci.* 86(10):2609–2618.

Liu, F., Cottrell, J. J., Furness, J. B., Rivera, L. R., Kelly, F. W., Wijesiriwardana, U., Pustovit, R. V., Fothergill, L. J., Bravo, D. M., Celi, P., Leury, B. J., Gabler, N. K. and Dunshea, F. R. 2016. Selenium and vitamin E together improve intestinal epithelial barrier function and alleviate oxidative stress in heat-stressed pigs. *Exp. Physiol.* 101(7):801–810.

Liu, F., Celi, P., Cottrell, J. J., Chauhan, S. S., Leury, B. J. and Dunshea, F. R. 2018. Effects of a short-term supranutritional selenium supplementation on redox balance, physiology and insulin-related metabolism in heat-stressed pigs. *J. Anim. Physiol. Anim. Nutr.* 102(1):276–285.

Liu, F., Zhao, W., Le, H. H., Cottrell, J. J., Green, M. P., Leury, B. J., Dunshea, F. R. and Bell, A. W. 2022. Review: what have we learned about the effects of heat stress on the pig industry? *Animal* 16:100349.

Liu, L., Chen, D., Yu, B., Yin, H., Huang, Z., Luo, Y., Zheng, P., Mao, X., Yu, J., Luo, J., Yan, H. and He, J. 2020. Fructooligosaccharides improve growth performance and intestinal epithelium function in weaned pigs exposed to enterotoxigenic *Escherichia coli*. *Food Funct.* 11(11):9599–9612.

Liu, Y., Song, M., Che, T. M., Almeida, J. A. S., Lee, J. J., Bravo, D., Maddox, C. W. and Pettigrew, J. E. 2013. Dietary plant extracts alleviate diarrhea and alter immune responses of weaned pigs experimentally infected with a pathogenic *Escherichia coli*. *J. Anim. Sci.* 91(11):5294–5306.

Liu, Y., Espinosa, C. D., Abelilla, J. J., Casas, G. A., Lagos, L. V., Lee, S. A., Kwon, W. B., Mathai, J. K., Navarro, D. M. D. L., Jaworski, N. W. and Stein, H. H. 2018. Non-antibiotic feed additives in diets for pigs: a review. *Anim. Nutr.* 4(2):113–125.

López-Vergé, S., Gasa, J., Temple, D., Bonet, J., Coma, J. and Solà-Oriol, D. 2018. Strategies to improve the growth and homogeneity of growing-finishing pigs: feeder space and feeding management. *Porc. Health Manag.* 4:14.

Lu, L., Liao, X. and Luo, X. 2017. Nutritional strategies for reducing nitrogen, phosphorus and trace mineral excretions of livestock and poultry. *J. Integr. Agric.* 16(12):2815–2833.

Marin, S., Ramos, A. J., Cano-Sancho, G. and Sanchis, V. 2013. Mycotoxins: occurrence, toxicology, and exposure assessment. *Food Chem. Toxicol.* 60:218–237.

McPherson, R. L., Ji, F., Wu, G., BlantonJr., J. R. and Kim, S. W. 2004. Growth and compositional changes of fetal tissues in pigs. *J. Anim. Sci.* 82(9):2534–2540.

Menegat, M. B., Goodband, R. D., DeRouchey, J. M., Tokach, M. D., Woodworth, J. C. and Dritz, S. S. 2019. Kansas State University Swine Nutrition Guide: specialty ingredients in nursery diets. Available at: https://www.asi.k-state.edu/extension/swine/

swinenutritionguide/pdf/KSU%20Specialty%20Ingredients%20in%20Nursery%20Diets%20fact%20sheet.pdf

Menegat, M. B., Dritz, S. S., Tokach, M. D., Woodworth, J. C., DeRouchey, J. M. and Goodband, R. D. 2020. Phase-feeding strategies based on lysine specifications for grow-finish pigs. *J. Anim. Sci.* 98(1):skz366.

Meunier-Salaün, M. C., Edwards, S. A. and Robert, S. 2001. Effect of dietary fibre on the behaviour and health of the restricted fed sow. *Anim. Feed Sci. Technol.* 90(1-2):53-69.

Mroz, Z., Jongbloed, A. W. and Kemme, P. A. 1994. Apparent digestibility and retention of nutrients bound to phytate complexes as influenced by microbial phytase and feeding regimen in pigs. *J. Anim. Sci.* 72(1):126-132.

Noblet, J., Fortune, H., Shi, X. S. and Dubois, S. 1994. Prediction of net energy value of feeds for growing pigs. *J. Anim. Sci.* 72(2):344-354.

NRC 2012. *Nutrient Requirements of Swine.* 11th Rev. ed. Washington, DC: National Academic Press.

Nyachoti, C. M., Kiarie, E., Bhandari, S. K., Zhang, G. and Krause, D. O. 2012. Weaned pig responses to *Escherichia coli* K88 oral challenge when receiving a lysozyme supplement. *J. Anim. Sci.* 90(1):252-260.

Oliphant, K. and Allen-Vercoe, E. 2019. Macronutrient metabolism by the human gut microbiome: major fermentation by-products and their impact on host health. *Microbiome* 7(1):91.

Osunbami, O. T. and Adeola, O. 2022. Energy value of hydrolyzed feather meal and flash-dried poultry protein for broiler chickens and pigs. *J. Anim. Sci.* 100(3):skac073.

Patience, J. F., Rossoni-Serão, M. C. and Gutiérrez, N. A. 2015. A review of feed efficiency in swine: biology and application. *J. Anim. Sci. Biotechnol.* 6(1):33.

Peña, S. T., Stone, F., Gummow, B., Parker, A. J., Paris, D. B. B. P., Peña, S. T., Stone, F., Gummow, B., Parker, A. J. and Paris, D. B. B. P. 2019. Tropical summer induces DNA fragmentation in boar spermatozoa: implications for evaluating seasonal infertility. *Reprod. Fertil. Dev.* 31(3):590-601.

PIC Nutrition and Feeding Guidelines 2021. Available at: https://www.pic.com/wp-content/uploads/sites/3/2021/03/PIC-Nutrition-Manual_English-Imperial.pdf

Pieper, R., Boudry, C., Bindelle, J., Vahjen, W. and Zentek, J. 2014. Interaction between dietary protein content and the source of carbohydrates along the gastrointestinal tract of weaned piglets. *Arch. Anim. Nutr.* 68(4):263-280.

Pieper, R., Tudela, C. V., Taciak, M., Bindelle, J., Pérez, J. F. and Zentek, J. 2016. Health relevance of intestinal protein fermentation in young pigs. *Anim. Health Res. Rev.* 17(2):137-147.

Pierron, A., Alassane-Kpembi, I. and Oswald, I. P. 2016. Impact of mycotoxin on immune response and consequences for pig health. *Anim. Nutr.* 2(2):63-68.

Pomar, C. and Remus, A. 2019. Precision pig feeding: a breakthrough toward sustainability. *Anim. Front.* 9(2):52-59.

Pomar, C., Andretta, I. and Remus, A. 2021. Feeding strategies to reduce nutrient losses and improve the sustainability of growing pigs. *Front. Vet. Sci.* 8:742220.

Rao, Z. X., Tokach, M. D., Woodworth, J. C., DeRouchey, J. M., Goodband, R. D. and Gebhardt, J. T. 2021. Evaluation of nutritional strategies to slow growth rate then induce compensatory growth in 90-kg finishing pigs. *Transl. Anim. Sci.* 5(3):txab037.

Renaudeau, D., Collin, A., Yahav, S., de Basilio, V., Gourdine, J. L. and Collier, R. J. 2012. Adaptation to hot climate and strategies to alleviate heat stress in livestock production. *Animal* 6(5):707–728.

Rhoads, R. P., Baumgard, L. H., Suagee, J. K. and Sanders, S. R. 2013. Nutritional interventions to alleviate the negative consequences of heat stress. *Adv. Nutr.* 4(3):267–276.

Rosero, D. S., Boyd, R. D., Odle, J. and van Heugten, E. 2016. Optimizing dietary lipid use to improve essential fatty acid status and reproductive performance of the modern lactating sow: a review. *J. Anim. Sci. Biotechnol.* 7:34.

Shurson, G. C. and Urriola, P. E. 2022. Sustainable swine feeding programs require the convergence of multiple dimensions of circular agriculture and food systems with One Health. *Anim. Front.* 12(6):30–40.

Shurson, G. C., Pelton, R. E. O., Yang, Z., Urriola, P. E. and Schmitt, J. 2022. Environmental impacts of eco-nutrition swine feeding programs in spatially explicit geographic regions of the United States. *J. Anim. Sci.* 100(12):skac356.

Sodhi, S., Sharma, A., Brar, A. P. S. and Brar, R. S. 2008. Effect of α tocopherol and selenium on antioxidant status, lipid peroxidation and hepatopathy induced by Malathion in chicks. *Pestic. Biochem. Physiol.* 90(2):82–86.

Speight, S. M., Estienne, M. J., Harper, A. F., Crawford, R. J., Knight, J. W. and Whitaker, B. D. 2012. Effects of dietary supplementation with an organic source of selenium on characteristics of semen quality and *in vitro* fertility in boars. *J. Anim. Sci.* 90(3):761–770.

Strathe, A. V., Bruun, T. S., Geertsen, N., Zerrahn, J.-E. and Hansen, C. F. 2017. Increased dietary protein levels during lactation improved sow and litter performance. *Anim. Feed Sci. Technol.* 232:169–181.

Sung, J. Y. and Kim, B. G. 2021. Prediction equations for digestible and metabolizable energy concentrations in feed ingredients and diets for pigs based on chemical composition. *Anim. Biosci.* 34(2):306–311.

Sung, J. Y., Ji, S. Y. and Kim, B. G. 2022a. Additivity of digestible energy and nutrient concentrations in hatchery byproducts fed to nursery pigs. *Anim. Biosci.* 35(3):453–460.

Sung, J. Y., Wiltafsky-Martin, M. K., González-Vega, J. C. and Adeola, O. 2022b. Autoclaving time-related reduction in metabolizable energy of poultry meal is greater in growing pigs compared with broiler chickens. *J. Anim. Sci.* 100(5):skac117.

Sung, J. Y., Johnson, T. A., Ragland, D. and Adeola, O. 2023. Impact of ileal indigestible protein on fecal nitrogen excretion and fecal microbiota may be greater compared with total protein concentration of diets in growing pigs. *J. Anim. Sci.* 101:skac409.

Temba, B. A., Sultanbawa, Y., Kriticos, D. J., Fox, G. P., Harvey, J. J. W. and Fletcher, M. T. 2016. Tools for Defusing a Major Global Food and Feed Safety Risk: nonbiological postharvest procedures to decontaminate mycotoxins in foods and feeds. *J. Agric. Food Chem.* 64(47):8959–8972.

Theil, P. K., Farmer, C. and Feyera, T. 2022. Physiology and nutrition of late gestating and transition sows. *J. Anim. Sci.* 100(6):skac176.

Tian, Q. Y. and Piao, X. S. 2019. Essential oil blend could decrease diarrhea prevalence by improving antioxidative capability for weaned pigs. *Animals* 9(10):847.

Tolosa, J., Rodríguez-Carrasco, Y., Ruiz, M. J. and Vila-Donat, P. 2021. Multi-mycotoxin occurrence in feed, metabolism and carry-over to animal-derived food products: a review. *Food Chem. Toxicol.* 158:112661.

Tran, T. H. T., Everaert, N. and Bindelle, J. 2018. Review on the effects of potential prebiotics on controlling intestinal enteropathogens Salmonella and *Escherichia coli* in pig production. *J. Anim. Physiol. Anim. Nutr.* 102(1):17–32.

Tso, K. H., Ju, J. C., Fan, Y. K. and Chiang, H. I. 2019. Enzyme degradation reagents effectively remove mycotoxins deoxynivalenol and zearalenone from pig and poultry artificial digestive juices. *Toxins* 11(10):599.

Vallet, J. L., Calderón-Díaz, J. A., Stalder, K. J., Phillips, C., Cushman, R. A., Miles, J. R., Rempel, L. A., Rohrer, G. A., Lents, C. A., Freking, B. A. and Nonneman, D. J. 2016. Litter-of-origin trait effects on gilt development. *J. Anim. Sci.* 94(1):96–105.

Van Boeckel, T. P., Brower, C., Gilbert, M., Grenfell, B. T., Levin, S. A., Robinson, T. P., Teillant, A. and Laxminarayan, R. 2015. Global trends in antimicrobial use in food animals. *Proc. Natl. Acad. Sci. U. S. A.* 112(18):5649–5654.

Vier, C. M., Dritz, S. S., Wu, F., Tokach, M. D., DeRouchey, J. M., Goodband, R. D., Gonçalves, M. A. D., Orlando, U. A. D., Chitakasempornkul, K. and Woodworth, J. C. 2019. Standardized total tract digestible phosphorus requirement of 24- to 130-kg pigs. *J. Anim. Sci.* 97(10):4023–4031.

Wähner, M., Geyer, M., Hallfarth, G. and Hühn, U. 2004. Der einflfluss von Zulagen einer vitaminemulsion mit L-Carnititin auf die spermaeigenschaftften von besamungsebern. *Zuchtungskunde* 76:196–207.

Wambacq, E., Vanhoutte, I., Audenaert, K., De Gelder, L. and Haesaert, G. 2016. Occurrence, prevention and remediation of toxigenic fungi and mycotoxins in silage: a review. *J. Sci. Food Agric.* 96(7):2284–2302.

Wang, H., Xu, R., Zhang, H., Su, Y. and Zhu, W. 2020. Swine gut microbiota and its interaction with host nutrient metabolism. *Anim. Nutr.* 6(4):410–420.

Wang, Y., Zhou, J., Wang, G., Cai, S., Zeng, X. and Qiao, S. 2018. Advances in low-protein diets for swine. *J. Anim. Sci. Biotechnol.* 9:60.

Woyengo, T. A., Beltranena, E. and Zijlstra, R. T. 2014. Nonruminant nutrition symposium: Controlling feed cost by including alternative ingredients into pig diets: A review. *J. Anim. Sci.* 92(4):1293–1305.

Woyengo, T. A., Beltranena, E. and Zijlstra, R. T. 2017. Effect of anti-nutritional factors of oilseed co-products on feed intake of pigs and poultry. *Anim. Feed Sci. Technol.* 233:76–86.

Xu, Y. T., Liu, L. I., Long, S. F., Pan, L. and Piao, X. S. 2018. Effect of organic acids and essential oils on performance, intestinal health and digestive enzyme activities of weaned pigs. *Anim. Feed Sci. Technol.* 235:110–119.

Zhai, H., Adeola, O. and Liu, J. 2022. Phosphorus nutrition of growing pigs. *Anim. Nutr.* 9:127–137.

Zhang, M., Dunshea, F. R., Warner, R. D., DiGiacomo, K., Osei-Amponsah, R. and Chauhan, S. S. 2020. Impacts of heat stress on meat quality and strategies for amelioration: a review. *Int. J. Biometeorol.* 64(9):1613–1628.

Zhang, Z., Xu, J., Zhang, X., Wang, J., Xie, H., Sun, Y., Zhang, Q., Chang, Z. and Liu, Y. 2022. Protective effect of SeMet on liver injury induced by ochratoxin A in rabbits. *Toxins* 14(9):628.

Zhu, C., Yang, J., Nie, X., Wu, Q., Wang, L. and Jiang, Z. 2022. Influences of dietary vitamin E, selenium-enriched yeast, and soy isoflavone supplementation on growth performance, antioxidant capacity, carcass traits, meat quality and gut microbiota in finishing pigs. *Antioxidants* 11(8):1510.

Chapter 4

Modelling feed requirements for pigs to optimize feed efficiency

Charlotte Gaillard, Ludovic Brossard, Florence Garcia-Launay and Jaap van Milgen, Institut Agro, France

1 Introduction

Animal production is confronted with the ongoing challenge of sustainability. According to FAO (2011), global animal production is projected to rise by approximately 70% in the coming decades to meet the growing demand for animal protein. To ensure sustainable practices, it is crucial to enhance the efficiency of animal production. This necessitates integrating various dimensions such as animal health and welfare, food quality and security, environment, and meeting consumer and citizen expectations. In pig production, feed and feeding play a significant role in controlling performance, with immediate and reversible effects. Feed constitutes a substantial portion of production costs (typically 60–70%) and significantly impacts economic outcomes. Moreover, feed also influences other aspects of sustainability, including performance, animal welfare and health, product quality, and environmental impact. The efficiency of converting dietary inputs into animal products is relatively low for most nutrients. Protein efficiency rarely surpasses 50%, while phosphorus and energy efficiencies are even lower. This highlights the importance of understanding nutritional requirements and their availability in feed ingredients to develop feeding strategies that enhance feed efficiency.

http://dx.doi.org/10.19103/AS.2024.0140.07

The nutrient requirement for specific production purposes, such as growth, protein deposition, milk production, or maintenance, can be determined when all other nutrients are provided at adequate levels (Fuller, 2004). Factors related to the animal (e.g. genetics, age, weight, sex, social status, and health), the feed (e.g. feed allowance, nutrient composition, and digestibility), and housing conditions (e.g. ambient temperature and space allowance) affect nutrient requirements (Noblet and Quiniou, 1999). Two commonly used methods for determining pig nutrient requirements are empirical and factorial methods (Patience et al., 1995). A comparison of these methods has been presented by Hauschild et al. (2010) and Pomar et al. (2013).

In the empirical approach, groups of pigs are fed with increasing levels of a nutrient while measuring one or more performance traits during a specific time period. The nutrient requirement in this method corresponds to the population requirement for the given performance and time interval. The estimated requirement may vary depending on the measured performance trait and the statistical model used for estimation (Pomar et al., 2013). However, differences in pig characteristics (e.g. adiposity in growing pigs and prolificacy in sows) and environmental effects limit the generalizability of results to other production scenarios (Pomar et al., 2003). Hence, the factorial approach is typically preferred as a reference method.

In the factorial approach, the estimated nutrient requirement includes maintenance and production requirements, as well as the efficiency of nutrient utilization (Fuller and Chamberlain, 1982; van Milgen and Noblet, 2003). The requirement is determined as the amount of the specific nutrient that enables the animal to perform its necessary functions without limiting growth (Pomar et al., 2013). For instance, the requirement for lysine is calculated by summing the lysine requirements for specific functions (e.g. basal endogenous losses and daily gain), expressed as the quantity of the nutrient needed per unit of feed intake, body weight (BW), or daily gain, as demonstrated in Pomar et al. (2013). While the empirical approach provides requirements at the population level, the factorial approach allows estimating individual animal requirements at a particular stage. Accounting for variability among pigs in a population is necessary when applying the factorial method to determine population requirements. Using the requirement of an average pig to feed the population implies that half of the population will receive insufficient nutrients, resulting in lower-than-expected performance (Pomar et al., 2003; Brossard et al., 2009; Hauschild et al., 2010).

Modelling approaches based on the factorial method have been developed since the 1970s to predict how growing pigs or sows respond to nutrient supply. These models simulate the performance of individual animals and serve as decision support tools for assessing nutrient requirements and identifying appropriate feeding strategies (e.g. van Milgen et al., 2008; NRC,

2012). However, the variation among animals has been found to impact population response, nutrient utilization efficiency, and ultimately the optimal nutrient supply for the population (Leclercq and Beaumont, 2000; Pomar et al., 2003; Brossard et al., 2009). To address this, stochasticity has been incorporated into models to account for variability and simulate group-level pig responses (e.g. Ferguson et al., 1997; Knap, 1999; Pomar et al., 2003; Vautier et al., 2013). These approaches have demonstrated the importance of considering variation among animals to improve nutrient efficiency on-farm. Monitoring devices that enable the characterization of each animal on-farm and feeding devices controlled by real-time decision support tools play a crucial role in assessing inter-animal variation.

In this chapter, we present briefly the existing modelling approaches developed to predict nutrient requirements of a single individual (growing pig or sow) in terms of protein/amino acids, energy, and minerals. We then present how variability among animals can be integrated in modelling nutrient requirements. We also illustrate how these models can be used in the development of precision feeding systems that allow improvement of feed efficiency for gestating sows.

2 Understanding feed efficiency

2.1 Definitions

Commonly, feed efficiency (FE) is defined as the ratio of mean daily BW gain to mean daily feed consumption over a given period. It can also be expressed as the ratio of growth to energy intake or inversely, and in that case it represents the efficiency of converting feed into BW gain (Gilbert, 2015). The animals with the lowest feed conversion ratio (FCR) tend to be the most efficient (Bouquet, 2013). Genetic improvement has reduced the pigs FCR (Bouquet, 2013) but recently, the idea of selecting animals with a low residual feed intake (RFI) arise. Indeed pigs with high RFI are less energy efficient because they produce more heat, mainly due to increased physical activity and basal metabolic rate (Barea et al., 2010). However, recent studies at IFIP (French Pork and Pig Institute) showed that selecting for FCR results is still more economic than selecting for RFI (Bouquet, 2013).

For fattening pigs, FCR is usually calculated from 10 weeks of age to slaughter but can also be calculated from weaning to around 10 weeks of age, or from weaning to slaughter (Gilbert, 2015). For lactating sows, improved feed efficiency translates into increased litter weight when sows eat more and therefore produce more milk. For pregnant sows, feed efficiency remains to be defined and could be calculated in two ways: (1) according to the BW of the sow and the back fat thickness at farrowing, in order to assess whether the objectives defined at insemination are achieved, or (2) according to the number of alive piglets at birth and possibly the litter weight at birth. In all cases, feed efficiency can be improved depending on the quantity and quality of feed distributed, its texture, and density, which also reduces environmental loads.

2.2 Factors

The review of Gaillard et al. (2020a) describe in details the factors affecting feed efficiency. They are listed next and their effect briefly summarized.

Spillage: Increased spillage increases the FCR. Spillage can be reduced by selecting feeder type (i.e. a linear dump feeder should be preferred to other types, Pierozan et al., 2016), feeder characteristics (individual shoulder protection, number of places per pig, and water supply), feeder localization in the pen (Averós et al., 2012), and by feeding the pigs restrictively (Patience et al., 2015) or choosing the right type of feed (i.e. pellets instead of mash diets, Vukmirović et al., 2017).

Feed intake, growth, and maintenance: Over their growth, pigs become less efficient at converting feed into BW gain (Patience et al., 2015) due to the increase in maintenance requirements and the decrease in the muscle lean-to-fat ratio. Their ingestion increased above the protein deposition capacity; therefore, fat deposition increases which decreases feed efficiency. This effect is dependent to the sex due to differences in feed intake and growth rate (Noblet et al., 1994), but also to the age of the animal. For example, during lactation, sows of parity 1 or 2 consume about 15% less feed than older sows (Koketsu et al., 1996) linked to the increase in maintenance energy requirements and increase in BW (O'Grady et al., 1985). Sows' mean daily feed intake also increases as litter size increases (i.e. mean daily feed intake of 4 kg lower with a small litter (<7 piglets) compared to larger litters, Koketsu et al., 1996). Finally, environmental factors such as space allowance, group size, number of feeders, flooring conditions, enrichment, temperature, ventilation rate, relative humidity, pathogens, and stressors influence FCR (Averós et al., 2010, 2012) due to variation in feed intake.

Digestibility of feed and digestive efficiency of pigs: FCR increases as the indigestible portion of the feed increases, which depends greatly on its composition (Sauvant et al., 2004) and on animal digestive capacity (Noblet et al., 2013). For example, a greater proportion of fibre in the diet is a challenge for digestion and results in high variability in pig performance. The digestibility of energy also depends on heritable genetic variability (Noblet et al., 2013) so selecting pigs for digestive efficiency would improve nutrient efficiency (Kyriazakis, 2011).

Composition of weight gain (lean-to-fat ratio): The chemical and tissue compositions of BW gain influence FCR. The chemical composition of empty BW gain in minerals and protein (around 3% and 16–17% of empty BW, respectively) is similar among breeds and sexes (Noblet et al., 1994) while the percentages of water and lipids are related to genotype and sex, which influence the energy content of gain. Adipose tissue yields around four times as much energy as muscle tissue because it contains less water. This results in large differences in the energy content of BW gain among genotypes (e.g. 11.2 MJ/kg in Piétrain males vs. 21.1 MJ/kg in Meishan castrates) and between sexes (12.3 MJ/kg, 13.8 MJ/kg, and 15.6 MJ/kg for Large-White males, females, and castrated males, respectively) (Noblet et al., 1994). Selection for fast growing and lean animals has increased the potential for protein deposition and reduced the amount of energy required to improve FCR (Gilbert, 2015). For a lean animal, a restriction in energy supply reduced similarly the deposition of proteins and lipids while for a fat animal, fat deposition is reduced in priority without increasing protein deposition (Gilbert, 2015).

Nutrient utilization: As described earlier, the FCR is affected by the efficiency of energy utilization, which depends on the energy content of BW gain and the effect of maintenance requirements. Similarly, to the energy, the efficiency of utilizing nutrients, such as amino acids (AA) and minerals, is also affected by nutrient digestibility and maintenance requirements. In case of nutrient undersupply, growth rate decreases resulting in an increased FCR, while in case of oversupply excretion is increased and efficiency of retention reduced. Improving nutrients digestibility and providing them over time close to individual requirements will improve their efficiency of use and reduce excretion.

2.3 Strategies to improve feed efficiency

Feed efficiency and related traits (like FCR and RFI) are a major target to select the most efficient animals. The development of new technology and automatic data recording allows for the search of new traits like feed intake and feeding behaviour (daily occupation time, daily feeder visit, and daily feeding rate), easily recorded on farm at an individual level (Reyer et al., 2017). Genetic factors are also influencing the abundance of distinct bacterial species (Benson et al., 2010) linked with FE (McCormack et al., 2019; Tan et al., 2017) but methods to phenotype the microbiota should be developed to use this information on farms. Finally, robust indicators need to be developed to quantify the sensitivity of animals to variations in the environment (Gilbert et al., 2017).

The stressors may have additive effects, so reducing or removing one or more, will improve pig performance (Black et al., 2001). For example, structural changes in buildings (i.e. addition of cooling systems, floor type, improving cleanliness, etc.) can be done to reduce climate stress and microbial load (Black et al., 2001; Renaudeau et al., 2011; Le Floc'h et al., 2006). The type of feeder (design, access, location) also has an effect on FCR (Rantanen et al., 1994).

Another way to improve FE is to adjust closely the nutrient supplies to the animals' requirements. Therefore as a first step, there is a need to estimate accurately individual requirements over time thanks to updated models, described thereafter (van Milgen et al., 2008; Gaillard et al., 2019). As a second step, automatic feeders and precision feeding strategies should be developed to refine the adjustment of nutrient supply to requirements.

3 Estimating feed requirements

3.1 Models

As mentioned earlier, the factorial approach allows for the determination of nutrient requirements for a specific performance level in a given environment. However, this performance level is influenced by various factors, including pig characteristics (e.g. age, weight, sex, health status, and genetic potential), feeding level, feed quality, and housing conditions. Additionally, this method does not consider the potential interactions among these factors. Achieving more precise nutrient requirement definitions involves establishing response curves that depict the relationship between nutritional supply and performance based on the pig's physiological status. Modelling approaches enable the integration of these response curves to predict performance (outputs from the model) using information such as feed quantity and quality, animal characteristics, and housing conditions (inputs to the model) through a set of equations. Different types of models are available (France and Thornley, 1984; Sauvant, 2005). Empirical models, also known as 'black box' models, establish relationships between inputs and outputs without relying on underlying biological mechanisms. Mechanistic models incorporate biological mechanisms to calculate outputs from inputs, making them more complex and involving multiple equations and parameters. Models that predict outputs at a specific point in time or space are referred to as static models, whereas dynamic models describe the dynamic nature of the response over a given period (e.g. the fattening period). Deterministic models utilize a fixed set of parameters and do not account for variability among animals. In contrast, stochastic models incorporate variability in model parameters, allowing for the consideration of variation among animals within a population.

3.1.1 Growing pigs

Since the 1970s, different models, most of which are dynamic, deterministic, and more or less mechanistic, have been developed to simulate pig growth and to determine nutrient requirements. They are often based on the association of empirical equations and mechanistic description of physiological or biochemical functions of the animal. The concepts used in these models have been summarized and compared by several authors (Bastianelli and Sauvant, 1997; Ferguson, 2006; Kyriazakis and Sandberg, 2006; Luiting and Knap, 2006; van Milgen et al., 2012). Whittemore and Fawcett (1974, 1976) proposed one of the first nutritional models describing pig growth, with the objective of predicting body protein and lipid deposition from the energy and protein intake. In this semi-mechanistic model, feed intake is a model input; the energy not used for maintenance or protein deposition is used for lipid deposition and is therefore considered as an energy sink. An important concept used in this model was the linear-plateau relationship between ingested energy and protein deposition, with protein deposition being dependent on the minimum lipid-to-protein deposition ratio (the linear part) or on the energy supply or maximum of protein deposition (PDmax, the plateau).

The concepts of Whittemore and Fawcett (1974, 1976) have been used and developed further in several models such as those of Black et al. (1986), Moughan et al. (1987), Pomar et al. (1991a), Ferguson et al. (1994), de Lange (1995), Emmans (1997), Wellock et al. (2003), van Milgen et al. (2008), Yoosuk et al. (2011), NRC (2012), and Strathe et al. (2015). These models are often based on the nutrient partitioning concepts of Whittemore and Fawcett (1974, 1976), which are summarized by de Lange (1995) and van Milgen et al. (2012). Ingested nutrients are partitioned between maintenance requirements (including also endogenous losses and a minimum protein turnover rate) and requirements for protein and lipid deposition and their corresponding efficiencies. Body weight and carcass traits are determined from these depositions. Compared to the approach of Whittemore and Fawcett, various modifications have been proposed, for instance concerning the change in PDmax and the relationship between ingested energy and protein deposition during growth (van Milgen et al., 2008; NRC, 2012). In these models, feed intake can be considered as an input ('push' approach) or an output ('pull' approach). In the 'push' approach, ad libitum feed intake is typically modelled as a function of BW through linear, exponential, power functions or as a Gamma function describing feed intake as multiples of maintenance (e.g. Moughan et al., 1987; Pomar et al., 1991a; de Lange, 1995; van Milgen et al., 2008; NRC, 2012). Because lipid deposition is considered as an energy sink, energy not used for maintenance and protein deposition is deposited as lipid. Unless a specific control of energy intake is included in the model (e.g. with the Gamma function where animals eat for maintenance when they approach maturity), there will be no constraint on the way lipid deposition will evolve during growth. In the models of Black et al. (1986), Ferguson et al. (1994), Emmans (1997), Wellock et al. (2003) and Yoosuk et al. (2011), a 'pull' approach was used where feed intake is calculated to satisfy the requirements of maintenance and of the potential protein and lipid deposition. In these models, explicit response curves (e.g. Gompertz functions) are used to describe the change in the potential protein and lipid deposition during growth. Actual intake is then calculated from predicted feed intake by applying constraints such as ingestion capacity of animal, feed characteristics (e.g. nutrient composition, density and water holding capacity) or environmental conditions (e.g. temperature). In this approach, when the animal tends to maturity, protein and lipid deposition tends towards zero and feed intake corresponds to the energy expenditure for maintenance.

The different models presented here predict growth, as affected by nutrient supply, animal characteristics and, in some cases, housing conditions. These models allow one to calculate nutrient requirements to achieve this growth and identify possible limiting factors. Most of these models depend on a large number of parameters that are difficult to obtain or estimate under practical conditions. Although these models are used in a research context, practical application of models requires that these models are easily accessible and user-friendly (e.g. as a dedicated software tool) and that required model inputs can be provided by the user (e.g. TMV by Werkgroep, 1991; NRC, 2012) and InraPorc by Dourmad et al., 2008; van Milgen et al., 2008). For example, the InraPorc software summarizes the phenotypic potential of a growing pig (i.e. growth potential and ad libitum feed intake) by five model parameters. These parameters can be estimated through a statistical routine using on-farm recorded data of BW and feed intake. Tools such as InraPorc can be used to estimate nutrient and energy requirements and utilization, evaluate the consequence of different feeding strategies, and identify feeding strategies that allow one to improve performance traits such as feed efficiency. An example of the change in nutrient and energy utilization during growth obtained from InraPorc is given in Fig. 1 for the utilization of digestible energy (Fig. 1a) and standardized ileal digestible (SID) lysine (Fig. 1b) as a function of BW. Van Milgen et al. (2009) illustrated the use of InraPorc to adapt feeding strategies and they simulated the performance of a pig between 27 kg and 100 kg BW having an average daily feed intake of 2.74 kg/d and an average daily gain (ADG) of 1.10 kg/d. Diets were formulated on a least-cost basis using feed ingredient prices of May 2008 in France and a net energy content of 10 MJ/kg. A first diet (D1) was formulated to contain 9.2 g SID lysine, a level 10% higher than the requirement of the animal at the start of the studied period to account for the requirement of the population. Two other diets (D2 and D3) were formulated with lower SID lysine levels corresponding to the requirement at around 60 kg BW and at the end of the studied period. Three feeding strategies were then defined: a single phase strategy with D1 (SP), a two-phase strategy with D1 offered until 60 kg BW and D2 for higher body weights (TP), and a multiphase strategy (MP) with a progressive replacement of D1 by D3 to follow as much as possible the changes in the SID lysine requirement. Growth performance obtained for the

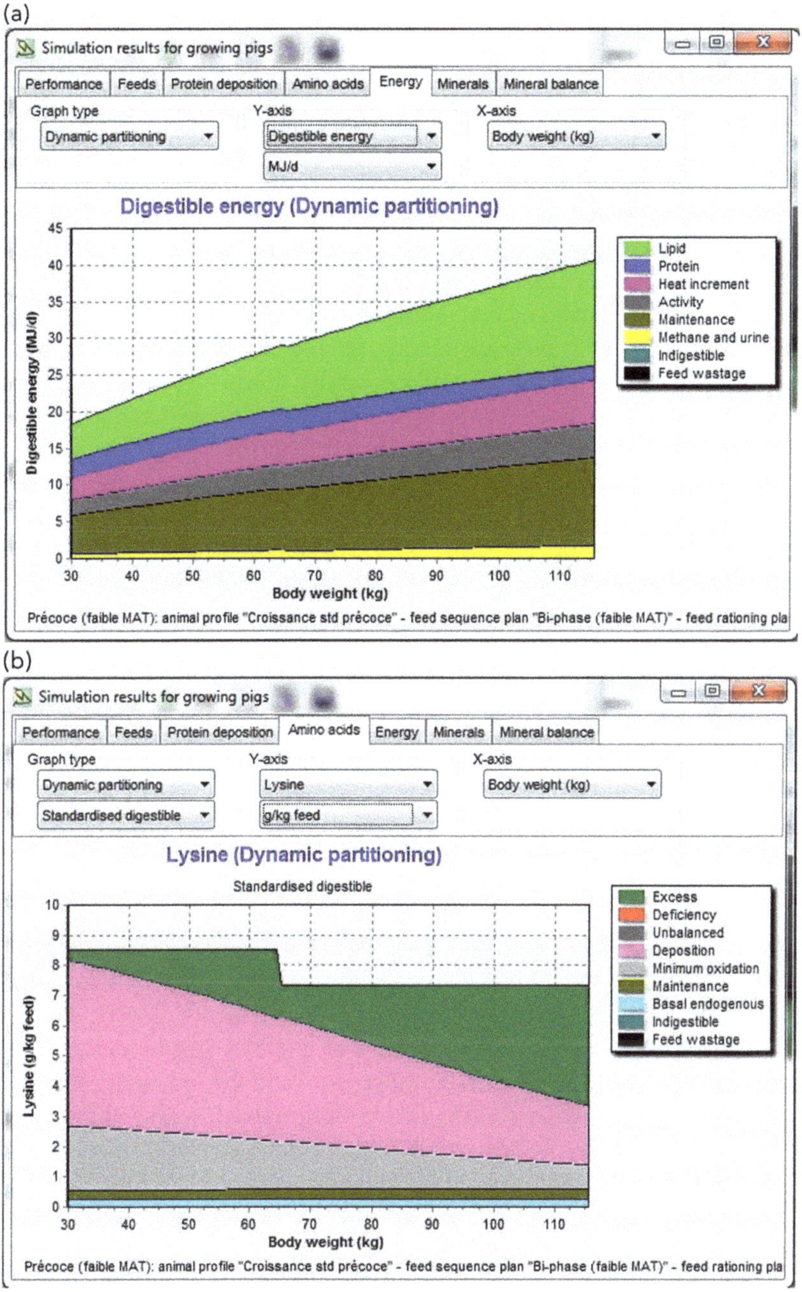

Figure 1 Simulation of changes in digestible energy (a) and lysine requirements (b) depending on body weight for animals with an average daily gain of 763 g/d and a feed to gain ratio of 2.95 kg/kg.

three strategies was identical as the diets were formulated to avoid deficiencies. However, total nitrogen (N) intake decreased from 4.79 kg for SP to 3.86 kg for MP. Consequently, N excretion decreased from 3.01 kg for SP to 2.08 for MP, and the efficiency of N utilization was increased. In addition, better adapting the nutrient supply to the requirement reduced feed costs (in comparison with the SP strategy) by 5.4% for TP and by 10.8% for MP. This example shows the interest of using modelling nutrient requirements to improve feed efficiency.

Many current growth models for pigs consider that BW, water and ash deposition, and fat and lean weight are determined from body protein and lipid weight, as suggested by Whittemore and Fawcett (1976). Van Milgen et al. (2008) and NRC (2012) estimate empty BW and BW from protein and lipid body mass without explicitly determining water and ash mass. As indicated by Létourneau-Montminy et al. (2015), these approaches imply that body ash (i.e. all body minerals) mass and growth are driven by growth of body protein and/or lipid. Létourneau-Montminy et al. (2015) observed that pigs fed deficient diets modulated growth of soft tissues and bone mineralization independently (Pomar et al., 2006; Rousseau, 2013). Consequently, total body ash can be reduced without necessarily reducing mineral deposition and growth of soft tissues.

Minerals play a crucial role in pig nutrition, particularly phosphorus (P) which is essential for bone development and metabolism in growing pigs. However, dietary phosphorus from plant sources is poorly digested, necessitating the use of expensive inorganic supplements or phytase enzymes to enhance digestion. This inefficient utilization of phosphorus also contributes to environmental issues like eutrophication. To address these challenges, precise modelling approaches are advocated to determine optimal mineral requirements, minimizing excess excretion.

Factorial models incorporating mineral requirements have been developed for growing pigs and sows. Notably, Jondreville and Dourmad (2005) estimated P requirements for various physiological stages, and these principles have been integrated into models like InraPorc for growing pigs (van Milgen et al., 2008) and sows (Dourmad et al., 2008). These models consider factors such as apparent digestible P, diet form, and the influence of phytase addition on digestibility. By adjusting dietary digestible P supply based on pig performance and physiological status, the proposed approach allows for the evaluation of performance level effects on apparent digestible P requirements. For example, a slight improvement in feed conversion ratio necessitates a corresponding increase in dietary P concentration.

Although estimating mineral requirements based on performance has limitations, specific models have been developed, particularly for phosphorus. These models enable independent variation of body mineral content, simulate phases of deficiency (e.g. bone weakening with maintained performance, followed by growth reduction), and compensatory bone mineralization. They consider the mechanisms of P intake, digestion, retention, and excretion, as well as other minerals and phytase addition. Steps of mineral digestion in the stomach, and in the small and large intestine, can be differentiated (e.g. Symeou et al., 2014a,b) as well as excretion in the urine and feces. While primarily used for research purposes, these models are intended to be integrated into decision support tools and feeding strategies to optimize growth performance, bone mineralization, and P retention while minimizing excretion.

3.1.2 Gestating sows models

In contrast to models for growing pigs, only few models describing nutrient utilization in reproductive sows have been developed (Williams et al., 1985; Dourmad, 1987; Pomar et al., 1991b; Pettigrew et al., 1992; Dourmad et al., 2008; NRC, 2012; Hansen et al., 2014), and most of these are research models. Some of these are more of calculation models to determine nutrient requirements (e.g. NRC, 2012). More mechanistic models also exist, such as InraPorc (Dourmad et al., 2008) and Hansen et al. (2014). These two models describe energy and nutrient partitioning on a daily basis using a similar structure. Recent updates have been done on the InraPorc nutritional model to answer the question of predicting requirements at the individual level of gestating and lactating sows, producing two models, one for the gestating sows and one for the lactating sows, described in two publications (Gaillard et al., 2019; Gauthier et al., 2019). These models calculate the daily nutritional requirements for each sow and each day and can be incorporated into tailor-made feeding equipment for sows. The nutritional requirements of sows are modelled by a factorial approach, which makes it possible to evaluate the requirements according to the use of nutrients by the different body compartments. This approach consists in determining for each nutrient (energy, AA, minerals) the expenses related to maintenance and production. Maintenance requirements cover basal metabolism, moderate physical activity, and thermoregulation in the thermal comfort zone. Production requirements are specific to each physiological stage. During lactation, they are due to the synthesis and secretion of milk while during gestation they are caused by the development of the uterus and the udder. The use of nutrients between different body compartments therefore changes during the reproductive cycle. In addition, the body reserves of the sow play a buffer role. The energy and excess nutrients in the feed help to build up the sow's body reserves; and in case of deficit, these reserves are mobilized.

These two models take into account the information available on farm for each sow such as their genotype, litter rank, prolificacy (number of piglets and litter weight), body condition of the sow (weight and backfat thickness) at different times (i.e. insemination, farrowing), physical activity (averaged and fixed), and type of housing (individual or group). Some information needed as inputs for the models are not available and must be estimated. For example, in the gestating sow model, the target weight to be reached at the end of gestation, and the size and the litter weight are estimated from historical data from the farm and few individual characteristics (rank of litter, backfat weight, and thickness at insemination, Gaillard et al., 2019). For lactating sows, the model requires daily milk production values and feed intake, but there is no automatic system to measure a sow's milk production and lactating sows are fed *ad libitum*. Algorithms have therefore been developed to estimate litter weight at weaning (Gauthier et al., 2020), a proxy for milk production, and the daily intake of the sow.

The two models were then evaluated with large databases from different farms (Gaillard et al., 2021, 2020b, 2019; Gauthier et al., 2019). The effects of the week of gestation or lactation, the litter rank, and the farm were significant on the daily nutrient requirements of sows and their performance (Figs 2 and 3). Simulations were then ran, still using databases, to compare a precision feeding strategy to a conventional one (fixed digestible lysine content over the gestation) for gestating and lactating sows. The precision feeding was simulated by 'mixing', every day and for each sow, two feeds that differ in their

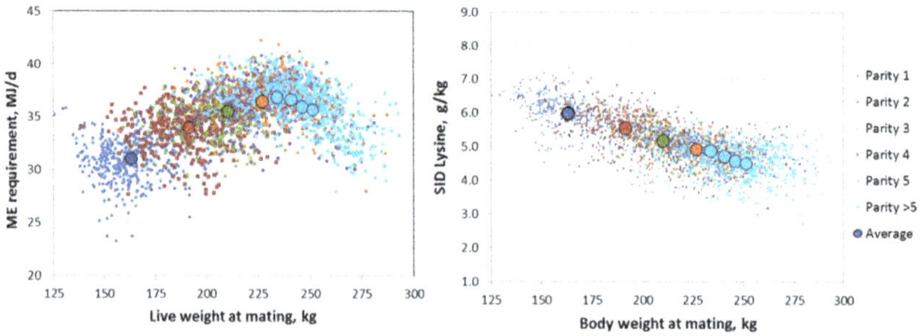

Figure 2 Variation of the energy (MJ ME/d, on the left) and lysine requirements (g/kg, on the right) between the gestating sows of different parity. Source: Gaillard et al. (2019).

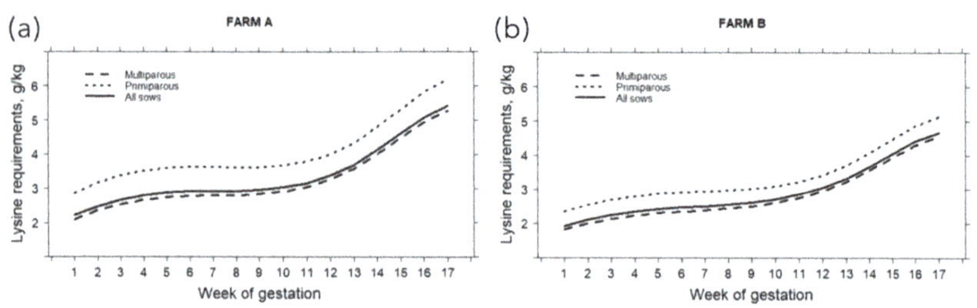

Figure 3 Variation of the digestible lysine requirements (g/kg) during sows' gestation for two farms (a and b). Source: Gaillard et al. (2020b).

nutrient content (a feed rich in lysine and a feed low in lysine, lysine being a performance-limiting amino acid). The results of these simulations showed that for gestating sows, a precision feeding strategy could reduce protein intake by approximately 25%, nitrogen excretion by 17%, phosphorus excretion by 15%, and feed cost by5% compared to a conventional feeding strategy. In addition, a higher proportion of sows would thus have their requirements covered at the end of gestation, especially the youngest, while a lower proportion would be overfed at the start of gestation, especially among the oldest. A farm effect was also observed regarding nitrogen and phosphorus excretion partly due to the fact that the sows on one farm had on average lower nutrient requirements per kilogram of feed resulting in a higher frequency of overfed animals. This highlights the importance of adjusting the nutrient content of low lysine feed for each farm. For the lactating sows, the results were similar to those simulated during gestation: digestible lysine requirements were covered for a greater proportion of sows on a precision feeding strategy compared to a conventional feeding strategy.

However, a precision feeding strategy based on adjusting lysine intake allows a limited reduction in excess phosphorus compared to nitrogen. In fact, mineral needs (i.e. phosphorus, calcium) are just as variable between sows and over time as those for lysine, but with a slightly different temporal dynamic. About 60–80% of phosphorus is stored in the bones and the remaining 20% is present in the cells and tissues of the body (Favus et al., 2006). Inadequate phosphorus intake occurs mainly at the end of the gestation and during the lactation, a period during which metabolism is intensified (NRC, 2012). This can lead to reduced efficiency of feed protein assimilation and health issues such as stunted growth in young pigs or low mineralization resulting in leg weakness for growing pigs and posterior paralysis for sows (van Riet et al., 2013). As for growing pigs, models or equations predicting mineral requirements need to be revised first; and then precision feeding adjusting feed supply based on energy, amino acids and minerals requirements simultaneously need to be develop. This will require also some technological modification, with for example feeders able to mix at least three diets instead of two. Equations for the prediction of phosphorus and calcium requirements have been recently updated and summarized in Quiniou et al. (2021), based on the physiological stage and the level of production of the sows. The authors specified that phosphorus can be supplied on a digestible basis with reference to tables of nutritive values of feedstuffs, but this is presently not the case for calcium. Consequently, different hypotheses for calcium digestibility were used and need to be updated.

3.2 Data collection and uncertainties

Data quality is crucial in utilizing sensor-generated data, which produces large volumes of data on a daily basis. To extract the relevant information from each measurement, data cleaning and sorting are necessary, often involving smoothing techniques based on the objective (Friggens and Robert, 2016). These cleaned data serve as inputs for models that predict animal nutrient requirements, necessitating strict cleaning rules and thresholds for repeatability and reliability. Additionally, precision in feed delivery is important in practical applications. However, recent experiments discovered limitations with sow feeders designed for precision feeding using two diets. The feeders required a minimum feed proportion of 50 g to ensure adequate precision, even in cases where the calculated requirements for one of the diets were lower (Gaillard et al., 2021).

To enhance the accuracy of predicting average sow energy requirements during gestation, Gaillard et al. (2019) proposed a model that initially calculated the maternal BW objective at farrowing based on the sow's age and a farm-adjusted growth curve. However, to address inter-individual variability, it would be beneficial to incorporate the actual progression of sow BW, and when possible, sow BT, into the model to adjust energy supply throughout gestation. Automated weighing scales can be utilized for collecting daily individual BW data on farms, while software advancements enable the automatic and real-time estimation of BW through video recordings (Kashiha et al., 2014; Pezzuolo et al., 2018). This practical solution eliminates the need for traditional scales and their calibration uncertainties, and it may also provide additional information on body condition. Furthermore, considering 3-D image analyses, similar to those used in dairy cows, could be valuable for evaluating sow body condition (Le Cozler et al., 2019). However, this technology has not yet been evaluated specifically in sows.

In precision feeding studies by Gaillard et al. (2020b, 2021), two diets were mixed to achieve precise nutrient supply, with a primary focus on improving amino acid (AA) availability. However, the efficiency of precision feeding in ensuring adequate phosphorus (P) supply was lower compared to AA, as the dynamics of P and AA requirements differ. To enhance precision feeding further, future research could explore formulating diets based on simultaneous AA and mineral requirements, potentially involving the mixing of three diets.

To enhance the precision of nutritional models used in precision feeding, it would be beneficial to incorporate behavioral and health measurements if their relationship with feed intake and energy/nutrient requirements is established. Recent advancements in technology have facilitated high-throughput measurements of individual sow behaviors, offering new opportunities for improved precision in feeding models.

3.3 Estimation from groups to individuals

Variability among animals plays a significant role in determining nutritional requirements, but most existing models focus on average animals and lack consideration for individual variation. Incorporating variability in models is essential for various reasons. Variability can greatly impact the profitability of production systems and the evaluation of different feeding strategies. Additionally, it is crucial for meaningful statistical testing and studying relationships between performance criteria. Recent studies have emphasized the importance of accounting for between-animal variation in evaluating biological responses and developing effective nutritional programs. Pomar et al. (2003) demonstrated the influence of genetic variance on protein deposition using a growth model, showing that increased variability led to curvilinear responses and higher nutrient requirements. Similarly, Brossard et al. (2009) highlighted the variation in lysine requirement among pigs and its dependence on feeding strategies and growth periods. Considering variability in models enables a more comprehensive understanding of nutrient utilization efficiency and optimal nutrient levels.

New technologies offer the opportunity to collect individual data in farm animals, enabling precision feeding and improved feed efficiency. Radiofrequency identification (RFID) and sensors play a vital role in automated data collection and can support real-time decision-making. Various sensors, such as electronic feeding stations and weighing systems, provide valuable information on feed intake, body weight, body composition, activity levels, and temperature. However, data processing is essential to extract meaningful information from the collected data. Cleaning and smoothing techniques are applied to eliminate abnormal values and extract the desired biological characteristics. The processed data can then be utilized as inputs for models that predict animals' nutrient requirements.

4 Improving models: the example of gestating sows

Environmental conditions and health issues or discomforts can potentially affect pigs' behaviors, and therefore their nutrient requirements (mostly energy requirement). Therefore, some factors, listed in the review of Gaillard et al. (2021), are briefly described thereafter as well as their impact on nutrient requirements.

4.1 Behavior and health: potential factors affecting nutrient requirements

4.1.1 Level of activity

Physical activity is a source of variability of energy requirements but often a unique value for all the animals is integrated into the estimation of nutritional requirements (i.e. 240 min standing per day for the gestating sows, Gaillard et al., 2019). However, it has been show that instantaneous heat production will double during standing compared to lying down (Noblet et al., 1993), and double again when walking (Marcon et al., 2020). Several factors will also affects physical activity like the sows' characteristics, the farms' management and environmental conditions (Cariolet and Dantzer, 1984). Accelerometer fixed as an ear tag on the ear of the sows can be used to record the number of movements per hour of its owner for three different positions (lying down, standing, and walking; Cornou and Lundbye-Christensen, 2012). However, they are expensive, a bit invasive for the animal and not very resistant. Standstill systems fixed on the building of the animals (Oliviero et al., 2008) or cameras with associated software (Yang et al., 2020; Durand et al., 2023) can be used instead to estimate their real-time activity of each animal.

4.1.2 Feeding and drinking behaviors

Feeding represents about 15% of total daily heat production due to eating and standing (Noblet et al., 1993). Therefore, feeding duration, which depends on the type of diet, is an interesting measurement to take into account in the calculation of the energy requirements. Other feeding behaviors like the number of non-nutritive visits to the feeder could be valuable indicators of illness or discomfort. This would require the definition of a 'baseline' number of non-feeding visits and time spent at the feeders to be able to detect variations due to an event or disease. Similarly, a lack of water supply has been linked with a decrease in feed intake, feed efficiency, and growth rate (Valros, 2018). In practice, a decrease in water consumption can be used as an indicator of reduced feed consumption, or as a predictor of diseases, especially in barns were feed intake is not recorded individually. However, water intake is highly fluctuating among sows, between days, with the feeding system, the type of diet ambient temperature, which makes it difficult to identify individual baseline and perturbations as noted for the activity level.

4.1.3 Body condition

Body weight and backfat thickness measured at insemination are used as inputs in the nutritional sow model as well as the historical herd BW for each parity, to estimate the BW curve during the gestation. To update this model and avoid using historical data, it would be useful to get individual daily body weights. To record body weight automatically and individually, foreleg weighing systems (Ramaekers et al., 1995), walk-through weighing of pigs using machine vision (Wang et al., 2008), photogrammetry to determine the three-dimensional shapes of pigs (Wu et al., 2004), and analysis of images based on videos (Kashiha et al., 2014; Pezzuolo et al., 2018) can be used. To record backfat thickness a small ultrasound scanner is used manually. It takes time and the measurements are not so precise. Automatic measurement methods are being developed like for example one based on three-dimensional (3D) shape analysis able to estimate the body condition of dairy cows (Le Cozler et al., 2019). Cariolet (1997) showed that the frequency of body lesions, stereotypies, and time spent standing after eating decreased for gestating sows when their body condition score increased. However, high body weight and backfat thickness have also been associated with locomotion and farrowing problems. Therefore, an appropriate management of body condition and consequently of feed supplies seems to be of major importance for sow welfare and performance.

4.1.4 Automatic detection of behaviors

The automatic detection of aggressive sequences in fattening pigs is possible through computer vision-based algorithms (Chen et al., 2017) with a mean accuracy of 97% for high aggression and 95.8% for medium aggression (Jensen and Yngvesson, 1998). This distinction between medium- and high-level aggressions is promising and should be integrated in nutrition models, as they will affect energy loss. To our knowledge, no automatic system of visual detection of aggression in group-housed sows exists. Aggressions can also be evaluated based on the identification of the screams emitted by the pigs (Illmann et al., 2018; Briefer et al., 2019). Indeed, vocal parameters (e.g. duration of the screams, peak frequency, fundamental frequency, noise ratio) are specific to situations encountered by animals (Briefer, 2012). Software able to automatically extract screams from background noise in growing pigs have been developed at the group level with a sensitivity of 72% and a precision of 82% (Vandermeulen et al., 2015). Here again, the difficulty would be to obtain individual data.

4.2 Environmental factors are strongly linked with nutrient requirements

Environmental parameters like the external temperature, humidity, housing conditions can affect the average daily gain, stress, posture, and eating habits of an animal. Therefore, an appropriate control of these environmental factors, could contribute to the improvement of precision farming and pig welfare.

4.2.1 Temperature and humidity

For sows kept indoors, it is recommended to maintain a room temperature between 16°C and 20°C (Verstegen and Curtis, 1988; Black et al., 1993). For growing and finishing pigs fed *ad libitum*, lower critical temperatures have been estimated at 13–14°C and 10–11°C, respectively. Temperatures outside these ranges may induce heat or cold stress with negative effects on animal health, welfare (Wegner et al., 2014a,b), and feed intake which will therefore lead to under or over nutrients supply. This range of temperature varies with the housing conditions (type of floor, humidity of the room, etc.), feeding and animals characteristics, which therefore underline the need to adjust individually feed allowances with the temperature. For example, Holmes and Close (1977) calculated that a thin sow weighing 140 kg should consume additional feed (12 kJ ME/d) at the rate of 59 g d^{-1} $°C^{-1}$ of coldness, while a fat sow of the same BW should consume additional feed at the rate 34 g d^{-1} $°C^{-1}$. When considering heat stress, barrows and gilts ate about 60–100 g less feed each day per °C of heat stress (Heitman and Hughes, 1949; Heitman et al., 1958). However, gestating sows are fed restrictively and heat stress should therefore rarely affect their feed intake. A temperature-humidity index (THI; Thom, 1959) is often used to define the level of heat stress in farm animals (Berman et al., 2016; Wegner et al., 2014b, 2016), taking into account the air temperature and the relative humidity. This THI index could easily be added to the nutritional models to adjust daily supply based on room conditions. Finally, air quality and temperature should be controlled simultaneously and continuously to optimize the need for a ventilation system and guarantee the health and productivity of the pigs as high levels of noxious gases could decrease appetite and average daily body weight gain at temperatures from 17°C to 28°C for growing-finishing pigs (Massabie et al., 1997). At 24°C and 28°C, increasing air velocity to renew the air and avoid increasing gas concentrations increased the growth rate and feed consumption, but decreased feed efficiency, due to higher fat deposition.

4.2.2 Floor and bedding

The type of floor is an important insulation factor that can affect the low critical temperature (LCT) of the animals. For example, an increase in LCT has been reported when the pigs were housed on slatted floor (+3–4°C, Verstegen and van der Hel, 1974; Mount, 1975) or wet floor (Mount, 1975) compared to a well insulated concrete floor. The presence of bedding should also be carefully considered. For example, straw bedding decreases LCT of 4°C (Verstegen and Curtis, 1988) and under cold conditions the straw eaten by the animals significantly contributes to thermoregulation (Noblet et al., 1989). Moreover, bedding often allow the sows to express their natural behaviour, and to spend less time and energy for abnormal behaviour (Cronin et al., 1986; Croney and Millman, 2007).

4.2.3 Light and noise

In pig farms, lighting requirements in the EU mandate a minimum of 40 lux for 8 h per day, as light intensity has been associated with productive benefits. Stevenson et al. (1983) found that litter weight increased by 141 ± 6 g with every 10 lux increase in light intensity. Increased duration of light exposure (14-16 hours per day) in pigs, as observed in Martelli et al. (2005, 2015), resulted in more resting time and less time spent on other behaviours, potentially reducing energy consumption and waste. Noise levels in European pig farms should not exceed 85 dB for extended periods, with avoidance of sudden loud noises as recommended by the EU Directive (2008). Noise variations in barns from equipment and management practices, can lead to behavioral and physiological changes. Prolonged or intermittent exposure to noise levels of 90 dB has been associated with decreased growth rates in pigs (Otten et al., 2004) and can induce increased activity and elevated stress hormone levels (Bowles, 1995; Kanitz et al., 2005; Turner et al., 2005).

Environmental, behavioural, and health measurements could be recorded automatically in the aim to be integrated into the nutritional models to adjust feed intake or nutrient supplies individually. However, new tools should be developed to extract knowledge from these data with the objectives to characterize the recorded behaviors (normal/abnormal) and to understand the interactions between animals, and the effects of disturbing events.

5 Case study: optimising feed efficiency in gestating sows

Using modelling approaches to predict individual nutrient requirements and propose feeding strategies from groups to individuals has been shown by simulation to improve gestating sows feed efficiency (Gaillard et al., 2019). A data set of 2511 gestations reporting sow characteristics at insemination and their farrowing performance was used as an input for a Python model, adapted from InraPorc, predicting nutrient requirement during gestation. Litter size, gestation weeks, and parity have been reported to significantly affect total metabolizable energy (ME) requirement, standardized ileal digestible lysine (SID Lys) requirement, standardized total tract digestible phosphorus (STTD-P) and total calcium (Total-Ca) requirements. These results underline the need for different diets that vary in amino acid and mineral composition, according to gestation stage and parity, showing the interest of a multiphase feeding strategy which could easily be set up in practice by grouping the sows according to their parity and gestation stage (early, late), and moving them to the feed line

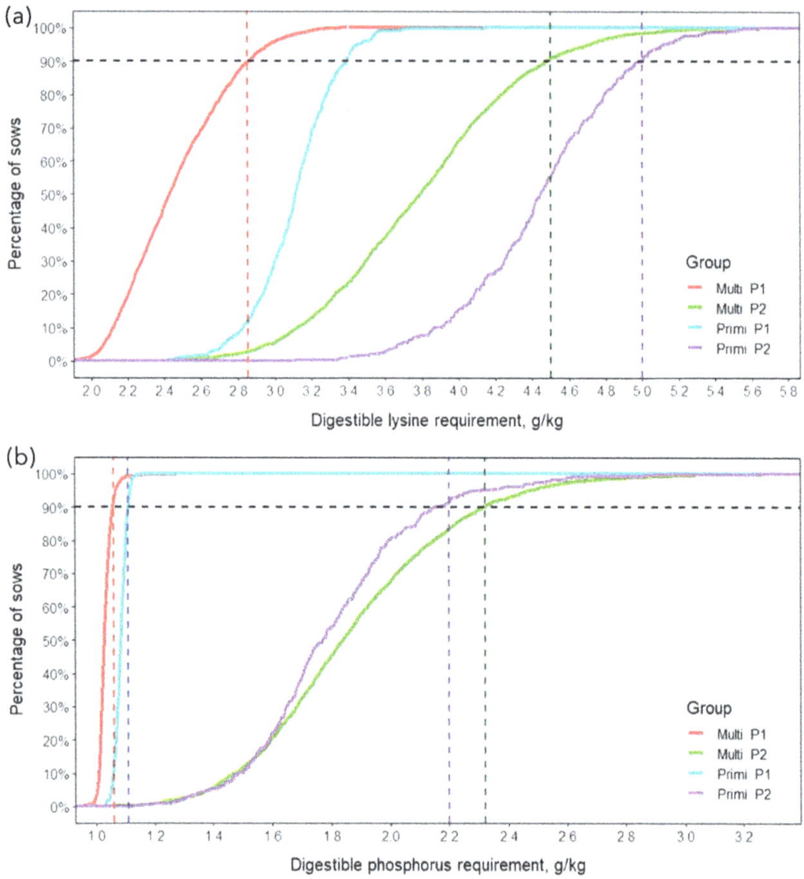

Figure 4 Cumulative distribution of the SID lysine requirement per kg of feed (a) and the digestible phosphorus requirement per kg of feed (b) according to group (multiparous in P1, multiparous in P2, primiparous in P1, primiparous in P2 with P1 being the period from weeks 0 to 11, and P2 from weeks 12 to 17). These plots can be used to determine the concentrations of SID lysine and digestible phosphorus (STTD-P) needed to satisfy the requirements of 90% of the sows (vertical dotted lines). Source: Gaillard et al. (2019).

that carries the appropriate ration. Therefore, based on empirical cumulative distribution functions, a 4-diets strategy, varying in SID Lys and STTD-P content according to parity (primiparous vs. multiparous), and gestation period (early vs. late gestation period) may be put forward to meet the requirements of 90% of the sows (Fig. 4). Two diets were proposed for multiparous sows, one diet in early gestation with low SID Lys and STTD-P contents (2.8 g/kg and 1.1 g/kg, respectively), and one diet with higher SID Lys and STTD-P contents (4.5 g/kg and 2.3 g/kg, respectively). Similar diets were proposed for primiparous cows with slightly higher SID Lys contents (early: 3.4 g SID Lys/kg and 1.1 g STTD-P/

kg; late: 5.0 g SID Lys/kg, 2.2 g STTD-P/kg). Nevertheless, looking at the huge variability among sows of the same parity group, this strategy would only constitute a first step towards precision feeding. The advantage of this 'group–multiphase feeding strategy' being that it can be easily put in place on farm without deep renovations or costs required for individual precision feeding strategies (i.e. cost of smart feeders). As a second step, a precision feeding (PF) with individualized feeding curves adjusted regarding lysine requirements has been simulated and compared to a conventional feeding (CF) strategy in two farms (A and B) for sows during gestation (Gaillard et al., 2020b). The calculated SID Lys of the PF strategy was 27% and 32% lower than CF for farms A and B, respectively. The simulated SID Lys retention was higher in PF compared with CF sows (45.7% vs. 33.3% respectively), with a greater effect in farm B than in farm A. The simulated deposition of SID Lys increased with gestation stage from an average of 3.48 (farm A) and 3.93 (farm B) g/kg over the first 8 weeks of gestation, to an average of 8.21 (farm A) and 8.80 (farm B) g/kg during the last 2 weeks of gestation. The PF strategy reduced the proportion of sows receiving lysine below their requirements compared to CF, especially in the last 2 weeks of gestation (1.05% and 43.6% in farm A, respectively, and 0.24% and 13.6% in farm B, respectively). The proportion of sows fed in excess of lysine was also reduced with the PF strategy, with a more marked effect in primiparous sows.

6 Conclusion and future trends

A precise knowledge of nutrient requirements in growing pigs and reproductive sows is required to better adapt feeding strategies and thus increase feed efficiency and economic results, and reduce the environmental impact of pig production. Modelling approaches have been developed to integrate this knowledge and allow taking into account the possible interactions between the pig and production conditions. Recent development to integrate the effect of individual variability on requirement estimations allowed better accounting for these interactions. Even though additional research is still required, modelling approaches appear to be interesting alternatives and complements to experimentation.

One could argue that most nutritional models for pigs that have been developed during the last 50 years are conceptually very similar to the first model proposed by Whittemore and Fawcett. What can and will bring new life to modelling is the massive availability of data that occurred during the last decade. Rather than predicting growth from some equation describing feed intake as a function of body weight (or the other way around in a pull approach), we can now measure these traits frequently and on an individual basis. Because both feed intake and growth can be observed, it allows addressing (phenotypic or genotypic) variation in nutritional models. Why do two animal that eat the same quantity of feed growth differently? Is it about because of differences in physical activity (i.e. energy expenditure), difference in body composition (i.e. energy partitioning), or the efficiency of nutrient and energy utilization?

Precision livestock feeding calls upon models to integrate different sources of information and then predict an expected outcome. Mechanistic models have been used for this purpose but the availability of data and the development of machine-learning techniques nowadays provide an alternative for mechanistic models. The question therefore arises if we need to understand the underlying biological mechanisms to make accurate predictions. At first sight, one may say no, because it is likely that machine-learning will outperform mechanistic models in predictive ability. However, the inputs that are now used in precision livestock feeding are on a rather aggregated scale (e.g. body weight, feed intake, images). This may change if measurements can be made on lower biological levels of 'aggregation' (e.g. body temperature, saliva, urine, feces). Which nutrient should be targeted if not only growth but also the saliva cortisol or urinary creatinine contents differs between pigs? Understanding of the biological mechanisms are then important to identify the appropriate nutritional (or other) levers. This because all the more important when precision feeding becomes more and more 'precise' and nutritional safety margins are reduced. This argues in favor of 'grey box' models combining advantages of mechanistic models and 'black box' models originating from machine learning process. This grey box type also contributes to answer questions arising about explicability of machine learning models; this is important to make them applicable to farm situations and acceptable by users. Furthermore, consideration about sensors are to be integrated. In one side, we can expect the development of sensors to obtain information at lower biological levels; complexity of some criteria (health indicators for instance) requires also to develop analysis to obtain information by combining data from different sensors taking into account time dynamic of parameters. On the other side, there are concerns about the price and the environmental impact of these sensors. So there are compromises to find between the number of sensors (to reduce the cost and the impact of their production), the frequency of data collection (to extend service life of sensors) and the type of data to collect depending of our ability to combine this data to obtain the expected information. This data availability will influence the way the models will be set up to adapt to data available. Other applications than precision feeding can be imagined for nutritional models. For instance, they can be integrated in models of farm functioning. Threshold effects of level of nutrient supply depending on feeding strategies can indeed be observed when variability among animals is considered. So nutritional models can help to explore animal performance, economic results, and environmental impacts at batch, building or farm level, depending on farmer strategies about feeding or batch management for instance. In a context of climate change and its impact on temperature at some periods (e.g. heat waves), and of competition for resources (impacting feed composition in terms of fibres for instance), better taking into account nutritional impact of strategies to cope with these aspects in farm-level models can help to develop decision-making tool. For instance, development on integration of temperature effect in nutritional model to be integrating in farm level tool is in progress (ThermiPig model, Brossard et al., 2019; Quiniou et al., 2021).

In conclusion, nutritional models will still be developed and improved to follow knowledge and technical evolutions as well as users' expectations that are diversified.

7 Where to look for further information

- A good introduction to the subject for non-specialists is provided by the article: Noblet et al. (2016), 'Etat des lieux et perspectives sur la nutrition du porc', Journées Rech. Porcine, 48, 85-96.
- Some books exist on feed efficiency and modelling: 'Mechanistic modelling in pig and poultry production' (2006, Eds. R. Gous, T. Morris and C. Fisher, CABI Publishing, Wallingford, UK); 'Mathematical modelling in animal nutrition' (2008, Eds. J. France and E. Kebreab, CABI Publishing, Wallingford, UK); 'Feed Efficiency in Swine' (2012, Ed. J. F. Patience, Wageningen Academic Publishers, Wageningen, The Netherlands); 'Nutritional modelling for pigs and poultry' (2015, Eds. N. K. Sakomura, R. M. Gous, I. Kyriazakis and L. Hauschild, Cabi Publication, Wallingford, UK).
- Every year, the European Federation of Animal Science (EAAP, www.eaap .org) organizes its annual meeting with sessions on pigs with a focus on feed use or feed efficiency, and also since 2016, a dedicated commission for PLF (precision livestock farming). Every 5 years, the international Workshop 'Modelling Nutrient Digestion and Utilization in Farm Animals' (Modnut) is organized.

A list of research centres readers can investigate, for example, for possible collaboration as well as to keep up with research trends:

a) French National Institute for Agricultural Research (INRA) (www.inra.fr/);
b) Agriculture and Agri-Food Canada (AAC) (www.agr.gc.ca/);
c) Wageningen University and Research (WUR) (https://www.wur.nl/); and
d) Newcastle University (www.ncl.ac.uk/).

8 References

Averós, X., Brossard, L., Dourmad, J. Y., de Greef, K. H., Edge, H. L., Edwards, S. A. and Meunier-Salaun, M. C. (2010). A meta-analysis of the combined effect of housing and environmental enrichment characteristics on the behaviour and performance of pigs, *Appl. Anim. Behav. Sci.* 127(3–4), 73–85. https://doi.org/10.1016/j.applanim.2010.09.010.

Averós, X., Brossard, L., Dourmad, J. Y., de Greef, K. H., Edwards, S. A. and Meunier-Salaün, M. C. (2012). Meta-analysis on the effects of the physical environment, animal traits, feeder and feed characteristics on the feeding behaviour and performance of growing-finishing pigs, *Animal* 6(8), 1275–1289. https://doi.org/10.1017/S1751731112000328.

Barea, R., Dubois, S., Gilbert, H., Sellier, P., van Milgen, J. and Noblet, J. (2010). Energy utilization in pigs selected for high and low residual feed intake, *J. Anim. Sci.* 88(6), 2062–2072. https://doi.org/10.2527/jas.2009-2395.

Bastianelli, D. and Sauvant, D. (1997). Modelling the mechanisms of pig growth, *Livest. Prod. Sci.* 51(1–3), 97–107.

Benson, A. K., Kelly, S. A., Legge, R., Ma, F., Low, S. J., Kim, J., Zhang, M., Oh, P. L., Nehrenberg, D., Hua, K., Kachman, S. D., Moriyama, E. N., Walter, J., Peterson, D. A. and Pomp, D. (2010). Individuality in gut microbiota composition is a complex polygenic trait shaped by multiple environmental and host genetic factors, *Proc. Natl. Acad. Sci. U. S. A.* 107(44), 18933–18938.

Berman, A., Horovitz, T., Kaim, M. and Gacitua, H. (2016). A comparison of THI indices leads to a sensible heat-based heat stress index for shaded cattle that aligns temperature and humidity stress, *Int. J. Biometeorol.* 60(10), 1453–1462.

Black, J. L., Campbell, R. G., Williams, I. H., James, K. J. and Davies, G. T. (1986). Simulation of energy and amino acid utilisation in the pig, *Res. Dev. Agric.* 3(3), 121–145.

Black, J. L., Mullan, B. P., Lorschy, M. L. and Giles, L. R. (1993). Lactation in the sow during heat stress, *Livest. Prod. Sci.* 35(1–2), 153–170.

Black, J. L., Giles, L. R., Wynn, P. C., Knowles, A. G., Kerr, C. A., Jones, M. R., Storm, A. D., Gallagher, N. L. and Eamens, G. J. (2001). A review. Factors limiting the performance of growing pigs in commercial environments. In 8th Biennial Conf. Aust. Pig Sci. Assoc. Adelaide, pp. 9–36.

Bouquet, A. (2013). Vers une sélection d'animaux encore plus efficaces!, *Tech.Porc* 14, 30–31.

Bowles, A. E. (1995). Responses of wildlife to noise. In *Wildlife and Recreationists: Coexistence through Management and Research* (Eds. Knight, R., Gutzwiller, K.), Island Press, Covelo, CA, pp. 109–156.

Briefer, E. F. (2012). Vocal expression of emotions in mammals: Mechanisms of production and evidence, *J. Zool.* 288(1), 1–20. https://doi.org/10.1111/j.1469-7998.2012.00920.x.

Briefer, E., Linhart, P., Policht, R., Spinka, M., Leliveld, L. M. C., Düpjan, S., Puppe, B., Padilla de la Torre, M., Janczak, A. M., Bourguet, C., Deiss, V., Boissy, A., Guérin, C., Read, E., Coulon, M., Hillmann, E. and Tallet, C. (2019). Vocal expression of emotional valence in pigs across multiple call types and contexts, *Peer J Prepr.* 7, e27934v27931.

Brossard, L., Dourmad, J. Y., Rivest, J. and van Milgen, J. (2009). Modelling the variation in performance of a population of growing pig as affected by lysine supply and feeding strategy, *Animal* 3(8), 1114–1123.

Brossard, L., Cadero, A., Dourmad, J.-Y., Renaudeau, D., Garcia-Launay, F., Marcon, M. and Quiniou, N. (2019). Combining a bioclimatic and a growth model to assess the effect of management practices and building ambiance on growing pig performances at the batch level. 9. Workshop on Modelling Nutrient Digestion and Utilization in Farm Animals (Modnut), 2019-09-14, Itamambuca, Ubatuba, Brazil.

Cariolet, R. (1997). Evaluation du bien être chez la truie gestante bloquée. Relation entre le bien être et la productivité numérique, *Journée Rech, Porcine* 29, 149-160.

Cariolet, R. and Dantzer, R. (1984). Motor activity in tethered sows during pregnancy, *Ann. Rech. Vet.* 15(2), 257-261.

Chen, C., Zhu, W., Ma, C., Guo, Y., Huang, W. and Ruan, C. (2017). Image motion feature extraction for recognition of aggressive behaviors among group-housed pigs, *Comput. Electron. Agri.* 142, 380-387.

Cornou, C. and Lundbye-Christensen, S. (2012). Modeling of sows diurnal activity pattern and detection of parturition using acceleration measurements, *Comput. Elec Agri.* 80, 97-104.

Croney, C. C. and Millman, S. T. (2007). Board-invited review: The ethical and behavioral bases for farm animal welfare legislation, *J. Anim. Sci.* 85(2), 556-565.

Cronin, G. M., van Tartwijk, J. M. F. M., van der Hel, W. and Verstegen, M. W. A. (1986). The influence of degree of adaptation to tether-housing by sows in relation to behaviour and energy metabolism, *Anim. Sci.* 42(2), 257-268.

Dourmad, J. Y. (1987). Composition du gain de poids de la truie gestante: prévision en fonction des apports énergétiques et protéiques, *Journées Rech, Porcine* 19, 203-214.

Dourmad, J. Y., Étienne, M., Valancogne, A., Dubois, S., van Milgen, J. and Noblet, J. (2008). InraPorc: A model and decision support tool for the nutrition of sows, *Anim. Feed Sci. Technol.* 143(1-4), 372-386.

Durand, M., Dourmad, J. Y. and Gaillard, C. (2023). Prédiction de l'activité physique journalière de truies gestantes à partir de créneaux ponctuels d'analyse vidéo manuelle, *Journées Rech. Porcine* 55, 2 pages.

Emmans, G. C. (1997). A method to predict the food intake of domestic animals from birth to maturity as a function of time, *J. Theor. Biol.* 186(2), 189-199.

FAO. (2011). *World Livestock 2011. Livestock in Food Security*, Food and Agriculture Organization of the United Nations, Rome.

Favus, M. J., Bushinsky, D. A. and Lemann, J. Jr. (2006). Chapter 13. Regulation of calcium, magnesium and phosphate metabolism. In *Primer on the Metabolic Bone Diseases and Disorders of Bone Metabolism* (Eds. Favus, M. J.), American Society for Bone and Mineral Research, Washington, DC, pp. 76-83.

Ferguson, N. S. (2006). Basic concepts describing animal growth and feed intake. In *Mechanistic Modelling in Pig and Poultry Production* (Eds. Gous, R., Morris, T., Fisher, C.), CAB International Publishing, Wallingford, pp. 22-53.

Ferguson, N. S., Gous, R. M. and Emmans, G. C. (1994). Preferred components for the construction of a new simulation-model of growth, feed intake and nutrient-requirements of growing pigs, *S. Afr. J. Anim. Sci.* 24(1), 10-17.

France, J. and Thornley, J. M. H. (1984). *Mathematical Models in Agriculture*, Butterworths, London.

Friggens, N. and Robert, P. E. (2016). Chapter 2 - Faire émerger les informations clés des données de l'élevage de précision. In *Elevage de précision*, Editions France Agricole, Paris, 12-28.

Fuller, M. F. (2004). *The Encyclopedia of Farm Animal Production*, CABI Publishing, Wallingford, Oxon.

Fuller, M. F. and Chamberlain, A. G. (1982). Protein requirements of pigs. In *Recent Advances in Animal Nutrition* (Ed. Haresign, W.), Butterwoths, London, pp. 175–186.

Gaillard, C., Gauthier, R., Cloutier, L. and Dourmad, J. Y. (2019). Exploration of individual variability to better predict the nutrient requirements of gestating sows, *J. Anim. Sci.* 97(12), 4934-4945.

Gaillard, C., Brossard, L. and Dourmad, J. Y. (2020a). Review: Improvement of feed and nutrient efficiency in pig production through precision feeding, *Anim. Feed Sci. Technol.* 268, 114611.

Gaillard, C., Quiniou, N., Gauthier, R., Cloutier, L. and Dourmad, J. Y. (2020b). Evaluation of a decision support system for precision feeding of gestating sows, *J. Anim. Sci.* 98(9), skaa255.

Gaillard, C., Durand, M., Largouët, C., Dourmad, J. Y. and Tallet, C. (2021). Review - Effects of the environment and animal behavior on nutrient requirements for gestating sows: Future improvements in precision feeding, *Anim. Feed Sci. Technol.* 279, 115034. https://doi.org/10.1016/j.anifeedsci.2021.115034.

Gauthier, R., Largouët, C., Gaillard, C., Cloutier, L., Guay, F. and Dourmad, J. Y. (2019). Dynamic modeling of nutrient use and individual requirements of lactating sows, *J. Anim. Sci.* 97(7), 2822–2836.

Gauthier, R., Largouët, C. and Dourmad, J. Y. (2020). Machine learning algorithms to predict litter weight gain from birth to weaning in swine. 71. Annual Meeting of the European Federation of Animal Science (EAAP), Virtual meeting, Portugal.

Gilbert, H. (2015). Sélection pour l'efficacité alimentaire chez le porc en croissance: opportunités et challenges, *Journées Rech, Porcine* 47, 19–30.

Hansen, A. V., Strathe, A. B., Theil, P. K. and Kebreab, E. (2014). Energy and nutrient deposition and excretion in the reproducing sow: Model development and evaluation, *J. Anim. Sci.* 92(6), 2458–2472.

Hauschild, L., Pomar, C. and Lovatto, P. A. (2010). Systematic comparison of the empirical and factorial methods used to estimate the nutrient requirements of growing pigs, *Animal* 4(5), 714–723.

Heitman, H. and Hughes, E. H. (1949). The effects of air temperature and relative humidity on the physiological well being of swine, *J. Anim. Sci.* 8(2), 171–181.

Heitman, H., Kelley, C. F. and Bond, T. E. (1958). Ambient air temperature and weight gain in swine, *J. Anim. Sci.* 17(1), 62–67.

Holmes, C. W. and Close, W. H. (1977). The influence of climatic variables on energy metabolism and associated aspects of productivity in pigs. In *Nutrition and the Climatic Environment* (Eds. Haresign, W., Swan, H., Lewis, D.), Butterworths, London, pp. 51–74.

Illmann, G., Leszkowov, I. and Simeckova, M. (2018). Do sows respond to sibling competition at the udder Day 1 post-partum?, *Appl. Anim. Behav. Sci.* 200, 51–55. https://doi.org/10.1016/j.applanim.2017.11.009.

Jensen, P. and Yngvesson, J. (1998). Aggression between unacquainted pigs—sequential assessment and effects of familiarity and weight, *Appl. Anim. Behav. Sci.* 58(1-2), 49–61. https://doi.org/10.1016/S0168-1591(97)00097-X.

Jondreville, C. and Dourmad, J. Y. (2005). Le phosphore dans la nutrition des porcs, INRA (Prods.), *Animal* 18(3), 183–192.

Kanitz, E., Otten, W. and Tuchscherer, M. (2005). Central and peripheral effects of repeated noise stress on hypothalamic-pituitary-adrenocortical axis in pigs, *Livest. Prod. Sci.* 94(3), 213–224.

Kashiha, M. A., Bahr, C., Ott, S., Moons, C. P. H., Niewold, T. A., Tuyttens, F. and Berckmans, D. (2014). Automatic monitoring of pig locomotion using image analysis, *Livest. Sci.* 159, 141–148. https://doi.org/10.1016/j.livsci.2013.11.007.

Knap, P. W. (1999). Simulation of growth in pigs: Evaluation of a model to relate thermoregulation to body protein and lipid content and deposition, *Anim. Sci.* 68(4), 655–679.

Koketsu, Y., Dial, G. D., Pettigrew, J. E., Marsh, W. E. and King, V. L. (1996). Characterization of feed intake patterns during lactation in commercial swine herds, *J. Anim. Sci.* 74(6), 1202–1210. https://doi.org/10.2527/1996.7461202x.

Kyriazakis, I. (2011). Opportunities to improve nutrient efficiency in pigs and poultry through breeding, *Animal* 5(6), 821–832. https://doi.org/10.1017/S1751731110002545.

Kyriazakis, I. and Sandberg, F. B. (2006). The problem of predicting the partitioning of scarce resources during sickness and health in pigs. In *Mechanistic Modelling in Pig and Poultry Production* (Eds. Gous, R., Morris, T., Fisher, C.), CAB International Publishing, Wallingford, pp. 117–142.

de Lange, C. F. M. (1995). Framework for a simplified model to demonstrate principles of nutrient partitioning for growth in the pig. In *Modelling Growth in the Pig* (Eds. Moughan, P. J., Verstegen, M. W. A., Visser-Reynevelds, M. I.), Wageningen Pers, Wageningen, pp. 71–85.

Leclercq, B. and Beaumont, C. (2000). Etude par simulation de la réponse des troupeaux de volailles aux apports d'acides aminés et de protéines, *INRA Prod. Anim.* 13(1), 47–59.

Le Cozler, Y., Allain, C., Caillot, A., Delouard, J. M., Delattre, L., Luginbuhl, T. and Faverdin, P. (2019). High-precision scanning system for complete 3D cow body shape imaging and analysis of morphological traits, *Comput. Electron. Agric.* 157, 447–453. https://doi.org/10.1016/j.compag.2019.01.019.

Le Floc'h, N., Jondreville, C., Matte, J. J. and Seve, B. (2006). Importance of sanitary environment for growth performance and plasma nutrient homeostasis during the post-weaning period in piglets, *Archiv. Anim. Nutri.* 60, 23–34. https://doi.org/10.1080/17450390500467810.

Létourneau-Montminy, M. P., Narcy, A., Dourmad, J. Y., Crenshaw, T. D. and Pomar, C. (2015). Modeling the metabolic fate of dietary phosphorus and calcium and the dynamics of body ash content in growing pigs, *J. Anim. Sci.* 93(3), 1200–1217.

Luiting, P. and Knap, P. W. (2006). Comparison of pig growth models – the genetic point of view. In *Mechanistic Modelling in Pig and Poultry Production* (Eds. Gous, R., Morris, T., Fisher, C.), CAB International Publishing, Wallingford, pp. 260–281.

Marcon, M., Quiniou, N., Courboulay, V., Rousselière, Y., Melot, G., Meunier-Salaün, M. C., Labussière, E., Ramonet, Y., Cherel, P. and Le Mer, M. (2020). Améliorer les performances et le bien être des truies gravides par la mobilisation de nouvelles technologies pour une alimentation de précision et la détection de signaux comportementaux, *Innov, Agron. INRAE* 79, 245–256.

Martelli, G., Scalabrin, M., Scipioni, R. and Sardi, L. (2005). The effects of the duration of the artificial photoperiod on the growth parameters and behaviour of heavy pigs, *Vet. Res. Commun.* 29(Suppl. 2), 367–369.

Martelli, G., Nannoni, E., Grandi, M., Bonaldo, A., Zaghini, G., Vitali, M., Biagi, G. and Sardi, L. (2015). Growth parameters, behavior, and meat and ham quality of heavy pigs subjected to photoperiods of different duration, *J. Anim. Sci.* 93(2), 758–766.

Massabie, P., Grainer, R. and Dividich, S. L. (1997). Effects on environment conditions on the performance of growing-finishing pig. In Proceedings of the 5th International Symposium on Livestock Environment, American Society of Agricultural & Biological Engineers (ASABE), St. Joseph, MI, pp. 1010–1016.

McCormack, U. M., Curiao, T., Metzler-Zebeli, B. U., Magowan, E., Berry, D. P., Reyer, H., Prieto, M. L., Buzoianu, S. G., Harrison, M., Rebeiz, N., Crispie, F., Cotter, P. D., O'Sullivan, O., Gardiner, G. E. and Lawlor, P. G. (2019). Porcine feed efficiency associated intestinal microbiota and physiological traits: Finding consistent cross-locational biomarkers for residual feed intake, *Systems* 4(4), 324–318. https://doi.org/10.1128/mSystems.00324-18.

Moughan, P. J., Smith, W. C. and Pearson, G. (1987). Description and validation of a model simulating growth in the pig (20–90 kg liveweight), *New Zeal, J. Agric. Res.* 30(4), 481–489.

Mount, L. E. (1975). The assessment of thermal environment in relation to pig production, *Livest. Prod. Sci.* 2(4), 381–392.

Noblet, J. and Quiniou, N. (1999). Principaux facteurs de variation du besoin en acides aminés du porc en croissance, *Techni.-Porc* 22(4), 9–16.

Noblet, J., Dourmad, J. Y., Le Dividich, J. and Dubois, S. (1989). Effect of ambient temperature and addition of straw or alfalfa in the diet on energy metabolism in pregnant sows, *Livest. Prod. Sci.* 21(4), 309–324. https://doi.org/10.1016/0301-6226(89)90091-2.

Noblet, J., Shi, X. S. and Dubois, S. (1993). Energy cost of standing activity in sows, *Livest. Prod. Sci.* 34(1–2), 127–136.

Noblet, J., Karege, C. and Dubois, S. (1994). 'Prise en compte de la variabilité de la composition corporelle pour la prévision du besoin énergétique et de l'efficacité alimentaire chez le porc en croissance, *Journées Rech, Porcine* 26, 267–276.

Noblet, J., Gilbert, H., Jaguelin-Peyraud, Y. and Lebrun, T. (2013). Evidence of genetic variability for digestive efficiency in the growing pig fed a fibrous diet, *Animal* 7(8), 1259–1264. https://doi.org/10.1017/S1751731113000463.

NRC. (2012). *Nutrient Requirements of Swine* (11th edn.), National Academy Press, Washington, DC.

O'Grady, J. F., Lynch, P. B. and Kearney, P. A. (1985). Voluntary feed intake by lactating sows, *Livest. Prod. Sci.* 12(4), 355-365. https://doi.org/10.1016/0301-6226(85)90134-4.

Oliviero, C., Pastell, M., Heinonen, M., Heikkonen, J., Valros, A., Ahokas, J., Vainio, O. and Peltoniemi, O. A. T. (2008). Using movement sensors to detect the onset of farrowing, *Biosyst. Eng.* 100(2), 281–285.

Otten, W., Kanitz, E., Puppe, B., Tuchscherer, M., Brüssow, K. P., Nürnberg, G. and Stabenow, B. (2004). Acute and long term effects of chronic intermittent noise stress on hypothalamic-pituitary-adrenocortical and sympatho-adrenomedullary axis in pigs, *Anim. Sci.* 78(2), 271–283. https://doi.org/10.1017/S1357729800054060.

Patience, J. F., Rossoni-Serão, M. C. and Gutiérrez, N. A. (2015). A review of feed efficiency in swine: Biology and application, *J. Anim. Sci. Biotechnol.* 6(1), 33. https://doi.org/10.1186/s40104-015-0031-2.

Patience, J. F., Thacker, P. A. and de Lange, C. F. M. (1995). *Swine Nutrition Guide* (2nd edn.), University of Saskatchewan, Prairie Swine Center, Saskatoon.

Pettigrew, J. E., Gill, M., France, J. and Close, W. H. (1992). A mathematical integration of energy and amino acid metabolism of lactating sows, *J. Anim. Sci.* 70(12), 3742-3761.

Pezzuolo, A., Guarino, M., Sartori, L., González, L. A. and Marinello, F. (2018). On-barn pig weight stimation based on body measurements by a Kinect v1 depth camera, *Comput. Electron. Agri.* 148, 29-36. https://doi.org/10.1016/j.compag.2018.03.003.

Pierozan, C. R., Agostini, P. S., Gasa, J., Novais, A. K., Dias, C. P., Santos, R. S. K., Pereira Jr., M., Nagi, J. G., Alves, J. B. and Silva, C. A. (2016). Factors affecting the daily feed intake and feed conversion ratio of pigs in grow-finishing units: The case of a company, *Porc., Health Manag.* 2, 7. https://doi.org/10.1186/s40813-016-0023-4.

Pomar, C., Harris, D. L. and Minvielle, F. (1991a). Computer-simulation model of swine production systems. 1. Modeling the growth of young-pigs, *J. Anim. Sci.* 69(4), 1468–1488.

Pomar, C., Harris, D. L. and Minvielle, F. (1991b). Computer simulation model of swine production systems: II. Modeling body composition and weight of female pigs, fetal development, milk production, and growth of suckling pigs, *J. Anim. Sci.* 69(4), 1489–1502.

Pomar, C., Kyriazakis, I., Emmans, G. C. and Knap, P. W. (2003). Modeling stochasticity: Dealing with populations rather than individual pigs, *J. Anim. Sci.* 81(Suppl. 2), E178–186.

Pomar, C., Jondreville, C., Dourmad, J.-Y. and Bernier, J. (2006). Influence du niveau de phosphore des aliments sur les performances zootechniques et la rétention corporelle de calcium, phosphore, potassium, sodium, magnésium, fer et zinc chez le porc de 20 à 100 kg de poids vif, *Journées Rech, Porcine* 38, 209–216.

Pomar, C., Pomar, J., Rivest, J., Cloutier, L., Létourneau-Montminy, M.-P., Andretta, I. and Hauschild, L. (2013). Estimating real-time individual amino acid requirements in growing finishing pigs: Towards a new definition of nutrient requirements in growing-finishing pigs? In *Nutritional Modelling for Pigs and Poultry* (Eds. Sakomura, N. K., Gous, R. M., Kyriazakis, I., Hauschild, L.), GBR, CABI Publishing, Wallingford, pp. 157–174.

Quiniou, N., Boudon, A., Dourmad, J. Y., Moinecourt, M., Priymenko, N. and Narcy, A. (2021). Modélisation du besoin en calcium de la truie reproductrice et variation du rapport phosphocalcique des aliments selon le niveau de performance, *INRAE Prod. Anim.* 34(1), 61–78. https://doi.org/10.20870/productions-animales.2021.34.1.4723.

Quiniou, N., Cadero, A., Marcon, M. and Brossard, L. (2021). Simulating with the ThermiPig model the impact of fattening room characteristics under different climates on the performance of pigs, *53, Journées Rech. Porcine*, 2021-02-01.

Ramaekers, P. J. L., Huiskes, J. H., Verstegen, M. W. A., den Hartog, L. A., Vesseur, P. C. and Swinkels, J. W. G. M. (1995). Estimating individual body weights of group-housed growing-finishing pigs using a forelegs weighing system, *Comput. Electron. Agri.* 13(1), 1–12. https://doi.org/10.1016/0168-1699(95)00009-S.

Rantanen, M. M., Richert, B. T., Goodband, R. D., Hines, R. H., Nelssen, J. L. and Tokach, M. D. (1994). Effect of feeder design on finishing pig growth performance, *Kansas Agricultural Experiment Station Research Reports* (10). https://doi.org/10.4148/2378-5977.6435.

Renaudeau, D., Gourdine, J. L. and St-Pierre, N. R. (2011). A meta-analysis of the effects of high ambient temperature on growth performance of growing-finishing pigs, *J. Anim. Sci.* 89(7), 2220–2230. https://doi.org/10.2527/jas.2010-3329.

Reyer, H., Shirali, M., Ponsuksili, S., Murani, E., Varley, P. F., Jensen, J. and Wimmers, K. (2017). Exploring the genetics of feed efficiency and feeding behaviour traits in a pig line highly selected for performance characteristics, *Mol. Genet. Genomics* 292(5), 1001–1011. https://doi.org/10.1007/s00438-017-1325-1.

Rousseau, X. (2013). Optimisation de l'utilisation de phosphore alimentaire chez le porc et le poulet en croissance, PhD Thesis, Université François-Rabelais, Tours.

Sauvant, D. (2005). La modélisation de la réponse des porcs aux pratiques alimentaires et d'élevage, *Journées Rech, Porcine* 37, 283–290.

Sauvant, D., Perez, J.-M. and Tran, G. (2004). *Tables of Composition and Nutritive Value of Feed Materials: Pigs, Poultry, Cattle, Sheep, Goats, Rabbits, Horses, Fish*. INRA Editions, Wageningen Academic Publishers, Versailles.

Stevenson, J. S., Pollmann, D. S., Davis, D. L. and Murphy, J. P. (1983). Influence of supplemental light on sow performance during and after lactation, *J. Anim. Sci.* 56(6), 1282–1286. https://doi.org/10.2527/jas1983.5661282x.

Strathe, A. B., Danfaer, A., Jorgensen, H. and Kebreab, E. (2015). A dynamic growth model for prediction of nutrient partitioning and manure production in growing-finishing pigs: Model development and evaluation, *J. Anim. Sci.* 93(3), 1061–1073.

Symeou, V., Leinonen, I. and Kyriazakis, I. (2014a). Modelling phosphorus intake, digestion, retention and excretion in growing and finishing pig: Model description, *Animal* 8(10), 1612–1621.

Symeou, V., Leinonen, I. and Kyriazakis, I. (2014b). Modelling phosphorus intake, digestion, retention and excretion in growing and finishing pig: Model evaluation, *Animal* 8(10), 1622–1631.

Tan, Z., Yang, T., Wang, Y., Xing, K., Zhang, F., Zhao, X., Ao, H., Chen, S., Liu, J. and Wang, C. (2017). Metagenomic analysis of cecal microbiome identified microbiota and functional capacities associated with feed efficiency in Landrace finishing pigs, *Front. Microbiol.* 8, 1546. https://doi.org/10.3389/fmicb.2017.01546.

Thom, E. C. (1959). The discomfort index, *Weatherwise* 12(2), 57.

Turner, J. G., Parrish, J. L., Hughes, L. F., Toth, L. A. and Caspary, D. M. (2005). Hearing in laboratory animals: Strain differences and non-auditory effects of noise, *Comp. Med.* 55(1), 12–23.

Valros, A. (2018). Chapter 5 - Tail biting. In *Advances in Pig Welfare* (Eds. Spinka, M.), Woodhead Publishing, pp. 137–166. Available at: https://linkinghub.elsevier.com/retrieve/pii/B9780081010129000046.

Vandermeulen, J., Bahr, C., Tullo, E., Fontana, I., Ott, S., Kashiha, M., Guarino, M., Moons, C. P. H., Tuyttens, F. A. M., Niewold, T. A. and Berckmans, D. (2015). Discerning pig screams in production environments, *PLoS ONE* 10(4), e0123111. https://doi.org/10.1371/journal.pone.0123111.

van Milgen, J. and Noblet, J. (2003). Partitioning of energy intake to heat, protein, and fat in growing pigs, *J. Anim. Sci.* 81(E), E86–93.

van Milgen, J., Valancogne, A., Dubois, S., Dourmad, J. Y., Seve, B. and Noblet, J. (2008). InraPorc: A model and decision support tool for the nutrition of growing pigs, *Anim. Feed Sci. Technol.* 143(1–4), 387–405. https://doi.org/10.1016/j.anifeedsci.2007.05.020.

van Milgen, J., Noblet, J., Dourmad, J. Y., Labussière, E., Garcia-Launay, F. and Brossard, L. (2012). Precision pork production: Predicting the impact of nutritional strategies on carcass quality, *Meat Sci.* 92(3), 182–187.

van Riet, M. M. J., Millet, S., Aluwé, M. and Janssens, G. P. J. (2013). Impact of nutrition on lameness and claw health in sows, *Livest. Sci.* 156(1-3), 24-35.

Vautier, B., Quiniou, N., van Milgen, J. and Brossard, L. (2013). Accounting for variability among individual pigs in deterministic growth models, *Animal* 7(8), 1265-1273.

Verstegen, M. W. A. and van der Hel, W. (1974). Effects of temperature and type of floor on metabolic rate and effective critical temperature in groups of growing pigs, *Anim. Sci.* 18(1), 1-11.

Verstegen, M. W. and Curtis, S. E. (1988). Energetics of sows and gilts in gestation crates in the cold, *J. Anim. Sci.* 66(11), 2865-2875.

Vukmirović, Đ., Čolović, R., Rakita, S., Brlek, T., Duragić, O. and Solà-Oriol, D. (2017). Importance of feed structure (particle size) and feed form (mash vs. pellets) in pig nutrition - a review, *Anim. Feed Sci. Technol.* 233, 133-144. https://doi.org/10.1016/j.anifeedsci.2017.06.016.

Wang, Y., Yang, W., Winter, P. and Walker, L. (2008). Walk-through weighing of pigs using machine vision and an artificial neural network, *Biosyst. Eng.* 100(1), 117-125. https://doi.org/10.1016/j.biosystemseng.2007.08.008.

Wegner, K., Lambertz, C., Das, G. and Gauly, M. (2014a). Climatic conditions in sow barns in Northern Germany, *Zuchtungskunde* 86, 200-211.

Wegner, K., Lambertz, C., Das, G., Reiner, G. and Gauly, M. (2014b). Climatic effects on sow fertility and piglet survival under influence of a moderate climate, *Animal* 8(9), 1526-1533.

Wegner, K., Lambertz, C., Das, G., Reiner, G. and Gauly, M. (2016). Effects of temperature and temperature-humidity index on the reproductive performance of sows during summer months under a temperate climate, *Anim. Sci. J.* 87(11), 1334-1339. https://doi.org/10.1111/asj.12569.

Wellock, I. J., Emmans, G. C. and Kyriazakis, I. (2003). Modelling the effects of thermal environment and dietary composition on pig performance: Model logic and concepts, *Anim. Sci.* 77(2), 255-266.

Williams, I. H., Close, W. H. and Cole, D. J. A. (1985). Strategies for sow nutrition: Predicting the response of pregnant animals to protein and energy intake. In *Recent Advances in Animal Nutrition* (Eds. Haresign, W., Cole, D. J. A.), Nottingham University Press, Nottingham, pp. 133-147.

Whittemore, C. T. and Fawcett, R. H. (1974). Model responses of the growing pig to the dietary intake of energy and protein, *Anim. Sci.* 19(2), 221-231.

Whittemore, C. T. and Fawcett, R. H. (1976). Theoretical aspects of a flexible model to simulate protein and lipid growth in pigs, *Anim. Prod.* 22(1), 87-96.

Wu, J., Tillett, R., McFarlane, N., Ju, X., Siebert, J. P. and Schofield, P. (2004). Extracting the three-dimensional shape of live pigs using stereo photogrammetry, *Comput. Electron. Agri.* 44(3), 203-222. https://doi.org/10.1016/j.compag.2004.05.003.

Yang, A., Huang, H., Zheng, B., Li, S., Gan, H., Chen, C., Yang, X. and Xue, Y. (2020). An automatic recognition framework for sow daily behaviours based on motion and image analyses, *Biosyst. Eng.* 192, 56-71. https://doi.org/10.1016/j.biosystemseng.2020.01.016.

Yoosuk, S., Ong, H. B., Roan, S. W., Morgan, C. A. and Whittemore, C. T. (2011). A simulation model for predicting the voluntary feed intake of a growing pig, *Acta Agric. Scan. A* 61(4), 168-186.

Chapter 5

New approaches for determining the nutritional value of pig feed

Gerald C. Shurson, University of Minnesota, USA

1 Introduction

The ultimate goal of feeding pigs is to maximize nutritional efficiency and minimize the environmental footprint in the production of safe, wholesome lean pork to feed an increasing number of consumers around the world. Accurate assessment of the nutritional value of pig feed is one of the main factors for achieving this goal. Measures used to determine the types and proportions of feed ingredients used in swine diet formulations affect cost, ability to achieve the genetic potential for productivity, nutritional efficiency, and subsequent environmental impacts of pig production. Nutritional efficiency is the proportion of dietary energy and nutrients consumed that are used for productive purposes such as lean growth, reproduction, and lactation and is dependent on using nutritional evaluation measures that most accurately predict biological responses from the chemical composition of the diet (Shurson et al., 2021a). Inaccurate determination of the true nutritional value of pig feed can lead to overfeeding energy and nutrients which result in excess cost, nutrient excretion, and greenhouse gas emissions that are detrimental to the environment. Likewise, inaccurate assessment of nutritional value can also result in underfeeding energy and nutrients which limit the pig's ability

http://dx.doi.org/10.19103/AS.2024.0140.09

to achieve its genetic potential for growth, reproduction, and lactation that subsequently reduces productivity and profitability and may lead to nutrient deficiencies as well as adverse animal health and well-being outcomes.

Historically, the determination of the nutritional value of feed ingredients has focused on estimating the amount of utilizable energy [digestible energy (DE); metabolizable energy (ME); and net energy (NE)] derived from carbohydrates, lipids, and excess protein while also quantifying the concentration and digestibility of amino acids and phosphorus because energy, protein, and phosphorus are the three most expensive components of swine diets. In contrast, limited analysis of fatty acids, vitamins, and trace minerals in raw materials is conducted by commercial feed manufacturers because analytical procedures are more complex and expensive, chemical composition is poorly associated with bioavailability, composition is highly variable within and among ingredient sources, and the relative ease and low cost of ensuring nutritional adequacy by adding supplemental lipid sources and vitamin-trace mineral premixes to swine diets. Except for mycotoxins, limited analysis is routinely conducted to assess anti-nutritional factors (lipid and protein oxidation compounds, biogenic amines, trypsin inhibitors, and glucosinolates) in feed ingredients for commercial swine feeds because analyses are expensive, require specialized equipment, and there are limited, if any, mitigation options. The discontinuation of using subtherapeutic growth-promoting levels of antimicrobials in many countries has led to increased interest in quantifying the concentrations of functional nutrients in various feed ingredients. However, as for anti-nutritional factors, specialized analytical equipment and methods are often required for their assessment in raw materials. Therefore, the future of analytics for swine feed evaluation must involve increasing our ability to predict the growth, reproductive, lactation, and health responses of pigs from the chemical composition of diets.

Fortunately, improved analytical measures, experimental methodologies, and diet formulation practices that more accurately quantify the biological and physiological value of nutrients and other chemical components of feed ingredients and complete feeds fed to swine are being developed. The emergence of liquid chromatography and mass spectroscopy along with other highly sensitive analytical equipment and methods are becoming more widely available and used to quantify many anti-nutritional factors, toxicants, functional nutrients, and other chemical constituents of interest in feed ingredients. As a result, these highly sensitive analytics are being used to determine biological mechanisms of nutrients, metabolites, and dietary additives on metabolic processes (Chen et al., 2018; Wang et al., 2018; Chen et al., 2019b; Guo et al., 2023), as well as assessing feed contaminants that are associated with human food safety and quality concerns (Malik et al., 2010; Wang et al., 2016; Fu et al., 2017). In addition, the use of several 'omics' techniques (i.e. genomics, transcriptomics, proteomics, metabolomics, and microbiome) is becoming a popular and more holistic molecular approach for evaluating and studying all aspects of animal biology including the role of diet and nutrition on animal health and well-being (Pinu et al., 2019).

Evaluation of the nutritional value of feed ingredients has also evolved to include the digestion kinetics of starch, fiber, and protein in pigs (Gerrits et al., 2021). As more is learned about nutrient digestion kinetics in various feed ingredients used in swine diets, more strategic diet formulation strategies studies will become possible for improving nutritional efficiency, minimizing the incidence of diarrhea in weaned pigs, and reducing the amount of nutrients excreted in manure. Considerations must also be given to the molecular structure of feeds (Yu and Prates, 2017) and the effects of various types of feed processing (Rojas and Stein, 2015; Salazar-Villanea, 2016; De Vries et al., 2012) on energy and nutrient digestibility, which have generally not been previously considered when formulating swine diets.

Precision feed formulation and feeding programs are an emerging approach for optimizing nutritional efficiency and productivity while minimizing the environmental footprint in pig production systems (Andretta et al., 2018; Pomar et al., 2021). Precision feeding is not possible without dynamic and accurate estimates of net energy and nutrient digestibility and bioavailability data in specific batches of each feed ingredient used to formulate diets fed to pigs. Multi-objective feed formulation is a new approach which is a key component of precision feeding that utilizes Life Cycle Assessment (LCA) data for several environmental impacts of various sources and types of feed ingredients used in swine diets (Mackenzie et al., 2016; Garcia-Launay et al., 2018). Low environmental impacts of feed ingredients are associated with production inputs and processes and are indirectly linked to their nutritional composition and digestibility affecting carbon, nitrogen, and phosphorus utilization efficiency and emissions. The Global Feed LCA Institute (GFLI; https://tools.blonkconsultants.nl/tool/16/) has developed the largest database of feed ingredients with the most LCA indicator variables (n = 18) using standardized guidelines and methodologies (LEAP, 2015). These LCA environmental impact data are part of nutritional value assessment because they are being used as constraints in formulating diets with reduced environmental impacts while maintaining productivity (de Quelen et al., 2021).

As a result, the types of measures used to evaluate feed ingredients for use in swine diets have greatly expanded from a historical perspective and must be considered along with traditional measures used to characterize energy and nutrient content and digestibility of feed ingredients to achieve increased sustainability of pig production. Therefore, the key considerations for determining the true nutritional value of feed ingredients and complete feeds for swine include the following:

1. Using dynamic estimates rather than static published 'book values' of nutrient composition and digestibility of feed ingredients to improve accuracy when formulating swine diets and to minimize overfeeding and underfeeding of energy and nutrients to pigs;
2. Understanding and using proper sampling procedures of ingredients for chemical analysis to ensure accurate nutrient quantification;
3. Understanding the limitations of analytical procedures, chemical measures, relative accuracy of chemical assays, and associated acceptable analytical variation when interpreting results;
4. Understanding of the benefits and limitations of experimental methodologies used to estimate the biologically useful (digestible and bioavailable) portions of energy, amino acids, phosphorus, and other nutrients in feed ingredients;
5. Understanding the positive and negative impacts of physical form and thermal processing on the nutritional value of feed ingredients;
6. Using validated models and prediction equations to dynamically estimate metabolizable energy, net energy, and digestible amino acid concentrations of feed ingredients;
7. Considering the detrimental effects of various anti-nutritional factors in feed ingredients on their nutritional value and quantifying those compounds so that constraints can be used in feed formulation to limit ingredient inclusion rates;
8. Considering the beneficial effects of functional nutrients in some ingredients to reduce reliance on using various health and growth-promoting feed additives;
9. Including environmental impact factors of feed ingredients when formulating swine diets to reduce the environmental footprint of pig production; and
10. Using precision feeding practices to increase nutritional efficiency and profitability while reducing environment impacts of pig production.

2 Methodology reviews and studies to determine energy, nutrient content and digestibility in feed ingredients

2.1 Feed sampling and analytical variation

Accurate quantification of the nutrient content of feed ingredients and complete feeds is dependent on sampling methods and understanding of acceptable analytical variation of assay procedures. This is especially important for trace minerals and vitamins because they represent a very small proportion of the total chemical components in feed and may not be uniformly distributed. For example, high variation in the analysis of zinc concentrations from the same batch of feed is commonly observed when subsamples are analyzed at different times, which is likely due to inadequate sampling procedures to obtain a representative sample (Holen et al., 2020). Results from one study have shown that a minimum of 34 feed samples were needed to achieve 95% confidence that the analyzed zinc concentration would be within 4 mg/kg of the actual value (Jones et al., 2017). However, these authors recognized that analyzing this number of samples per batch would be too costly and impractical and suggested that collecting and analyzing two to five subsamples to achieve the goal of obtaining results within 15 mg/kg of the expected zinc concentration may be acceptable under commercial conditions. Analytical variation associated with various nutrients should be considered when evaluating laboratory results from feed analysis for various nutrients (Menegat et al., 2019). Cromwell et al. (1999) evaluated variability in nutrient analyses among corn and soybean meal sources and university research laboratories and reported that except for crude protein and selenium, analytical variability among labs was as great as, and in some cases greater than, variability in nutrient composition among sources of corn and soybean meal. In addition, variation within laboratories tended to be less than variation among laboratories. Therefore, efforts to minimize causes of variation in nutrient analysis within and among laboratories are essential for the accurate determination of the nutrient content of feed ingredients.

2.2 Moisture and water activity

Traditional wet chemistry procedures continue to serve as the standard for determining the chemical composition of feed ingredients. Accurate determination of moisture content in feed ingredients is an essential part of feed ingredient evaluation because it affects pricing, shipping, storage, extent of nutrient dilution in diet formulation, and feed consumption. Thiex and Richardson (2003) reviewed the various methods used to determine moisture content and identified several challenges that cause high variability in results and bias. Chemical characteristics of various types of feed ingredients affect the accuracy of results from moisture content determinations. For example, Ahn et al. (2014) compared different oven-drying methods for determining moisture content of various types of feed ingredients and reported that drying samples at 135°C for 2 h is inappropriate for measuring moisture content in whey permeate, whey powder, distillers dried grains with solubles (DDGS), and diets containing these ingredients. In contrast to moisture content, water activity is rarely measured in feed ingredients used in swine diets. However, it provides important information for predicting the types of microorganisms that may cause spoilage, survival of viral pathogens, susceptibility to lipid and protein oxidation, enzyme activity, and vitamin stability (Shurson et al., 2021a). Because water activity is a better predictor than moisture content of these negative consequences on nutritional value of ingredients, it should be included as part of quality assurance protocols during feed ingredient manufacturing and storage.

2.3 Starch, fiber and lipid digestibility

Methods for determining digestibility of lipids (Stein, 2017), fiber (Agyekum and Nyachoti, 2017; Stein, 2017), and starch (Stein, 2017; Jeziemy et al., 2017) have been generally described but more comprehensive reviews are needed. Jeziemy et al. (2017) evaluated starch digestibility in grain legumes and showed that concentration and digestibility vary based on analytical method, and digestibility is greater when using polarimetric rather than enzymatic methods. Stein (2017) indicated that ileal digestibility must be determined for amino acids, starch, and lipids because nutrients not absorbed in the small intestine will be fermented or changed in the large intestine, resulting in inaccurate total tract digestibility estimates.

2.4 Digestible, metabolizable and net energy

Several reviews have described the differences and challenges of using various energy evaluation methods to determine DE, ME, and NE content of feed ingredients (de Lange and Birkett, 2005; Kil et al., 2013; Kong and Adeola, 2014; Velayudhan et al., 2015; Zhang and Adeola, 2017; Li et al., 2018; Noblet et al., 2022). The newest developments involve evaluating experimental protocols to determine energy values in high-fiber diets. Gao et al. (2023) reported that experimental protocols for determining DE and ME content in high-fiber ingredients should include a minimum 7-day adaptation period to allow gut microflora to adjust to high-fiber diets, especially if they contain high concentrations of insoluble fiber. However, Lyu et al. (2018) indicated that a 12-day adaptation period is necessary for NE determination of high-fiber ingredients regardless of a lack of effect of length of adaptation period on apparent total tract digestibility coefficients for most nutrients. Liu et al. (2020) compared using a 3-, 5-, and 7-day time-based total fecal collection period to determine DE and ME and reported that a 5-day collection period was adequate. Chen et al. (2014) developed a computer-controlled simulation digestion system to determine DE in feed ingredients for growing pigs and results showed high accuracy and repeatability when compared with *in vivo* determined DE values among ingredients representing a wide range in energy and nutrient composition.

2.5 Amino acid digestibility

Similar to energy, several reviews have summarized the benefits and limitations of using different methodologies to determine the digestibility of amino acids in feed ingredients (Kong and Adeola, 2014; Stein, 2017; Zhang and Adeola, 2017). Accurate estimates of standardized ileal digestibility (SID) of amino acids are dependent on accurate estimates of basal endogenous losses. Bloxham et al. (2023) compared total amino content and SID values in two databases and observed that digestibility values were affected by total amino acid content in the ingredients, which may be a result of inadequate methods to estimate basal endogenous losses of amino acids and statistical analysis. Park et al. (2013) developed prediction equations to estimate basal endogenous losses of protein and amino acids in pigs fed nitrogen-free diets using experimental data from 34 published studies and observed that initial pig body weight and the ratio of feed intake per maintenance feed intake were two main factors that reduced these losses.

2.6 Protein quality

Assessing the nutritional value of protein in feed ingredients involves more than determining indispensable amino acid digestibility and must include more holistic assessments of biological value to improve nitrogen utilization efficiency and reduce the environmental footprint of nitrogen in swine diets. Liebert (2017) reviewed the limitations of using historical measures of protein quality (e.g. protein efficiency ratio, nitrogen efficiency ratio, biological value, net protein utilization, and physiological value of protein) and proposed a modeling approach based on nitrogen excretion per unit of deposition by accounting for differences in pig performance and the quantity and quality of protein consumed. The *in vitro* pepsin digestibility assay has been used for rapid estimation of protein quality of high-protein ingredients for poultry, but results are often misleading due to the interactive effects of lipids and carbohydrates in feed samples (Bryan and Classen, 2020). Although these challenges can be overcome by using multi-enzyme assays, none of these assays account for the biological properties of ingredients in the gastrointestinal tract (Bryan and Classen, 2020).

2.7 Mineral content

Thermal combustion and inductively coupled plasma methodologies are commonly used to determine mineral content in feed ingredients, but direct comparisons of analytical results for various minerals have rarely been reported in the scientific literature. Kerr et al. (2008) conducted a study to compare the use of these methods on mineral composition in 45 feed ingredients used in swine feeds and observed high variability in mineral concentrations within and among sources, and some differences in sulfur concentrations between methods for a few feed ingredients.

2.8 Calcium and phosphorus digestibility

Methods for determining calcium and phosphorus digestibility of feed ingredients have been reviewed by Zhang and Adeola (2017) and Stein (2017). She et al. (2017) recommended that endogenous phosphorus losses need to be considered to improve the accuracy of phosphorus digestibility estimates and further evaluation is needed regarding the additivity of digestibility values for P in mixed diets. Shen et al. (2002) determined the endogenous fecal phosphorus excretion from feeding corn and reported that it was equivalent to 12.3% and 25.8% of daily total and available phosphorus requirements of pigs, respectively. Therefore, current apparent digestibility and availability estimates for corn underestimate the true utilization of P in growing pigs by about 35%, and results in overfeeding P and excessive P excretion. González-Vega and Stein (2014) indicated that calcium-phosphorus-phytase interactions must be considered in calcium digestibility studies because of the possible formation of insoluble calcium-phytate complexes or calcium-phosphorus complexes which may affect the efficacy of phytase and digestibility of calcium and phosphorus. Lautrou et al. (2021) reviewed the benefits and limitations of using total analyzed values, relative bioavailability, digestibility, along with mechanistic modeling and meta-analysis approaches for determining calcium and phosphorus utilization in feedstuffs as part of precision feeding strategies.

2.9 Near-infrared reflectance spectroscopy and other advanced technologies

Using wet chemistry procedures, the preparation of samples for analysis usually involves digestion and other decomposition processes that destroy the molecular structure of feed and may produce artifacts that affect the interpretation of results. Furthermore, the time involved in submitting samples and conducting laboratory analysis makes it difficult to use data from an ingredient batch in the diet formulation process. Therefore, near-infrared reflectance spectroscopy (NIRS) has become a popular alternative to wet chemistry analysis in the feed industry because it is a rapid, non-destructive, and relatively inexpensive method for determining the nutritional value of feed ingredients. Recent studies have shown that NIRS can be a useful method for estimating the available energy and most amino acids in corn-dried distillers grains with solubles (Zhou et al., 2012). Additional studies have demonstrated that protein and amino acid composition and digestibility (Noel et al., 2021), as well as the DE and ME values (Noel et al., 2022) of feed ingredients and complete feeds, can be accurately determined with NIRS. Paternostre et al. (2021) evaluated the usefulness of NIRS in predicting the nutrient content, digestibility, and net energy of pig feed using calibrations based on feed and feces. They observed that most chemical components could be accurately predicted in feed and feces, but net energy was more accurately predicted using feed spectra compared with feces spectra, while digestibility of nutrients was poorly estimated using either feed or feces spectra. Other rapid, non-invasive, and more advanced techniques that use synchrotron radiation infrared microspectroscopy and synchrotron *x*-ray methods are becoming available for use in determining the nutrient content and digestibility of feed ingredients without destroying the molecular structure (Yu and Prates, 2017).

2.10 Physicochemical properties of dietary fiber

Many feed ingredients derived from food and agro-industrial processes contain high amounts of dietary fiber that is highly variable in physicochemical properties. Therefore, it is important to measure these physicochemical properties in high-fiber ingredients to determine water solubility and hydration properties, such as swelling and water binding capacity, because these factors influence gastric emptying, rate of nutrient absorption in the small intestine, and fermentation in the large intestine of pigs (Bach Knudsen, 2001). Conventional methods for measuring hydration properties of dietary fiber utilize centrifugation, dialysis bags, and filtration but do not reflect *in vivo* digestion processes involving endogenous enzymes (Bach Knudsen, 2001). Zhou et al. (2018) compared the use of conventional analytical methods with an *in vitro* digestion assay and found that the *in vitro* method was more accurate for characterizing the hydration properties of dietary fiber than conventional methods.

2.11 Digestion kinetics nutrients in ingredients

Traditional feed evaluation systems used to determine the nutritional value of feed ingredients are based on dietary inclusion rates, constraints on nutrient concentrations, and nutrient digestibility data but do not account for differences in the rate, extent, and timing of nutrient digestion and fermentation in the gastrointestinal tract (Heyer et al., 2022). The complex interactions that occur within the lumen of the digestive tract affect the kinetics of digestion and the subsequent fate of nutrients after absorption (Gerrits et al., 2021). Because of the large variation in nutritional quality among feed ingredients, there is a tremendous need to develop new approaches for evaluating the digestion kinetics of feed ingredients. Current approaches evaluate the extent of digestion potential of various feed ingredients in the digestive tract by combining *in vitro* digestibility determinations with *in silico* digestive processes including digesta transit, nutrient hydrolysis, and absorption using dynamic, mechanistic models (Gerrits et al., 2021). Additional factors such as characterizing digesta physicochemical properties and rate of transit and *in vitro* digestion kinetics data of individual feed ingredients are needed to develop mechanistic models that accurately predict *in vivo* nutrient absorption kinetics of various ingredients (Gerrits et al., 2021).

Initial studies were conducted to determine the accuracy of predicting the *in vivo* glucose response from *in vitro* starch digestion using a modified Englyst assay (Van Kempen et al., 2010). These researchers showed that *in vitro* starch digestion kinetics can accurately predict portal glucose appearance up to 8 h postprandially if *in vitro* data are corrected for gastric emptying time. Amaral et al. (2015) determined the kinetics of starch digestion of corn ground to different particle sizes and showed that small particle sizes caused the most rapid starch digestion compared with larger particle sizes. Cadogan and Choct (2015) evaluated the pattern of nonstarch polysaccharide (NSP) digestion of two types of wheat, with or without xylanase supplementation, in the gastrointestinal tract of pigs and observed that the structures of NSPs, rather than their concentration and solubility, are important determinants of the anti-nutritional properties of NSPs in pig diets.

Schop (2020) conducted several studies to evaluate the impact of starch, dietary fiber, and protein digestion kinetics on absorption of nutrients in pigs and subsequently developed a computer model (SNAPIG) based on these results. Specifically, the digestive transit of solids and liquids along with the impact of diet viscosity, nutrient solubility, kinetics of protein hydrolysis, and feed intake level of pigs were quantified. These data were then used to develop SNAPIG, which is an *in silico* dynamic digestion model that can be used to predict the absorption of nutrients by simulating digestion kinetics in pigs fed different diets containing various ingredients with different physicochemical properties.

Zijlstra et al. (2012) developed *in vitro* methodologies to characterize the kinetics of digestion and fermentation and their connections with *in vivo* models to determine the extent, kinetics, and site of digestion or fermentation of starch and fiber in feed ingredients fed to pigs. Results from their studies showed that slowly digestible starch is partially degraded in the distal small intestine and large intestine, and subsequently fermented to produce volatile fatty acids which reduced insulin responses and whole-body energy use. Highly viscous purified fiber reduced glycemic responses and digesta passage rate resulting in increased ileal digestion of nutrients. However, fiber in feed grains reduced digestibility. Therefore, the development of more complex methodologies that combine *in vitro* and *in vivo* measures would be useful for determining mechanisms that connect physicochemical properties of starch and dietary fiber with whole-body nutrient use and intestinal health of growing pigs.

Heyer et al. (2022) evaluated the effects of fiber fermentation and protein digestion kinetics on mineral digestion in pigs and indicated that total tract digestibility of phosphorus is underestimated in diets containing fermentable ingredients because of greater endogenous losses of phosphorus. These authors suggested that quantification of fiber type, NSP fractions, and kinetics, along with their effect on digesta pH and solubility related to fermentation, is needed in future studies.

Chen et al. (2019a) reported that substantial differences in *in vitro* digestion kinetics exist among protein sources based on nitrogen solubilization, along with the release of free amino acids and low-molecular-weight peptides, and suggested that the kinetics of release of low-molecular-weight peptides may be useful in predicting *in vivo* protein digestion kinetics in pigs. Ye et al. (2022) fed diets containing low, medium, and highly degradable protein to lactating sows to determine protein digestion kinetics and reported that although kinetics of protein digestion had no effect on litter weight gain, milk composition, and nitrogen excretion, feeding protein sources with a high proportion of slowly degradable protein reduced body weight loss and protein mobilization of maternal body tissue during lactation.

2.12 Physical characteristics and feed processing effects on nutritional value

Most grains, oilseeds, and animal- and plant-derived by-products used in swine diets are subjected to particle size reduction and thermal processing before chemical analysis is performed. As a result, physical characteristics and time and temperature of exposure during heat treatment can have significant effects on energy and nutrient content and digestibility of feed ingredients. Heat treatment conditions of feed ingredients and complete feeds are also important for inactivating pathogenic bacteria, viruses, and parasites that may be present (Shurson et al., 2021b).

The particle size of feed ingredients and complete feeds should be expressed as the geometric mean diameter by weight along with the geometric standard deviation using the current standard method ANSI/ASAE S319.4 (ASABE, 2008). However, this method allows for the use of different sieve times, sieve agitators, and dispersion agents. Unfortunately, the number, type, and arrangement of agitators on sieves, and the use of dispersion agents, affect particle size determinations but are often not reported, which makes it difficult to compare results among laboratories and research studies (Goodband et al., 2006). Kalivoda et al. (2017) used different variations of the approved method to determine the combination of factors that most accurately estimate the particle size of corn, sorghum, and wheat samples. Grinding corn to reduce particle size from 865 to 339 μm linearly increased DE and ME concentration and starch digestibility but did not affect protein, amino acid, and phosphorus digestibility (Rojas and Stein, 2015). Furthermore, Lyu et al. (2021) reported that nutrient composition and physical characteristics are different between different particle size fractions in corn and soybean meal and may contribute to differences in pig growth responses.

The molecular structure of fiber and protein in feed ingredients affects the accessibility of enzymes for digestion and subsequent digestibility values of feed ingredients. Salazar-Villanea et al. (2016) reviewed studies that involved modifications of the protein structure of grain-based ingredients from processing (i.e. heat, shear, and moisture) to identify potential mechanisms that affect digestibility. These researchers indicated that the association between conformational changes of protein and protein digestibility is unclear, which is partially due to the confounding effects of protease inhibitors. Because of the multifactorial complexity of processing technologies used in feed production, model systems need to be developed to separate and explain the effects of various processing factors on structural changes and the nutritional value of proteins (Salazar-Villanea et al., 2016). The effects of feed processing on NSP in high-fiber ingredients were reviewed by de Vries et al. (2012), and they reported that results are challenging to interpret because the use of crude fiber, neutral detergent fiber, and acid detergent fiber digestibility values are inadequate for understanding the influence of modifications of the NSP during processing on digestibility compared with more detailed information provided by measuring NSPs. However, grinding was effective in increasing apparent ileal digestibility (AID), hydrothermal processes were more effective in improving solubility, viscosity, and digestibility of fiber than dry thermal processes, and heat processing increased the effectiveness of feed enzymes by up to six times compared with non-heat processed feeds.

3 Models and equations to dynamically predict energy, nutrient content and digestibility in feed ingredients

In addition to the complex, mechanistic models that have been developed to evaluate protein quality, calcium and phosphorus utilization, and digestion kinetics of feed ingredients, simpler models and equations are also useful for dynamically estimating DE, ME, and NE content and amino acid, calcium, and phosphorus digestibility of feed ingredients. The use of validated prediction equations based on chemical analysis data provides rapid, inexpensive, accurate estimates of important nutritional components that can be used in precision feed formulations to improve nutritional efficiency. In general, large, robust data sets representing the entire range of nutrient composition of an individual ingredient are necessary to obtain the high degree of accuracy desired. Robust prediction equations have also been developed using meta-analysis approaches which enable systematically combining data into homogeneous pools to determine more precise overall estimates of various factors affecting response variables (Borenstein et al., 2009).

A summary of publications providing mathematical models and prediction equations for energy, total and digestible amino acids, and digestible calcium and phosphorus of various ingredients used in swine diets is shown in Table 1. Numerous studies provide prediction equations developed for DE and ME for many feed ingredients used in swine diets, but only a few studies have provided NE equations for grain-based ingredients (Liu et al., 2015; Li et al., 2018; Wu et al., 2016b; Aldenhoven et al., 2020). More studies are needed for determining the NE content of more types and sources of ingredients *in vivo* so that larger databases can be constructed to develop and validate NE prediction equations.

Similar to grain-based ingredients, several equations to predict DE and ME concentrations from feed fats and oils have been developed for swine (Table 1). Unfortunately, their accuracy in predicting DE concentrations from estimates obtained from *in vivo* experiments is generally poor (Kellner and Patience, 2017; Kerr et al., 2016; Kerr et al., 2018). For example, Kerr et al. (2016) compared the DE content of five sources of distillers corn oil (DCO) determined by an *in vivo* experiment with predicted DE values using equations from Wiseman et al. (1998) and reported that the predicted DE content of these lipids averaged about 300 kcal/kg more than the *in vivo* determined DE content for all DCO sources except for the source containing about 94% free fatty acids (FFAs), which was predicted to have 1100 kcal/kg more than was actually determined *in vivo*. Lipids that were used to develop the Wiseman et al. (1998) equations contained a maximum moisture, impurities, and insoluble (MIU) concentration of 20 g/kg, which may partially explain why these equations may overestimate DE content of lipids because MIU contains no energy. As a result, Wealleans et al. (2021) proposed using an adjustment factor that reflects the energy dilution of MIU and reported that calculated energy values of fats and oils were reduced by as much as 46% compared with values obtained from the original prediction equation. Similarly, Kellner and Patience (2017) determined the DE content of 14 different lipids fed to 13 kg and 50 kg pigs, and compared these *in vivo* values with predicted values using an equation from Powles et al. (1995) along with equations derived from the chemical composition of lipids evaluated in their study. Their results showed that the Powles et al. (1995) equation underestimated the negative effect of the FFA content of lipids on DE concentrations and was not accurate for estimating the DE content of more saturated lipid sources. Likewise, Kerr et al. (2018) determined the DE content of 10 different fats and oils fed to 19 kg pigs, and compared *in vivo* values with predicted values using the Powles et al. (1995) prediction equation, and reported that the DE content was underestimated by 243 kcal/kg (tallow) to

Table 1 Summary of publications providing models and equations to dynamically predict energy, nutrient content, and digestibility in feed ingredients and complete feeds

Nutritional measure	Feed ingredients	References
NE	Multiple ingredients	Li et al. (2018)
	Canola meal, corn, corn DDGS, cottonseed meal, soybean meal, wheat, and wheat bran	Liu et al. (2015a); Aldenhoven et al. (2020); Wu et al. (2016b)
	Lipids	Kellner and Patience (2017)
	Corn, soybean meal, and wheat bran	Yang et al. (2020); Yang et al. (2021)
DE and ME	Canola meal, copra meal, corn germ meal, dehulled soybean meal, high protein distillers dried grains, palm kernel expellers, perilla meal, sesame meal, and tapioca distillers dried grains	Son et al. (2017)
	Multiple ingredients	Sung and Kim (2021)
	Corn co-products	Anderson et al. (2012)
	DDGS	Urriola et al. (2014); Wu et al. (2016a)
	Corn	Smith et al. (2015); Bertol et al. (2017)
	Double-low rapeseed cake	Wang et al. (2021)
	Wheat milling by-products	Huang et al. (2014)
	Lipids	Rosero et al. (2015); Kellner and Patience (2017); Kerr et al. (2016); Kerr et al. (2018)
	Compound feeds	Grümpel-Schlüter et al. (2021); Sung and Kim (2021)
Amino acid content	Soybean meal	Lee et al. (2021)
	Faba bean, peas, lupin, soybean, wheat, barley, corn, sorghum, oats decorticated, corn DDGS, soybean meal, and rapeseed meal	Messad et al. (2015)
Amino acid digestibility	Faba bean, peas, lupin, soybean, wheat, barley, corn, sorghum, oats decorticated, corn DDGS, soybean meal, and rapeseed meal	Messad et al. (2015)
	DDGS	Zeng et al. (2017)
	Sunflower seed meal	Liu et al. (2015b)
Ca and P digestibility	Meat and bone meal	Sulabo and Stein (2013)
P digestibility	Multiple ingredients	Symeou et al. (2014a, b)

2276 kcal/kg (coconut oil), except for fish oil, which was overestimated by 289 kcal/kg. When Kerr et al. (2018) used the Kellner and Patience (2017) DE prediction equations in a similar comparison, the predicted DE content of lipid sources was also underestimated (440–967 kcal/kg), but generally to a lesser extent than the corresponding DE estimates derived from the Powles et al. (1995) equation, while the DE content of tallow was overestimated by 393 kcal/kg. Although the results from these comparisons are disappointing, they provide some insight for future research considerations when attempting to develop more accurate energy prediction equations for lipids. These results indicate that (1) most of the energy equations developed for lipids appear to be too simplistic and do not include measures of oxidation and (2) prediction equations for specific fats and oils must be derived from experiments that have directly determined DE, ME, or NE content of the specific types of lipids being modeled.

A review of several studies evaluating the accuracy of estimates for predicting amino acid concentrations from crude protein content of some ingredients using simple regression analyses showed poor correlations between crude protein and lysine content, but somewhat better correlations for estimating other indispensable amino acids from crude protein (Shurson et al., 2021a). Messad et al. (2015) recognized the limitations of using a simple regression analysis for developing equations to predict amino acid content from crude protein content and used a meta-analysis to evaluate nutrient composition data from 34 studies for 12 ingredients. Results from this study showed that although several of the models provided relatively accurate estimates of indispensable amino acid concentrations predicted from crude protein content for many of the ingredients, using this approach does not provide the accuracy necessary for use in precision swine feed formulation (Messad et al., 2015). In contrast, the development and use of prediction equations to estimate SID of amino acids in several feed ingredients based on chemical composition appears to be a more promising alternative compared with using various *in vitro* methods. Zeng et al. (2017) conducted a meta-analysis of published chemical composition digestibility data for corn DDGS and developed simple prediction equations that provided better estimates than those reported from previously published studies. Similarly, Messad et al. (2015) used a meta-analysis approach to predict AID and SID of amino acids for 12 common feed ingredients and resulted in a relatively good positive linear association (R^2 = 0.72–0.88) between total concentrations of indispensable amino acids and their AID. However, correlations varied between ingredients because of the negative effect of NDF content on AID in some ingredients. Although the prediction accuracy of these SID equations was improved compared with the original versions, these researchers recommended that they be validated before being used in practical swine diet formulations (Messad et al., 2015).

Limited studies have been conducted to predict digestible calcium and phosphorus content in feed ingredients used in swine diets. Symeou et al. (2014a) developed a dynamic, deterministic model that estimates the proportion of total dietary phosphorus that is digested, retained, and the soluble and insoluble phosphorus excreted from pigs of different genotypes and fed diets of different composition. A sensitivity analysis of the model predictions was conducted and showed that the model satisfactorily predicted P digested, retained, and excreted using variable amounts of dietary inorganic P, Ca, and phytase supplementation in conventional European feed ingredients, but poor results were obtained when dried distillers grains with solubles and canola meal were evaluated (Symeou et al., 2014b). Therefore, improvements in model assumptions are needed to increase the accuracy of predictions when applied to more diverse feed ingredients used in swine diets around the world. Sulabo and Stein (2013) evaluated eight different sources of meat and bone meal to determine Ca and P digestibility and develop prediction equations. Their results showed a high positive correlation between the apparent total tract digestibility of Ca and the standardized total tract digestibility of P, a negative correlation between digestibility of Ca and P and with total Ca and P concentrations, and indicated that prediction equations using total P concentration can reasonably estimate Ca and P digestibility in meat and bone meal fed to pigs.

4 Anti-nutritional factors and associated analytical methods

In addition to accurate measurement of energy and nutrients in feed ingredients, it is equally important to determine the concentrations of various types of anti-nutritional factors that are naturally present in several feed ingredients because they can interfere with energy and nutrient digestion and absorption, reduce nutritional efficiency, and at high concentrations, be detrimental to animal health. Therefore, accurate quantification of anti-nutritional factors in various feed ingredients is essential to limit diet inclusion rates when these compounds exceed maximum thresholds that may lead to detrimental nutritional and health effects. The most widely studied anti-nutritional factors include trypsin inhibitors in legume seeds such as soybeans, glucosinolates in rapeseed and canola meal, and mycotoxins in grain and grain by-products (Table 2). However, metabolites from lipid and protein oxidation, along with biogenic amines, should also be considered when assessing the nutritional value of feed ingredients containing high protein and lipid concentrations (Table 2).

4.1 Trypsin inhibitors and urease activity

Trypsin inhibitors and urease activity are the main indicators of the nutritional quality of soybean meal. Trypsin inhibitors interfere with trypsin

Table 2 Summary of key anti-nutritional factors in factors in feed ingredients and references on the nutritional effects and methods for measurement

Anti-nutritional factor	Feed ingredient	Reference
Trypsin inhibitors	Soybeans and camelina	Woyengo et al. (2017); Liu (2019); Liu et al. (2021); Chen et al. (2020)
Glucosinolates	Canola and camelina	Woyengo et al. (2017); Szmigielska et al. (2000)
Cyanogenic glycosides	Flaxseed	Woyengo et al. (2017)
Carbonyls and other protein oxidation metabolites	Grain, oilseed, and animal by-products	Levine et al. (1990); Soglia et al. (2016); Celi and Gabai (2015); Feng et al. (2015); Zhang et al. (2017); Frame et al. (2020); Frame et al. (2021)
Secondary aldehydes and other lipid oxidation metabolites	Lipids	Shurson et al. (2015); Wang et al. (2016); Lindblom et al. (2017); Hung et al. (2017); Silva-Guillen et al. (2020); Winkler-Moser et al. (2020); Abeyrathne et al. (2021); Bacou et al. (2021); Shurson et al. (2021a)
Biogenic amines	Fish meal and animal by-products	Learey et al. (2018); Feddern et al. (2019); Özogul and Özogul (2019); Kočar et al. (2021); Wang et al. (2021)
Regulated mycotoxins	Grain and grain by-products	Escrivá et al. (2015); Vanhoutte et al. (2016); Peng et al. (2018); Holanda and Kim (2021); Tkaczy and Jedziniak (2021); Liu et al. (2022); Xu et al. (2022)
Co-occurring mycotoxins	Grain and grain by-products	Borutova et al. (2012); Khoshal et al. (2019)
Emerging mycotoxins	Grain and grain by-products	Fraeyman et al. (2017); Streit et al. (2013); Novak et al. (2021); Khoshal et al. (2019)
Masked mycotoxins	Grain and grain by-products	Berthiller et al. (2013); Streit et al. (2013); Rychlik et al. (2014); Zhang et al. (2019)

and chymotrypsin enzymes secreted by the pancreas and reduce protein and amino acid digestibility and feed intake in pigs (Woyengo et al., 2017). Urease is a naturally occurring enzyme found in soybeans that catalyzes the breakdown of urea to ammonia and results in an increase in pH. Determining urease activity and trypsin inhibitor concentrations is essential for assessing the adequacy of heat treatment during soybean and other legume seed processing because these compounds are heat labile and inactivated during adequate thermal processing conditions. Measuring urease activity in soybean meal is relatively easy because it is a procedure based on pH, but measuring trypsin inhibitors is more challenging because multiple types of equipment and several reagents are required in a long and complex reaction process (Chen et al., 2020). Although urease and trypsin inhibitors are not biochemically or functionally associated, urease has been used as an indicator for trypsin inhibitor concentrations because they have been shown to be correlated.

However, Chen et al. (2020) evaluated 980 samples of soybean products (i.e. solvent-extracted soybean meal, raw soybeans, full-fat extruded soybean meal, heat-inactivated full-fat soybean meal, and expeller soybean meal) collected from feed mills in 21 countries to determine if the correlation between urease activity and trypsin inhibitor concentrations was sufficiently high to use urease activity as a valid and reliable indicator of soybean meal quality. Results from this study showed that urease activity was not highly correlated with trypsin inhibitor concentration, which prevents its use as a surrogate alternative assay for directly determining trypsin inhibitor concentrations (Chen et al., 2020). Liu (2019) and Liu et al. (2021) discussed improvement in the trypsin inhibitor assay and its applications in legumes, cereals, and related products.

4.2 Glucosinolates

Glucosinolates are the primary anti-nutritional factor present in canola and rapeseed meal, and dietary concentrations should be considered when formulating swine diets (Woyengo et al., 2017). Myrosinase degrades glucosinolates into several toxic compounds including isothiocyanates, nitriles, and thiocyanates that interfere with the synthesis of thyroid hormones resulting in decreased feed intake and disruption of normal metabolic processes (Woyengo et al., 2017). Szmigielska et al. (2000) evaluated a simpler and low-cost membrane extraction/HPLC method for determining glucosinolate concentrations in canola seeds compared with existing procedures, and reported very high correlations of results with the palladate colorimetric procedure and gas chromatography methods. However, because there are more than 130 structurally different glucosinolates, it is important to identify and quantify these individual compounds because their biological activity varies from having beneficial effects to detrimental effects (Grosser and van Dam, 2017). A targeted and robust HPLC method has been validated to analyze glucosinolate profiles in many diverse plant species, and results have been compared with other faster and more advanced methods of glucosinolate analysis (Grosser and van Dam, 2017). In addition, a new low-cost analytical method for quantifying glucosinolates in camelina has been validated against Association of Official Agricultural Chemists criteria and can be adopted as a new standard method for quality control of camelina in the feed industry (Meza et al., 2022).

4.3 Protein oxidation

Minimizing oxidative stress in pigs is essential for optimizing health, productivity, and nutritional efficiency (Bacou et al., 2021). A dietary imbalance of prooxidants and antioxidants can lead to oxidative stress along with the consumption of oxidized proteins and lipids. Exposure to heat can cause proteins to oxidize and produce reactive oxygen species including superoxide ions, peroxides, hydroxy radicals, and hydroxyl ions that can cause more oxidative damage to proteins, lipids, and DNA (Feng et al., 2015; Zhang et al., 2017; Frame et al., 2021). Protein oxidation of feed ingredients is typically measured by determining carbonyl concentrations (Levine et al., 1990; Soglia et al., 2016). Although limited studies have been conducted to evaluate the effect of feeding diets containing ingredients with high amounts of oxidized protein, results from initial studies have shown reduced energy and nutrient digestibility (Celi and Gabai, 2015; Frame et al., 2020, 2021). Because of these potential effects, evaluation of the extent of protein oxidation in high protein feed ingredients that undergo heat processing may be useful when assessing the protein quality of ingredients.

4.4 Lipid oxidation

Compared with protein oxidation, much more is known about the negative effects of feeding oxidized lipids to pigs on oxidative stress and nutritional efficiency (Shurson et al., 2015; Hung et al., 2017; Lindblom et al., 2017; Silva-Guillen et al., 2020). Unfortunately, there is no single measure of lipid oxidation that can be used to accurately characterize the extent of lipid damage that occurs in fats and oils because oxidation is a dynamic process with several compounds simultaneously produced and degraded over time (Shurson et al., 2015; Winkler-Moser et al., 2020). Peroxide value (PV) is useful for evaluating the concentrations of free radicals produced during the initiation phase of oxidation, while *p*-anisidine value (AV) is a useful indicator of secondary aldehydes produced during the propagation stage of oxidation (Shurson et al., 2015; Wang et al., 2016). Wang et al. (2016) proposed using 4-hydroxynonenal as an accurate single aldehyde biomarker in oxidized lipids because it is highly correlated with physiological responses in pigs, but its measurement is more difficult than traditional lipid oxidation measures such as PV and AV. Therefore, until more definitive models or methods are developed that more accurately characterize the extent of oxidation in feed fats and oils, the combination of PV and AV has been suggested as useful indicators for connecting the extent of lipid oxidation in animal diets with physiological responses in swine and poultry (Shurson et al., 2021a).

4.5 Biogenic amines

Biogenic amines are also useful measures for determining the protein and amino acid quality of high-protein feed ingredients. The presence of biogenic amines in human food or animal feed indicates that some bacterial spoilage has occurred, and concentrations are correlated with bacterial counts and freshness of food (Özogul and Özogul, 2019). Biogenic amines are produced from precursors of free amino acids in food and feed under conditions that allow for bacterial growth resulting in the production of amino acid decarboxylase enzymes which catalyze these reactions (Özogul and Özogul, 2019). Biogenic amines are low-molecular-weight organic bases that are formed and degraded during the normal metabolism of microorganisms, plants, and animals, and are involved in several important physiological roles in humans and animals such as regulation of body temperature, stomach volume, stomach pH, and brain activity (ten Brink et al., 1990). These compounds also serve as essential components of cell growth by promoting DNA, RNA, and protein synthesis, stabilizing ribosomes, and increasing amino acid uptake by cells (Smith, 1990). It is difficult to determine the specific threshold or maximum tolerable concentrations when biogenic amines become toxic because relative toxicity varies among biogenic amine compounds, and detoxification efficiency varies by age of animals and species. Several recent publications have described new and traditional methods for determining biogenic amines in food and feed (Learey et al., 2018; Özogul and Özogul, 2019; Kočar et al., 2021; Wang et al., 2021).

4.6 Mycotoxins

Mycotoxins can also be a major dietary contributor to oxidative stress (Bacou et al., 2021) and reduce nutritional efficiency, productivity, and health of animals (Xu et al., 2022). The most commonly measured mycotoxins in feed ingredients and complete feeds include deoxynivalenol, zearalenone, aflatoxins, ochratoxin A, patulin, fumonisins, and T-2 toxins as part of quality assurance programs in commercial feed mills. Many of these mycotoxins are regulated in various countries because of their prevalence in various feed ingredients and their toxicity when fed to animals. As a result, several scientific reviews have been published summarizing the benefits and limitations of using various analytical methods for mycotoxin detection and quantification, as well as the toxic effects of feeding diets contaminated with individual mycotoxins to animals (Escrivá et al., 2015; Vanhoutte et al., 2016; Peng et al., 2018; Holanda and Kim, 2021; Li et al., 2021; Tkaczy and Jedziniak, 2021; Liu et al., 2022; Xu et al., 2022). Pigs are more sensitive to dietary mycotoxin contamination than other food animal species, and detrimental health and performance responses have been observed when feeding diets containing concentrations of combinations of mycotoxins that individually are considered to be 'safe' because they are below the maximum recommended dietary threshold concentrations. As a result, a more holistic approach is needed for assessing mycotoxin contamination in feed ingredients and should include co-occurring mycotoxins, emerging mycotoxins, and masked mycotoxins.

4.6.1 Co-occurring mycotoxins

Multiple mycotoxins are usually present in the same feed sample because some fungal species produce several mycotoxins, and multiple fungal species can be found in the same ingredient. Borutova et al. (2012) evaluated 1428 feed samples in Asia-Oceania in which they observed that multiple mycotoxins were commonly found in the same feed ingredient sample, and it was rare to detect only one mycotoxin in a sample. These authors reported that the correlation for co-occurrence of aflatoxins and fumonisins was very high, but correlations between other *Fusarium* mycotoxins (deoxynivalenol and zearalenone) with ochratoxin A and aflatoxins were low. Although the toxic effects of individual mycotoxins are fairly well characterized, the toxic effects of combinations of co-occurring mycotoxins are poorly understood (Khoshal et al., 2019).

4.6.2 Emerging mycotoxins

Emerging mycotoxins, which are secondary metabolites of fungi, are also a major concern because they are not routinely measured and are not regulated but can have detrimental effects on pigs (Khoshal et al., 2019). Most of the common emerging mycotoxins are produced by *Fusarium* species and include enniatins, beauvericin, apicidin, aurofusarin, culmorin, butanolide, fusaric acid, moniliformin, and fusaproliferin (Khoshal et al., 2019). Aspergillus species can produce sterigmatocystin and emodin while penicillium mold produces mycophenolic acid. Multiple mycotoxin screening has led to the identification of 139 different secondary metabolites in feed and feed ingredients (Streit et al., 2013). Although limited information is available on the effects of these secondary metabolites on nutrient utilization, health, and performance of pigs, results from an initial study by Novak et al. (2021) showed that the combination of beauvericin, enniatins, and deoxynivalenol in diets fed to weaned pigs reduced growth rate, reduced gut microbial diversity and liver enzyme activity, and caused liver, jejunum, and lymph node lesions.

4.6.3 Masked mycotoxins

Masked mycotoxins are derivatives of commonly measured and regulated mycotoxins that are undetectable due to their modified chemical structure when using conventional analytical methods. Berthiller et al. (2013) indicated that the *Fusarium* mycotoxins (deoxynivalenol, zearalenone, fumonisins, nivalenol, fusarenon-X, T-2 toxin, HT-2 toxin, and fusaric acid) are prone to being metabolized and transformed into various chemical derivatives. The most studied masked mycotoxin is deoxynivalenol-3-glucoside, which can be found in corn, wheat, barley, and oats. Other masked mycotoxins that have been studied include zearalenone-14-glucoside, nivalenol-3-glucoside, T-2-toxin-3-glucoside, and Ht-2-toxin-3-glucoside. Little is known about the relative toxicity of masked mycotoxins because limited research has focused on these compounds, but several studies suggest that their toxicity is a concern for food safety.

5 Functional nutrients

Functional feed ingredients have been described as those that contain certain nutrients and bioactive compounds or properties that promote animal health due to their (1) antimicrobial properties, (2) antioxidant properties, (3) influence gut microbiome composition and function, (4) ability to reduce inflammation by enhancing immune signaling and responses, (5) effect on stimulating feed consumption, and (6) improvement in gut health (Shurson et al., 2021a). Functional nutrients may be a single chemical component of a functional ingredient (e.g. copper or linolenic acid) or a combination of functional nutrients (e.g. phenolic acids and non-starch polysaccharides) that provide specific health-promoting effects. Numerous scientific reviews have been published to describe the beneficial health effects of specific functional ingredients and functional nutrients when included in swine diets (Table 3). While some of these functional nutrients are easily measured using standard laboratory analysis procedures for various nutrients (i.e. amino acids, fatty acids, copper, and zinc), analytical procedures for other functional nutrients require the use of specialized equipment and methods. Quantifying functional nutrient concentrations in feed ingredients is important to ensure that adequate amounts are present in complete diets to increase the likelihood of achieving the growth and health benefits expected.

6 Environmental impact indicators of feed ingredients

Reducing the environmental footprint of pork production has become an urgent need and new criteria for selecting feed ingredients used in swine diets. Feed production contributes 50–85% of climate change impact, 64–97% of eutrophication potential, 70–96% of energy use, and about 100% of land occupation in monogastric animal production systems (Garcia-Launay et al., 2018). Therefore, because feed ingredients used in animal feed production have a significant influence on environmental footprint, one of the most effective strategies to reduce the environmental impact of pork production is to use multi-objective diet formulation approaches including low environmental impact feed ingredients based on LCA methodologies (Mackenzie et al., 2016;

Table 3 Summary of functional nutrients, ingredient sources, biological functions, and key references

Bioactive compounds	Sources	Functions	Reference
Glutamate, glutamine, glycine, proline, and arginine	High protein feed ingredients and crystalline amino acids	Signaling pathways regulating gene expression, intracellular turnover, nutrient metabolism, oxidative defense, and reducing cell damage	Wu (2010); Wu (2014); Yang and Liao (2019)
Immunoglobulins	Spray-dried animal plasma	Improve immune response and gut barrier function	Peace et al. (2011); van Dijk et al. (2001)
β-glucans	Barley, oats, and wheat	Prebiotic	Jha and Berrocoso (2015); O'Connell et al. (2005); Stein and Kil (2006); de Lange et al. (2010); Lombardi et al. (2020)
Copper	Copper sulfate and tribasic copper chloride	Antibacterial that alters gut microbiome	Poole (2017); Shannon and Hill (2019)
Linoleic and linolenic acid	Plant and fish oils	Improve reproductive performance in sows, involved in inflammatory reactions and immune responses from bacterial challenges and epithelial barrier function	Rosero et al. (2016a); Rosero et al. (2016b); Lauridsen (2020)
Caproic acid, caprylic acid, capric acid, lauric acid; Glycerol monocaproate, glycerol monocaprylate, glycerol monocaprate, and glycerol monolaurate	Purified forms from various lipids	Antibacterial, antiviral, immune modulation activity, and improved gut health in pigs; feed pathogen mitigation	Zentek et al. (2011); Cochrane et al. (2020); Cochrane et al. (2016); Hanczakowska (2017); Jackman et al. (2020); Gebhardt et al. (2020); Dahmer et al. (2020)
Various types of non-starch polysaccharides	Various high-fiber ingredients	Alter gut microbiome, prebiotic, production of short chain fatty acids, improve innate and adaptive immune responses, antioxidant and bactericidal properties	Charalampopoulos et al. (2002); Jha and Berrocoso (2015); Agyekum and Nyachoti (2017); Tran et al. (2018); Jha et al. (2019); McRorie and Fahey (2015); Lindberg (2014); Makki et al. (2018); Beukema et al. (2020)

(Continued)

Table 3 (*Continued*)

Bioactive compounds	Sources	Functions	Reference
Phenolic acids including ferulic acid, *p*-coumeric acid, vanillic acid, and sinapic acid	Barley, corn, oats, rye, wheat, and grape extract	Antioxidant properties	Fiesel et al. (2014); Verhelst et al. (2014); Gerasopoulos et al. (2015); Vitaglione et al. (2008)
Lactose	Lactose and dried milk products	Improves nutrient digestibility, prebiotic, fermentation to lactic acid and volatile fatty acids in young pigs	O'Doherty et al. (2010); Mahan (1992); O'Connell et al. (2005); Bach Knudsen (2012)
Isoflavones	Soybean meal and other soy ingredients	Anti-inflammatory, antioxidant, anti-viral, decrease intestinal epithelial permeability, enhance growth and immune responses from PRRSV infection	Smith and Dilger (2018); Schweer et al. (2018)
Saponins	Soybean meal and other soy ingredients	Antioxidants, potential enhancers of passive immunity, vaccine adjuvants to increase immune response	Smith and Dilger (2018); Schweer et al. (2018)
Vitamins	A (carotenoids), C, D, E, K, niacin, pyridoxine, and riboflavin	Antioxidants, improve gastrointestinal function and health	Sinbad et al. (2019); Lauridsen et al. (2021)
Zinc	Zinc oxide and other zinc sources	Antibacterial effects that alter gut microbiome	Poole (2017); Shannon and Hill (2019); Shurson et al. (2022b)

Garcia-Launay et al. (2018); de Quelen et al., 2021; Méda et al., 2021; Soleimani and Gilbert, 2021). LCA is a compilation and evaluation of the inputs, outputs, and environmental impacts of a system used to produce a product throughout its life cycle (van Middelaar et al., 2019). Standardized methodology and guidelines have been established for determining various LCA environmental impact indicators (LEAP, 2015), but most of the limited number of LCA feed ingredient databases include feed ingredients used in the European Union, and LCA values are not directly applicable to ingredients used in the United States and other countries. However, the Global Feed LCA Institute (GFLI; https://tools.blonkconsultants.nl/tool/16/) has developed the largest database of feed ingredients with the most LCA indicator variables (n = 18; Table 4) and the future goal of widespread global application.

Recent studies have shown that feeding growing-finishing swine diets formulated using the least cost and LCA environmental impact measures resulted in significant reductions in environmental impacts with no effects on growth performance and carcass characteristics (de Quelen et al., 2021; Shurson et al., 2022a). Therefore, the determination of accurate LCA data for all feed ingredients used in swine diets will continue to be a new and important focus for assessing the nutritional value of feed ingredients to improve the environmental sustainability of pork production globally.

7 Conclusion

Assessing the nutritional value of feed ingredients used in swine diets has evolved beyond determining energy and nutrient content and digestibility. The nutritional value of feed ingredients must include assessing the positive (functional nutrients) as well as the negative (anti-nutritional) compounds that influence health, efficiency, and productivity of pigs. Feed represents the greatest opportunity to reduce the environmental footprint of pig production. Therefore, assessment of the LCA environmental impacts of feed ingredients based on geographic origin is also becoming a new component of nutritional value for selecting and using sources of feed ingredients. Approaches for dynamically estimating nutrient content and digestibility of actual feed ingredients being fed improve precision in feed formulation compared with using average, static published values. Near-infrared spectroscopy technology and calibrations have become accepted as a less expensive, rapid, and accurate alternative to traditional wet chemistry procedures not only for quantifying energy and nutrient content of various feed ingredients and complete feeds, but also for estimating net energy and digestible amino acid concentrations. Validated prediction equations and models developed using robust nutrient composition data sets for individual ingredients can also be an effective way

Table 4 Summary of Global Feed LCA Institute environmental impact measures and their functional units that have been applied to feed ingredients

Environmental impact measure	Functional unit per kg ingredient	Description
Global warming with or without land use change	kg CO_2 equivalent	Indicator of potential global warming due to emissions of greenhouse gases to the air, using carbon dioxide as a standard, with or without a change in land use
Stratospheric ozone depletion	kg CFC11 equivalent	Indicator of emissions to air that cause destruction of the stratospheric ozone layer using chlorofluorocarbon-11 as a reference standard
Ionizing radiation	kBq Co-60 equivalent	Impact on radiation as measured by kilobecquerels of cobalt-60 radioactive isotope as a reference standard
Ozone formation, human health	kg Nox equivalent	Impact on nitrous oxide gases that affect the ozone and human health
Fine particulate matter formation	kg PM2.5 equivalent	Impact on air quality as atmospheric particulate matter with particles having a diameter of less than 2.5 micrometers
Ozone formation, terrestrial ecosystems	kg Nox equivalent	Impact on nitrous oxide gases that affect the ozone and human health
Terrestrial acidification	kg SO_2 equivalent	Indicator of the potential acidification of soil and water due to the release of nitrogen oxide and sulfur oxide gases
Freshwater eutrophication	kg P equivalent	Indicator of the potential for increased phosphorus emission to freshwater
Marine eutrophication	kg N equivalent	Indicator of the potential for increased nitrogen emission to freshwater
Terrestrial ecotoxicity	kg 1,4-DCB	Impact of toxic substances emitted to the environment on land organisms using 1,4-dichlorobenzene as a standard
Freshwater ecotoxicity	kg 1,4-DCB	Impact of toxic substances emitted to the environment on freshwater organisms using 1,4-dichlorobenzene as a standard
Marine ecotoxicity	kg 1,4-DCB	Impact of toxic substances emitted to the environment on seawater organisms using 1,4-dichlorobenzene as a standard
Human carcinogenic toxicity	kg 1,4-DCB	Impact of carcinogenic toxic substances on the environment using 1,4-dichlorobenzene as a standard
Human non-carcinogenic toxicity	kg 1,4-DCB	Impact of non-carcinogenic toxic substances on the environment using 1,4-dichlorobenzene as a standard
Land use	m^2a crop equivalent	Impact of converting non-agricultural land into agricultural use

(Continued)

Table 4 (*Continued*)

Environmental impact measure	Functional unit per kg ingredient	Description
Mineral resource scarcity	kg Cu equivalent	Indicator of depletion of natural inorganic mineral resources using copper as a standard
Fossil resource scarcity	kg oil equivalent	Indicator of the depletion of natural fossil fuel resources
Water consumption	m^3	Indicator of the amount of water (cubic meters) required to produce a kg of product

Source: Adapted from: Shurson and Urriola (2022).

of dynamically estimating DE, ME, NE, and digestible amino acid content of ingredients. However, more attention is needed for quantifying concentrations of anti-nutritional factors such as the extent of lipid and protein oxidation, trypsin inhibitors, glucosinolates, biogenic amines, and masked and emerging mycotoxins in feeding ingredients because of their detrimental effects on energy and nutrient utilization efficiency and animal health. In addition, greater consideration should be given toward measuring concentrations of various functional nutrients in selected ingredients that provide antioxidant, anti-inflammatory, antibacterial, antivirus, or prebiotic benefits in swine diets. Exploiting the benefits of functional nutrients may minimize the need for using various growth and health-promoting feed additives in swine diets. Finally, determining the LCA of feed ingredients and setting constraints in multi-objective feed formulations is rapidly becoming an important component of environmentally sustainable precision swine feeding programs.

8 Where to look for further information

8.1 Recommended review and original articles

8.1.1 General trends

- Pomar, C., Andretta, I. and Remus, A. (2021). Feeding strategies to reduce nutrient losses and improve sustainability of growing pigs, *Front. Vet Sci.* 8: 742220. doi:10.3389/fvets.2021.742220.
- Shurson, G. C. and Urriola, P. E. (2022). Sustainable swine feeding programs require the convergence of multiple dimensions of circular agriculture and food systems with One Health, *Anim. Front.* 12: 30–40. https://doi.org/10.1093/af/vfac077.
- Shurson, G. C., Hung, Y.-T., Jang, J. C. and Urriola, P. E. (2021). Measures matter – Determining the true nutri-physiological value of feed ingredients for swine, *Animals* 11: 1259. https://doi.org/10.3390/ani11051259.

- van der Poel, A. F. B., Abdollahi, M. R., Cheng, H., Colovic, R., den Hartog, L. A., Miladinovic, D., Page, G., Sijssens, K., Smillie, J. F., Thomas, M., Wang, W., Yu, P. and Hendriks, W. H. (2020). Future directions of animal feed technology research to meet the challenges of a changing world, *Anim. Feed Sci. Technol.* 270: 114692. https://doi.org/10.1016/j.anifeedsci.2020.114692.

8.1.2 Net energy

- Noblet, J., Wu, S.-B. and Choct, M. (2022). Methodologies for energy evaluation of pig and poultry feeds: A review, *Anim. Nutr.* 8: 185–203. https://doi.org/10.1016/j.aninu.2021.06.015.
- Li, Z., Liu, H., Li, Y., Lv, Z., Liu, L., Lai, C., Wang, J., Wang, F., Li, D. and Zhang, S. (2018). Methodologies on estimating the energy requirements for maintenance and determining net energy contents of feed ingredients in swine: A review of recent work, *J. Anim. Sci. Biotechnol.* 9: 39. https://doi.org/10.1186/s40104-018-0254-0.

8.1.3 Protein quality

- Liebert, F. (2017). Invited review: Further progress is needed in procedures for the biological evaluation of dietary protein quality in pig and poultry feeds, *Arch. Anim. Breed.* 60: 259–270. https://doi.org/10.5194/aab-60-259-2017.

8.1.4 Minerals

- Kerr, B. J., Ziemer, C. J., Weber, T. E., Trabue, S. L., Bearson, B. L., Shurson, G. C. and Whitney, M. H. (2008). Comparative sulfur analysis using thermal combustion or inductively coupled plasma methodology and mineral composition of common livestock feedstuffs, *J. Anim. Sci.* 86: 2377–2384. doi:10.2527/jas.2007-0811.
- Lautrou, M., Narcy, A., Dourmad, J.-Y., Pomar, C., Schmidely, P. and Montminy, M.-P. L. (2021). Dietary phosphorus and calcium utilization in growing pigs: Requirements and improvements. *Front. Vet. Sci.* 8: 734365. doi:10.3389/fvets.2021.734365.

8.1.5 Vitamins

- Yang, P., Wang, H. K., Zhu, M., Li, L. X. and Ma, Y. X. (2021). Degradation kinetics of vitamins in premixes for pig: Effects of choline, high concentrations of copper and zinc, and storage time. *Anim. Biosci.* 34: 701–713. https://doi.org/10.5713/ajas.20.0026.

8.1.6 Near-infrared spectroscopy

- Noel, S. J., Jørgensen, H. J. H. and Bach Knudsen, K. E. (2022). The use of near-infrared spectroscopy (NIRS) to determine the energy value of individual feedstuffs and mixed diets for pigs, *Anim. Feed Sci. Technol.* 283: 115156. https://doi.org/10.1016/j.anifeedsci.2021.115156.
- Noel, S. J., Jørgensen, H. J. H. and Bach Knudsen, K. E. (2021). Prediction of protein and amino acid composition and digestibility in individual feedstuffs and mixed diets for pigs using near-infrared spectroscopy, *Anim. Nutr.* 7: 1242–1252. https://doi.org/10.1016/j.aninu.2021.07.004.
- Paternostre, L., Baeten, V., Ampe, B., Millet, S. and De Boever, J. (2021). The usefulness of NIRS calibrations based on feed and feces spectra to predict nutrient content, digestibility and net energy of pig feeds, *Anim. Feed Sci. Technol.* 281: 115091. https://doi.org/10.1016/j.anifeedsci.2021.115091.

8.1.7 Digestion kinetics

- Gerrits, W. J. J., Schop, M. T. A., de Vries, S. and Dijkstra, J. (2021). ASAS-NANP symposium: Digestion kinetics in pigs: The next step in feed evaluation and a ready-to-use modeling exercise. *J. Anim. Sci.* 99: 1-8. https://doi.org/10.1093/jas/skab020.

8.1.8 Anti-nutritional factors

- Bacou, E., Walk, C., Rider, S., Litta, G. and Perez-Calvo, E. (2021). Dietary oxidative distress: A review of nutritional challenges and models for poultry, swine and fish, *Antioxidants* 10: 525. https://doi.org/10.3390/antiox10040525.
- Berthiller, F., Crews, C., Dall'Asta, C., De Saeger, S., Haesaert, G., Karlovsky, P., Oswald, I. P., Seefelder, W., Speijers, G. and Stroka, J. (2013). Masked mycotoxins: A review, *Mol. Nutr. Food Res.* 57: 165–186. doi:10.1002/mnfr.201100764.
- Cowieson, A. J., Ruckebusch, J. P., Knap, I., Guggenbuhl, P. and Fru-Nji, F. (2016). Phytate-free nutrition: A new paradigm in monogastric animal production. *Anim. Feed Sci. Technol.* 222: 180–189. http://dx.doi.org/10.1016/j.anifeedsci.2016.10.016.
- Winkler-Moser, J. K., Hwang, H.-S. and Kerr, B. J. (2020). Changes in markers of lipid oxidation and thermal treatment in feed-grade fats and oils, *J. Sci. Food Agric.* 100: 3328–3340. doi:10.1002/jsfa.10364.
- Zhang, Z., Nie, D., Fan, K., Yang, J., Guo, W. and Meng, J. (2019). A systematic review of plant-conjugated masked mycotoxins: Occurrence, toxicology,

and metabolism, *Crit. Rev. Food Sci. Nutr.* 60: 1523–1537. https://doi.org/10.1080/10408398.2019.1578944.

8.1.9 Functional nutrients

- Wu, G. (2010). Functional amino acids in growth, reproduction, and health, *Adv. Nutr.* 1: 31–37. doi:10.3945/an.110.1008.
- McRorie, J. W. and Fahey, G. C. (2015). Fiber supplements and clinically meaningful health benefits: Identifying the physiochemical characteristics of fiber that drive specific physiologic effects. In: Dietary Supplements in Health Promotion, T.C. Wallace, Ed. CRC Press, Boca Raton, FL, p. 161–206.

8.1.10 Environmental impacts

- Méda, B., Garcia-Launay, F., Dusart, L., Ponchant, P., Espagnol, S. and Wilfart, A. (2021). Reducing environmental impacts of feed using multiobjective formulation: What benefits at the farm gate for pig and broiler production? *Animal* 15: 100024. https://doi.org/10.1016/j.animal.2020.100024.
- Garcia-Launay, F., Dusart, L., Espagnol, S., Laisse-Redoux, S., Gaudré, D., Méda, B. and Wilfart, A. (2018). Multiobjective formulation is an effective method to reduce environmental impacts of livestock feeds, *Br. J. Nutr.* 120: 1298–1309. doi:10.1017/S0007114518002672.
- Mackenzie, S. G., Leinonen, I., Ferguson, N. and Kyriazakis, I. (2016). Towards a methodology to formulate sustainable diets for livestock: Accounting for environmental impact in diet formulation, *Br. J. Nutr.* 115: 1860–1874. doi:10.1017/S0007114516000763.

9 References

Abeyrathne, E. D. N. S., Nam, K. and Ahn, D. U. (2021). Analytical methods for lipid oxidation and antioxidant capacity in food systems, *Antioxidants* 10(10): 1587. 10.3390/antiox10101587.

Agyekum, A. K. and Nyachoti, C. M. (2017). Nutritional and metabolic consequences of feeding high-fiber diets to swine: a review, *Engineering* 3(5): 716–725. 10.1016/J.ENG.2017.03.010.

Ahn, J. Y., Kil, D. Y., Kong, C. and Kim, B. G. (2014). Comparison of oven-drying methods for determination of moisture content in feed ingredients, *Asian Australas. J. Anim. Sci.* 27(11): 1615–1622. 10.5713/ajas.2014.14305.

Aldenhoven, N., Gutierrez, N. A., Jaworski, N. W. and van Laar, H. (2020). 167 Analysis of variation in net energy prediction of feed ingredients fed to swine, *J. Anim. Sci.* 98 (Suppl. 3): 62. 10.1093/jas/skaa054.111.

Amaral, N. O., Amaral, L. G. M., Cantarelli, V. S., Fialho, E. T., Zangeronimo, M. G. and Rodrigues, P. B. (2015). Influence of maize particle size on the kinetics of starch digestion in the small intestine of growing pigs, *Anim. Prod. Sci.* 55(10): 1250–1254. 10.1071/AN14050.

Anderson, P. V., Kerr, B. J., Weber, T. E., Ziemer, C. J. and Shurson, G. C. (2012). Determination and prediction of digestible and metabolizable energy from chemical analysis of corn coproducts fed to finishing pigs, *J. Anim. Sci.* 90(4): 1242–1254. 10.2527/jas.2010-3605.

Andretta, I., Hauschild, L., Kipper, M., Pires, P. G. S. and Pomar, C. (2018). Environmental impacts of precision feeding programs applied to pig production, *Animal* 12(9): 1990–1998. 10.1017/S1751731117003159.

American Society of Agricultural and Biological Engineers. (2008). Standard ANSI/ASAE S319.4: method of determining and expressing fineness of feed materials by sieving. Revised in 2012, *Am. Soc. Agric., Biol. Eng.*, St. Joseph, MI.

Bach Knudsen, K. E. (2012). Lactose in diet influences the degradation of mixed linked β (1–3; 1–4)-D-glucan in the small intestine of pigs, *J. Anim. Sci.* 90 (Suppl. 4): 125–127. 10.2527/jas.53788.

Bach Knudsen, K. E. (2001). The nutritional significance of "dietary fibre" analysis, *Anim. Feed Sci. Technol.* 90(1–2): 3–20. 10.1016/S0377-8401(01)00193-6.

Bacou, E., Walk, C., Rider, S., Litta, G. and Perez-Calvo, E. (2021). Dietary oxidative distress: a review of nutritional challenges and models for poultry, swine and fish, *Antioxidants* 10(4): 525. 10.3390/antiox10040525.

Berthiller, F., Crews, C., Dall'Asta, C., De Saeger, S. D., Haesaert, G., Karlovsky, P., Oswald, I. P., Seefelder, W., Speijers, G. and Stroka, J. (2013). Masked mycotoxins: a review, *Mol. Nutr. Food Res.* 57(1): 165–186. 10.1002/mnfr.201100764.

Bertol, T. M., Zanotto, D. L., Coldebella, A. and Ludke, J. V. (2017). Development and validation of equations to predict the metabolizable energy value of corn for pigs, *J. Anim. Sci.* 95(1): 291–301. 10.2527/jas.2016.0832.

Beukema, M., Faas, M. M. and de Vos, P. (2020). The effects of different dietary fiber pectin structures on the gastrointestinal immune barrier: impact via gut microbiota and direct effects on immune cells, *Exp. Mol. Med.* 52(9): 1364–1376. 10.1038/s12276-020-0449-2.

Bloxham, D. J., Azain, M., Pesti, G. M. and Wu, S. B. (2023). Total amino acid level affects the results of standardized ileal digestibility assays for feed ingredients for swine, *Anim. Nutr.* 13: 19–30 10.1016/j.aninu.2022.08.018.

Borenstein, M., Hedges, L. V., Higgins, J. P. T. and Rothstein, H. R. (2009). *Introduction to Meta-analysis* (1st edn.). John Wiley and Sons Ltd., West Sussex, UK. 10.1002/9780470743386.

Borutova, R., Aragon, Y. A., Nährer, K. and Berthiller, F. (2012). Co-occurrence and statistical correlations between mycotoxins in feedstuffs collected in the Asia-Oceania in 2010, *Anim. Feed Sci. Technol.* 178(3–4): 190–197. 10.1016/j.anifeedsci.2012.09.015.

Bryan, D. D. S. L. and Classen, H. L. (2020). In vitro methods of assessing protein quality for poultry, *Animals (Basel)* 10(4): 551. 10.3390/ani10040551.

Cadogan, D. J. and Choct, M. (2015). Pattern of non-starch polysaccharide digestion along the gut of the pig; Contribution to available energy, *Anim. Nutr.* 1(3): 160–165. 10.1016/j.aninu.2015.08.011.

Celi, P. and Gabai, G. (2015). Oxidant/antioxidant balance in animal nutrition and health: the role of protein oxidation, *Front. Vet. Sci.* 2: 48. 10.3389/fvets.2015.00048.

Charalampopoulos, D., Wang, R., Pandiella, S. S. and Webb, C. (2002). Application of cereals and cereal components in functional foods: a review, *Int. J. Food Microbiol.* 79(1-2): 131-141. 10.1016/s0168-1605(02)00187-3.

Chen, H., Wierenga, P. A., Hendriks, W. H. and Jansman, A. J. M. (2019a). *In vitro* protein digestion kinetics of protein sources for pigs, *Animal* 13(6): 1154-1164. 10.1017/S1751731118002811.

Chen, H., Peng, L., de Nanclares, M. P., Trudeau, M. P., Yao, D., Cheng, Z., Urriola, P. E., Mydland, L. T., Shurson, G. C., Overland, M. and Chen, C. (2019b). Identifcation of sinapine-derived choline from a rapeseed diet as a source of serum trimethylamine N-oxide in pigs, *J. Agric. Food Chem.* 67(27): 7748-7754. 10.1021/acs.jafc.9b02950.

Chen, J., Wedekind, K., Escobar, J. and Vazquez-Añón, M. (2020). Trypsin inhibitor and urease activity of soybean meal products from different countries and impact of trypsin inhibitor on ileal amino acid digestibility in pig, *J. Americ. Oil Chem. Soc.* 97(10): 1151-1163. 10.1002/aocs.12394.

Chen, L., Gao, L. X., Huang, Q. H., Lu, Q. P., Sa, R. N. and Zhang, H. F. (2014). Prediction of digestible energy of feed ingredients for growing pigs using a computer-controlled simulated digestion system, *J. Anim. Sci.* 92(9): 3887-3894. 10.2527/jas.2013-7092.

Chen, C., de Nanclares, M. P., Kurtz, J. F., Trudeau, M. P., Wang, L., Yao, D., Saqui-Salces, M., Urriola, P. E., Mydland, L. T., Shurson, G. C. and Overland, M. (2018). Identification of redox imbalance as a prominent metabolic response elicited by rapeseed feeding in swine metabolome, *J. Anim. Sci.* 96(5): 1757-1768. 10.1093/jas/sky080.

Cochrane, R. A., Dritz, S. S., Woodworth, J. C., Stark, C. R., Saensukjaroenphon, M., Gebhardt, J. T., Bai, J., Hesse, R. A., Poulsen, E. G., Chen, Q., Gauger, P. C., Derscheid, R. J., Zhang, J., Tokach, M. D., Main, R. G. and Jones, C. K. (2020). Assessing the effects of medium-chain fatty acids and fat sources on PEDV infectivity, *Transl. Anim. Sci.* 4(2): txz179. 10.1093/tas/txz179.

Cochrane, R. A., Huss, A. R., Aldrich, G. C., Stark, C. R. and Jones, C. K. (2016). Evaluating chemical mitigation of Salmonella typhimurium ATCC 14028 in animal feed ingredients, *J. Food Prot.* 79(4): 672-676. 10.4315/0362-028X.JFP-15-320.

Cromwell, G. L., Calvert, C. C., Cline, T. R., Crenshaw, J. D., Crenshaw, T. D., Easter, R. A., Ewan, R. C., Hamilton, C. R., Hill, G. M., Lewis, A. J., Mahan, D. C., Miller, E. R., Nelssen, J. L., Pettigrew, J. E., Tribble, L. F., Veum, T. L. and Yen, J. T. (1999). Variability among sources and laboratories in nutrient analyses of corn and soybean meal. NCR-42 Committee on Swine Nutrition. North Central Regional-42, *J. Anim. Sci.* 77(12): 3262-3273. 10.2527/1999.77123262x.

Dahmer, P. L., Leubcke, G. E., Lerner, A. B. and Jones, C. K. (2020). Effects of medium-chain fatty acids as alternatives to ZnO or antibiotics in nursery pig diets, *Transl. Anim. Sci.* 4(3): txaa151. 10.1093/tas/txaa151.

de Lange, C. F. M. and Birkett, H. (2005). Characterization of useful energy content in swine and poultry feed ingredients, *Can. J. Anim. Sci.* 85(3): 269-280.

de Lange, C. F. M., Pluske, J., Gong, J. and Nyachoti, C. M. (2010). Strategic use of feed ingredients and feed additives to stimulate gut health and development in young pigs, *Livest. Sci.* 134(1-3): 124-134. 10.1016/j.livsci.2010.06.117.

de Quelen, F., Brossard, L., Wilfart, A., Dourmad, J. Y. and Garcia-Launay, F. (2021). Eco-friendly feed formulation and on-farm feed production as ways to reduce the environmental impacts of pig production without consequences on animal performance, *Front. Vet. Sci.* 8: 689012. 10.3389/fvets.2021.689012.

de Vries, S., Pustjens, A. M., Schols, H. A., Hendriks, W. H. and Gerrits, W. J. J. (2012). Improving digestive utilization of fiber-rich feedstuffs in pigs and poultry by processing and enzyme technologies: a review, *Anim. Feed Sci. Technol.* 178(3-4): 123-138. 10.1016/j.anifeedsci.2012.10.004.

Escrivá, L., Font, G. and Manyes, L. (2015). *In vivo* toxicity studies of fusarium mycotoxins in the last decade: a review, *Food Chem. Toxicol.* 78: 185-206. 10.1016/j.fct.2015.02.005.

Feddern, V., Mazzuco, H., Fonseca, F. N. and de Lima, G. J. M. M. (2019). A review on biogenic amines in food and feed: toxicological aspects, impact on health and control measures, *Anim. Prod. Sci.* 59(4): 608-618. 10.1071/AN18076.

Feng, X., Li, C., Ullah, N., Cao, J., Lan, Y., Ge, W., Hackman, R. M., Li, Z. and Chen, L. (2015). Susceptibility of whey protein isolate to oxidation and changes in physicochemical, structure, and digestibility characteristics, *J. Dairy Sci.* 98(11): 7602-7613. 10.3168/jds.2015-9814.

Fiesel, A., Gessner, D. K., Erika Most, E. and Eder, K. (2014). Effects of dietary polyphenol-rich plant product from grape or hop on pro-inflammatory gene expression in the intestine, nutrient digestibility, and faecal microbiota of weaned pigs, *BMC Vet. Res.* 10: 196. 10.1186/s12917-014-0196-5.

Fraeyman, S., Croubels, S., Devreese, M. and Antonissen, G. (2017). Emerging *Fusarium* and *Alternaria* mycotoxins: occurrence, toxicity and toxicokinetics, *Toxins* 9(7): 228. 10.3390/toxins9070228.

Frame, C. A., Huff-Lonergan, E., Kerr, B. J. and Serao, M. R. (2021). Feeding oxidized chicken byproduct meal impacts digestibility more than performance and oxidative status in nursery pigs, *J. Anim. Sci.* 99(2): 1-9. 10.1093/jas/skab029.

Frame, C. A., Johnson, E., Kilburn, L., Huff-Lonergan, E., Kerr, B. J. and Serao, M. R. (2020). Impact of dietary oxidized protein on oxidative status and performance in growing pigs, *J. Anim. Sci.* 98(5): 1-7. 10.1093/jas/skaa097.

Fu, Y., Zhao, C., Lu, X. and Xu, G. (2017). Nontargeted screening of chemical contaminants and illegal additives in food based on liquid chromatography-high resolution mass spectrometry, *TrAC Trends Anal. Chem.* 96: 89-98. 10.1016/j.trac.2017.07.014.

Gao, Q., Liu, Z., Li, K., Bai, G., Liu, L., Zhong, R., Chen, L. and Zhang, H. (2023). Time-course effects of different fiber-rich ingredients on energy values, microbiota composition and SCFA profile in growing pigs, *Anim. Nutr.* 12: 263-275 10.1016/j.aninu.2022.10.003.

Garcia-Launay, F., Dusart, L., Espagnol, S., Laisse-Redoux, S., Gaudré, D., Méda, B. and Wilfart, A. (2018). Multiobjective formulation is an effective method to reduce environmental impacts of livestock feeds, *Br. J. Nutr.* 120(11): 1298-1309. 10.1017/S0007114518002672.

Gebhardt, J. T., Thomson, K. A., Woodworth, J. C., Dritz, S. S., Tokach, M. D., DeRouchey, J. M., Goodband, R. D., Jones, C. K., Cochrane, R. A., Niederwerder, M. C., Fernando, S., Abbas, W. and Burkey, T. E. (2020). Effect of dietary medium-chain fatty acids on nursery pig growth performance, fecal microbial composition, and mitigation properties against porcine epidemic diarrhea virus following storage, *J. Anim. Sci.* 98(1): 1-11. 10.1093/jas/skz358.

Gerasopoulos, K., Stagos, D., Petrotos, K., Kokkas, S., Kantas, D., Goulas, P. and Kouretas, D. (2015). Feed supplemented with polyphenolic by-product from olive mill wastewater processing improves redox status in blood and tissues of piglets, *Food Chem. Toxicol.* 86: 319-327. 10.1016/j.fct.2015.11.007.

Gerrits, W. J. J., Schop, M. T. A., de Vries, S. and Dijkstra, J. (2021). ASAS-NANP symposium: digestion kinetics in pigs: the next step in feed evaluation and a ready-to-use modeling exercise, *J. Anim. Sci.* 99(2): 1-8. 10.1093/jas/skab020.

González-Vega, J. C. and Stein, H. H. (2014). Invited review- calcium digestibility and metabolism in pigs, *Asian-Australas. J. Anim. Sci.* 27(1): 1-9. 10.5713/ajas.2014.r.01.

Goodband, R. D., Diederich, W., Dritz, S. S., Tokach, M. D., DeRouchey, J. M. and Nelssen, J. L. (2006). Comparison of particle size analysis of ground grain with, or without, the use of a flow agent. In: *Kansas State University Swine Day Report. Rep. Prog. No. WRP966*. Kansas State University, Manhattan, pp. 163–168.

Guo, Y., Wang, L., Hanson, A., Urriola, P. E., Shurson, G. C. and Chen, C. (2023). Identification of protective amino acid metabolism events in nursery pigs fed thermally oxidized corn oil, *Metabolites* 13(1): 103. 10.3390/metabo13010103.

Grosser, K. and van Dam, N. M. (2017). A straightforward method for glucosinolate extraction and analysis with high-pressure liquid chromatography (HPLC), *J. Vis. Exp.* (121): 55425. 10.3791/55425.

Grümpel-Schlüter, A., Berk, A., Schäffler, M., Spiekers, H. and Dänicke, S. (2021). Evaluation of an equation for predicting metabolizable energy concentration in compound feeds for pigs, *Arch. Anim. Nutr.* 75(4): 251–262. 10.1080/1745039X.2021.1947066.

Hanczakowska, E. (2017). The use of medium-chain fatty acids in piglet feeding – a review, *Ann. Anim. Sci.* 17(4): 967-977. 10.1515/aoas-2016-0099.

Heyer, C. M. E., Jaworski, N. W., Page, G. I. and Zijlstra, R. T. (2022). Effect of fiber fermentation and protein digestion kinetics on mineral digestion in pigs, *Animals* 12(16): 2053. 10.3390/ani12162053.

Holanda, D. M. and Kim, S. W. (2021). Mycotoxin occurrence, toxicity, and detoxifying agents in pig production with an emphasis on deoxynivalenol, *Toxins* 13(2): 171. 10.3390/toxins13020171.

Holen, J. P., Johnston, L. J., Urriola, P. E., Garrett, J. E. and Shurson, G. C. (2020). Comparative digestibility of polysaccharide-complexed zinc and zinc sulfate in diets for gestating and lactating sows, *J. Anim. Sci.* 98(4): skaa079.

Huang, Q., Shi, C. X., Su, Y. B., Liu, Z. Y., Li, D. F., Liu, L., Huang, C. F., Piao, X. S. and Lai, C. H. (2014). Prediction of the digestible and metabolizable energy content of wheat milling by-products for growing pigs from chemical composition, *Anim. Feed Sci. Technol.* 196: 107–116. 10.1016/j.anifeedsci.2014.06.009.

Hung, Y. T., Hanson, A. R., Shurson, G. C. and Urriola, P. E. (2017). Peroxidized lipids reduce growth performance of poultry and swine: A meta-analysis, *Anim. Feed Sci. Technol.* 231: 47–58. 10.1016/j.anifeedsci.2017.06.013.

Jackman, J. A., Boyd, R. D. and Elrod, C. C. (2020). Medium-chain fatty acids and monoglycerides as feed additives for pig production: towards gut health improvement and feed pathogen mitigation, *J. Anim. Sci. Biotechnol.* 11: 44. 10.1186/s40104-020-00446-1.

Jeziemy, D., Mosenthin, R., Sauer, N., Schwadorf, K. and Rosenfelder-Kuon, P. (2017). Methodological impact of starch determination of starch content and lieal digestibility of starch in grain legumes for growing pigs, *J. Anim. Sci. Biotechnol.* 8: 4. 10.1186/s40104-016-0131-7.

Jha, R. and Berrocoso, J. D. (2015). Review: dietary fiber utilization and its effects on physiological functions and gut health of swine, *Animal* 9(9): 1441–1452. 10.1017/S1751731115000919.

Jha, R., Fouhse, J. M., Tiwari, U. P., Li, L. and Willing, B. P. (2019). Dietary fiber and intestinal health of monogastric animals, *Front. Vet. Sci.* 6: 48. 10.3389/fvets.2019.00048.

Jones, A. M., Woodworth, J. C., Vahl, C., Dritz, S. S., Tokach, M. D., Goodband, R. D. and DeRouchey, J. M. (2017). Assessment of sampling technique from feeders on copper, zinc, calcium, and phosphorus analysis, Kans. Agric. exp. Stn, *Res. Rep.* 3: 5.

Kalivoda, J. R., Jones, C. K. and Stark, C. R. (2017). Impact of varying analytical methodologies on grain particle size determination, *J. Anim. Sci.* 95(1): 113–119. 10.2527/jas.2016.0966.

Kellner, T. A. and Patience, J. F. (2017). The digestible energy, metabolizable energy, and net energy content of dietary fat sources in thirteen- and fifty-kilogram pigs, *J. Anim. Sci.* 95(9): 3984–3995. 10.2527/jas2017.1824.

Kerr, B. J., Curry, S. M. and Lindblom, S. C. (2018). Digestibility of energy and caloric value in nursery pigs fed commercially available lipids, *Appl. Anim. Sci.* 35: 291–297. 10.15232/aas.2018-01835.

Kerr, B. J., Dozier, W. A., III and Shurson, G. C. (2016). Lipid digestibility and energy content of distillers' corn oil in swine and poultry, *J. Anim. Sci.* 94(7): 2900–2908. 10.2527/jas.2016-0440.

Kerr, B. J., Ziemer, C. J., Weber, T. E., Trabue, S. L., Bearson, B. L., Shurson, G. C. and Whitney, M. H. (2008). Comparative sulfur analysis using thermal combustion or inductively coupled plasma methodology and mineral composition of common livestock feedstuffs, *J. Anim. Sci.* 86(9): 2377–2384. 10.2527/jas.2007-0811.

Khoshal, A. K., Novak, B., Martin, P. G. P., Jenkins, T., Neves, M., Schatzmayr, G., Oswald, I. P. and Pinton, P. (2019). Co-occurrence of DON and emerging mycotoxins in worldwide finished pig feed and their combined toxicity in intestinal cells, *Toxins* 11(12): 727. 10.3390/toxins11120727.

Kil, D. Y., Kim, B. G. and Stein, H. H. (2013). Feed energy evaluation for growing pigs, *Asian Australas. J. Anim. Sci.* 26(9): 1205–1217. 10.5713/ajas.2013.r.02.

Kočar, D., Köse, S., Tufan, B., Ščavničar, A. and Pompe, M. (2021). Determination of biogenic amines in fresh fish and processed fish products using IC-MS/MS, *Foods* 10(8): 1746. 10.3390/foods10081746.

Kong, C. and Adeola, O. (2014). Evaluation of amino acid and energy utilization in feedstuff for swine and poultry diets, *Asian Australas. J. Anim. Sci.* 27(7): 917–925. 10.5713/ajas.2014.r.02.

Lauridsen, C. (2020). Effects of dietary fatty acids on gut health and function of pigs pre- and post-weaning, *J. Anim. Sci.* 98(4): 1–12. 10.1093/jas/skaa086.

Lauridsen, C., Matte, L. L., Lessard, M., Celi, P. and Litta, G. (2021). Role of vitamins for gastro-intestinal functionality and health in pigs, *Anim. Feed Sci. Technol.* 273: 114823. 10.1016/j.anifeedsci.2021.114823.

Livestock Environmental Assessment and Performance Partnership (2015). Environmental performance of animal feed supply chains: Guidelines for. assessment. In: *Livest. Environ. Assess. Perform.* Partnership, Food and Agriculture Organization, Rome, Italy.

Learey, J. J., Crawford-Clark, S., Bowen, B. J., Barrow, C. J. and Adcock, J. L. (2018). Detection of biogenic amines in pet food ingredients by RP-HPLC with automated dansyl chloride derivatization, *J. Sep. Sci.* 41(24): 4430–4436. 10.1002/jssc.201800455.

Lee, S. A., Park, C. S. and Kim, B. G. (2021). Novel two-slope equations to predict amino acid concentrations using crude protein concentration in soybean meal, *Agriculture* 11(4): 280. 10.3390/agriculture11040280.

Levine, R. L., Garland, D., Oliver, C. N., Amici, A., Climent, I., Lenz, A. G., Ahn, B. W., Shaltiel, S. and Stadtman, E. R. (1990). Determination of carbonyl content in oxidatively modified proteins, *Methods Enzymol.* 186: 464–478. 10.1016/0076-6879(90)86141-h.

Li, R., Wen, Y., Wang, F. and He, P. (2021). Recent advances in immunoassays and biosensors for mycotoxins detection in feedstuffs and foods, *J. Anim. Sci. Biotechnol.* 12(1): 108. 10.1186/s40104-021-00629-4.

Li, Z., Liu, H., Li, Y., Lv, Z., Liu, L., Lai, C., Wang, J., Wang, F., Li, D. and Zhang, S. (2018). Methodologies on estimating the energy requirements for maintenance and determining net energy contents of feed ingredients in swine: a review of recent work, *J. Anim. Sci. Biotechnol.* 9: 39. 10.1186/s40104-018-0254-0.

Liebert, F. (2017). Invited review: further progress is needed in procedures for the biological evaluation of dietary protein quality in pig and poultry feeds, *Arch. Anim. Breed.* 60(3): 259–270. 10.5194/aab-60-259-2017.

Lindblom, S. C., Dozier, W. A., Shurson, G. C. and Kerr, B. J. (2017). Digestibility of energy and lipids and oxidative stress in nursery pigs fed commercially available lipids, *J. Anim. Sci.* 95(1): 239–247. 10.2527/jas.2016.0915.

Lindberg, J. E. (2014). Fiber effects in nutrition and gut health in pigs, *J. Anim. Sci. Biotechnol.* 5(1): 15. Available at: http://www.jasbsci.com/content/5/1/15.

Liu, D. W., Liu, L., Li, D. F. and Wang, F. L. (2015a). Determination and prediction of the net energy content of seven feed ingredients fed to growing pigs based on chemical composition, *Anim. Prod. Sci.* 55(9): 1152–1163. 10.1071/AN14091.

Liu, J. D., Li, Q. Y., Zeng, Z. K., Li, P., Xu, X., Wang, H. L., Zhang, S. and Piao, X. S. (2015b). Determination and prediction of the amino acid digestibility of sunflower seed meals in growing pigs, *Asian Australas. J. Anim. Sci.* 28(1): 86–94. 10.5713/ajas.14.0109.

Liu, K. (2019). Soybean trypsin inhibitor assay: further improvement of the standard method approved and reapproved by American Oil Chemists' Society and American Association of Cereal Chemists International, *J. Am. Oil Chem. Soc.* 96(6): 635–645. 10.1002/aocs.12205.

Liu, K., Seegers, S., Cao, W., Wanasundara, J., Chen, J., da Silva, A. E., Ross, K., Franco, A. L., Vrijenhoek, B., P., Li, Y., Wu, X. and Bloomer, S. (2021). An international collaborative study on trypsin inhibitor assay for legumes, cereals, and related products, *JAOCS* 98: 375–390. 10.1002/aocs.12459.

Liu, M., Zhao, L., Gong, G., Zhang, L., Shi, L., Dai, J., Han, Y., Wu, Y., Khalil, M. M. and Sun, L. (2022). Invited review: remediation strategies for mycotoxin control in feed, *J. Anim. Sci. Biotechnol.* 13(1): 19. 10.1186/s40104-021-00661-4.

Liu, Z., Zhong, R., Chen, L., Xie, F., Li, K., Liu, L. and Zhang, H. (2020). Effects of collection durations on the determination of energy values and nutrient digestibility of high-fiber diets in growing pigs by total fecal collection method, *Animals (Basel)* 10(2): 228. 10.3390/ani10020228.

Lombardi, P., Musco, N., Calabrò, S., Tudisco, R., Mastellone, V., Vastolo, A., Infascelli, F. and Cutrignelli, M. I. (2020). Different carbohydrate sources affect swine performance and post-prandial glycaemic response, *Ital. J. Anim. Sci.* 19(1): 421–430. 10.1080/1828051X.2020.1749899.

Lyu, F., van der Poel, A. F. B., Hendriks, W. H. and Thomas, M. (2021). Particle size distribution of hammer-milled maize and soybean meal, its nutrient composition and in vitro digestion characteristics, *Anim. Feed Sci. Technol.* 281: 115095. 10.1016/j.anifeedsci.2021.115095.

Lyu, Z. Q., Huang, C. F., Li, Y. K., Li, P. L., Liu, H., Chen, Y. F., Li, D. F. and Lai, C. H. (2018). Adaptation duration for net energy determination of high fiber diets in growing pigs, *Anim. Feed Sci. Technol.* 241: 15–26. 10.1016/j.anifeedsci.2018.04.008.

Mackenzie, S. G., Leinonen, I., Ferguson, N. and Kyriazakis, I. (2016). Towards a methodology to formulate sustainable diets for livestock: accounting for environmental impact in diet formulation, *Br. J. Nutr.* 115(10): 1860–1874. 10.1017/S0007114516000763.

Mahan, D. C. (1992). Efficacy of dried whey and its lactalbumin and lactose components at two dietary lysine levels on postweaning pig performance and nitrogen balance, *J. Anim. Sci.* 70(7): 2182–2187. 10.2527/1992.7072182x.

Makki, K., Deehan, E. C., Walter, J. and Bäckhed, F. (2018). The impact of dietary fiber on gut microbiota in host health and disease, *Cell Host Microbe* 23(6): 705–715. 10.1016/j.chom.2018.05.012.

Malik, A. K., Blasco, C. and Picó, Y. (2010). Liquid chromatography-mass spectroscopy in food safety, *J. Chromatogr. A* 1217(25): 4018–4040. 10.1016/j.chroma.2010.03.015.

McRorie, J. W. and Fahey, G. C. (2015). Fiber supplements and clinically meaningful health benefits: identifying the physiochemical characteristics of fiber that drive specific physiologic effects. In: *Dietary Supplements in Health Promotion*, T. C. Wallace (Eds). CRC Press, Boca Raton, FL, pp. 161–206.

Méda, B., Garcia-Launay, F., Dusart, L., Ponchant, P., Espagnol, S. and Wilfart, A. (2021). Reducing environmental impacts of feed using multiobjective formulation: what benefits at the farm gate for pig and broiler production?, *Animal* 15(1): 100024. 10.1016/j.animal.2020.100024.

Menegat, M. B., Goodband, R. D., DeRouchey, J. M., Tokach, M. D., Woodworth, J. C. and Dritz, S. S. (2019). *Kansas State University Swine Nutrition Guide: Feed Sampling and Analysis*, pp. 1–5.

Messad, F., Létourneau-Montminy, M. P., Charbonneau, E., Sauvant, D. and Guay, F. (2015). Prediction of standardized ileal digestibility and essential amino acid content of ingredients in swine: A meta-analysis, *Anim. Feed Sci. Technol.* 207: 204–221. 10.1016/j.anifeedsci.2015.06.012.

Meza, S., Zhou, Y., Chastain, J., Yang, Y., Cheng, H. H., Iassonova, D., Rivest, J. and You, H. (2022). Eco-efficient quantification of glucosinolates in Camelina seed, oil, and defatted meal: optimization, development, and validation of a UPLC-DAD method, *Antioxidants* 11(12): 2241. 10.3390/antiox11122441.

Noblet, J., Wu, S. B. and Choct, M. (2022). Methodologies for energy evaluation of pig and poultry feeds: a review, *Anim. Nutr.* 8(1): 185–203. 10.1016/j.aninu.2021.06.015.

Noel, S. J., Jørgensen, H. J. H. and Bach Knudsen, K. E. B. (2022). The use of near-infrared spectroscopy (NIRS) to determine the energy value of individual feedstuffs and mixed diets for pigs, *Anim. Feed Sci. Technol.* 283: 115156. 10.1016/j.anifeedsci.2021.115156.

Noel, S. J., Jørgensen, H. J. H. and Bach Knudsen, K. E. (2021). Prediction of protein and amino acid composition and digestibility in individual feedstuffs and mixed diets for pigs using near-infrared spectroscopy, *Anim. Nutr.* 7(4): 1242–1252. 10.1016/j.aninu.2021.07.004.

Novak, B., Lopes Hasuda, A., Ghanbari, M., Mayumi Maruo, V., Bracarense, A. P. F. R. L., Neves, M., Emsenhuber, C., Wein, S., Oswald, I. P., Pinton, P. and Schatzmayr,

D. (2021). Effects of *Fusarium* metabolites beauvericin and enniatins alone or in mixture with deoxynivalenol on weanling piglets, *Food Chem. Toxicol.* 158: 112719. 10.1016/j.fct.2021.112719.

O'Connell, J. M., Callan, J. J. and O'Doherty, J. V. (2005). The interaction between cereal type and lactose level on piglet performance and diet digestibility post weaning, *Anim. Sci.* 81(2): 265–269. 10.1079/ASC42030265.

O'Doherty, J. V., Dillon, S., Figat, S., Callan, J. J. and Sweeney, T. (2010). The effects of lactose inclusion and seaweed extract derived from Laminaria spp. on performance, digestibility of diet components and microbial populations in newly weaned pigs, *Anim. Feed Sci. Technol.* 157(3-4): 173–180. 10.1016/j.anifeedsci.2010.03.004.

Özogul, Y. and Özogul, F. (2019), Chapter 1. Biogenic amines formation, toxicity, regulations in food. In:. *Biogenic Amines in Food*, pp. 1–17. 10.1039/9781788015813-00001.

Park, C. S., Oh, S. I. and Kim, B. G. (2013). Prediction of basal endogenous losses of amino acids based on body weight and feed intake in pigs fed nitrogen-free diets, *Rev. Colomb. Cienc. Pecu* 26: 186–192.

Paternostre, L., Baeten, V., Ampe, B., Millet, S. and De Boever, J. (2021). The usefulness of NIRS calibrations based on feed and feces spectra to predict nutrient content, digestibility and net energy of pig feeds, *Anim. Feed Sci. Technol.* 281: 115091. 10.1016/j.anifeedsci.2021.115091.

Peace, R. M., Campbell, J., Polo, J., Crenshaw, J., Russell, L. and Moeser, A. (2011). Spray-dried porcine plasma influences intestinal barrier function, inflammation, and diarrhea in weaned pigs, *J. Nutr.* 141(7): 1312–1317. 10.3945/jn.110.136796.

Peng, W.-X., Marchal, J. L. M. and van der Poel, A. F. B. (2018). Strategies to prevent and reduce mycotoxins for compound feed manufacturing, *Anim. Feed Sci. Technol.* 237: 129–153. 10.1016/j.anifeedsci.2018.01.017.

Pinu, F. R., Beale, D. J., Paten, A. M., Kouremenos, K., Swarup, S., Schirra, H. J. and Wishart, D. (2019). Systems biology and multi-omics integration: viewpoints from the metabolomics research community, *Metabolites* 9(4): 76. 10.3390/metabo9040076.

Pomar, C., Andretta, I. and Remus, A. (2021). Feeding strategies to reduce nutrient losses and improve sustainability of growing pigs, *Front. Vet. Sci.* 8: 742220. 10.3389/fvets.2021.742220.

Poole, K. (2017). At the nexus of antibiotics and metals: the impact of Cu and Zn on antibiotic activity and resistance, *Trends Microbiol.* 25(10): 820–832. 10.1016/j.tim.2017.04.010.

Powles, J., Wiseman, J., Cole, D. J. A. and Jagger, S. (1995). Prediction of the apparent digestible energy value of fats given to pigs, *Anim. Sci.* 61(1): 149–154. 10.1017/S1357729800013631.

Rojas, O. J. and Stein, H. H. (2015). Effects of reducing particle size of corn grain on the concentration of digestible and metabolizable energy and on the digestibility of energy and nutrients in corn grain fed to growing pigs, *Livest. Sci.* 181: 187–193. 10.1016/j.livsci.2015.09.013.

Rosero, D. S., Odle, J., Arellano, C., Boyd, R. D. and van Heugten, E. (2015). Development of prediction equations to estimate the apparent digestibility energy content of lipids when fed to lactating sows, *J. Anim. Sci.* 93(3): 1165–1176. 10.2527/jas.2014-8402.

Rosero, D. S., Boyd, R. D., Odle, J. and van Heugten, E. (2016a). Optimizing dietary lipid use to improve essential fatty acid status and reproductive performance of the modern lactating sows: a review, *J. Anim. Sci. Biotechnol.* 7: 34. 10.1186/s40104-016-0092-x.

Rosero, D. S., Boyd, R. D., McCulley, M., Odle, J. and van Heugten, E. (2016b). Essential fatty acid supplementation during lactation is required to maximize the subsequent reproductive performance of the modern sow, *Anim. Reprod. Sci.* 168: 151–163. 10.1016/j.anireprosci.2016.03.010.

Rychlik, M., Humpf, H. U., Marko, D., Dänicke, S., Mally, A., Berthiller, F., Klaffke, H. and Lorenz, N. (2014). Proposal of a comprehensive definition of modified and other forms of mycotoxins including "masked" mycotoxins, *Mycotoxin Res.* 30(4): 197–205. 10.1007/s12550-014-0203-5.

Salazar-Villanea, S., Hendriks, W. H., Bruininx, E. M. A. M., Gruppen, H. and van der Poel, A. F. B. (2016). Protein structural changes during processing of vegetable feed ingredients used in swine diets: implications for nutritional value, *Nutr. Res. Rev.* 29(1): 126–141. 10.1017/S0954422416000056.

Schop, M. (2020). Modeling digestion kinetics in pigs - predicting nutrient absorption based on diet and ingredient properties, Ph.D. thesis. Wageningen University, Wageningen, the Netherlands. 10.18174/507537.

Schweer, W. P., Patience, J. F., Burrough, E. R., Kerr, B. J. and Gabler, N. K. (2018). Impact of PRRSV infection and dietary soybean meal on ileal amino acids digestibility and endogenous amino acids losses in growing pigs, *J. Anim. Sci.* 96(5): 1846–1859. 10.1093/jas/sky093.

Shannon, M. C. and Hill, G. M. (2019). Trace mineral supplementation for the intestinal health of young monogastric animals, *Front. Vet. Sci.* 6: 73. 10.3389/fvets.2019.00073.

She, Y., Li, D. and Zhang, S. (2017). Methodological aspects of determining phosphorus digestibility in swine: a review, *Anim. Nutr.* 3(2): 97–102. 10.1016/j.aninu.2017.02.003.

Shen, Y., Fan, M. Z., Ajakiye, A. and Archibold, T. (2002). Use of the regression analysis technique to determine the true digestible phosphorus digestibility and the endogenous phosphorus output associated with corn in growing pigs, *J. Nutr.* 132(6): 1199–1206. 10.1093/jn/132.6.1199.

Shurson, G. C. and Urriola, P. E. (2022). Sustainable swine feeding programs require the convergence of multiple dimensions of circular agriculture and food systems with One Health, *Anim. Front.* 12(6): 30–40. 10.1093/af/vfac077.

Shurson, G. C., Pelton, R. E. O., Yang, Z., Urriola, P. E. and Schmitt, J. (2022a). Environmental impacts of eco-nutrition swine feeding programs in spatially explicit geographic regions of the United States, *J. Anim. Sci.* 100(12): 1–7. 10.1093/jas/skac356.

Shurson, G. C., Urriola, P. E. and Hung, Y. T. (2022b). Too much of a good thing: rethinking feed formulation and feeding practices for zinc in swine diets to achieve One Health and environmental sustainability, *Animals (Basel)* 12(23): 3374. 10.3390/ani12233374.

Shurson, G. C., Urriola, P. E. and van de Ligt, J. L. G. (2022c). Can we effectively manage parasites, prions, and pathogens in the global feed industry to achieve One Health?, *Transbound. Emerg. Dis.* 69(1): 4–30. 10.1111/tbed.14205:10.111.

Shurson, G. C., Hung, Y. T., Jang, J. C. and Urriola, P. E. (2021). Measures matter - determining the true nutri-physiological value of feed ingredients for swine, *Animals (Basel)* 11(5): 1259. 10.3390/ani11051259.

Shurson, G. C., Kerr, B. J. and Hanson, A. R. (2015). Evaluating the quality of feed fats and oils and their effects on pig growth performance, *J. Anim. Sci. Biotechnol.* 6(1): 10. 10.1186/s40104-015-0005-4.

Silva-Guillen, Y. V., Arellano, C., Boyd, R. D., Martinez, G. and van Heugten, E. (2020). Growth performance, oxidative stress and immune status of newly weaned pigs fed peroxidized lipids with or without supplemental vitamin E or polyphenols, *J. Anim. Sci. Biotechnol.* 11: 22. 10.1186/s40104-020-0431-9.

Sinbad, O. O., Folorunsho, A. A., Olabisi, O. L., Ayoola, O. A. and Temitope, E. J. (2019). Vitamins as antioxidants, *J. Food Sci. Nutr. Res.* 2: 214–235. 10.26502/jfsnr.2642-11000021.

Smith, T. K. (1990). Effect of dietary putrescine on whole body growth and polyamine metabolism, *Proc. Soc. Exp. Biol. Med.* 194(4): 332–336. 10.3181/00379727-194-43100.

Smith, B. N. and Dilger, R. N. (2018). Immunomodulatory potential of dietary soybean-derived isoflavones and saponins in pigs, *J. Anim. Sci.* 96(4): 1288–1304. 10.1093/jas/sky036.

Smith, B., Hassen, A., Hinds, M., Rice, D., Jones, D., Sauber, T., Iiams, C., Sevenich, D., Allen, R., Owens, F., McNaughton, J. and Parsons, C. (2015). Predicting the digestible energy of corn determined with growing swine from nutrient composition and cross-species measurements, *J. Anim. Sci.* 93(3): 1025–1038. 10.2527/jas.2014-7807.

Soglia, F., Petracci, M. and Ertbjerg, P. (2016). Novel DNPH-based method for determination of protein carbonylation in muscle and meat, *Food Chem.* 197(A): 670–675. 10.1016/j.foodchem.2015.11.038.

Soleimani, T. and Gilbert, H. (2021). An approach to achieve overall farm feed efficiency in pig production: environmental evaluation through individual life cycle assessment, *Int. J. Life Cycle Assess.* 26(3): 455–469 10.1007/s11367-020-01860-3.

Son, A. R., Park, C. S. and Kim, B. G. (2017). Determination and prediction of digestible and metabolizable energy concentrations in byproduct feed ingredients fed to growing pigs, *Asian-Australas. J. Anim. Sci.* 30(4): 546–553. 10.5713/ajas.16.0607.

Stein, H. H. (2017). Procedures for determining digestibility of amino acids, lipids, starch, fibre, phosphorus, and calcium in feed ingredients fed to pigs, *Anim. Prod. Sci.* 57(11): 2317–2324. 10.1071/AN17343.

Stein, H. H. and Kil, D. Y. (2006). Reduced use of antibiotic growth promoters in diets fed to weanling pigs: dietary tools, Part 2, *Anim. Biotechnol.* 17(2): 217–231. 10.1080/10495390600957191.

Streit, E., Schwab, C., Sulyok, M., Naehrer, K., Krska, R. and Schatzmayr, G. (2013). Multi-mycotoxin screening reveals the occurrence of 139 different secondary metabolites in feed and feed ingredients, *Toxins* 5(3): 504–523. 10.3390/toxins5030504.

Sulabo, R. C. and Stein, H. H. (2013). Digestibility of phosphorus and calcium in meat and bone meal fed to growing pigs, *J. Anim. Sci.* 91(3): 1285–1294. 10.2527/jas.2011-4632.

Sung, J. Y. and Kim, B. G. (2021). Prediction equations for digestible and metabolizable energy concentrations in feed ingredients and diets for pigs based on chemical composition, *Anim. Biosci.* 34(2): 306–311. 10.5713/ajas.20.0293.

Symeou, V., Leinonen, I. and Kyriazakis, I. (2014a). Modelling phosphorus intake, digestion, retention and excretion in growing and finishing pigs: model description, *Animal* 8(10): 1612–1621. 10.1017/S1751731114001402.

Symeou, V., Leinonen, I. and Kyriazakis, I. (2014b). Modelling phosphorus intake, digestion, retention and excretion in growing and finishing pigs: model evaluation, *Animal* 8(10): 1622–1631. 10.1017/S1751731114001414.

Szmigielska, A. M., Schoenau, J. J. and Levers, V. (2000). Determination of glucosinolates in canola seeds using anion exchange membrane extraction combined with the high-pressure liquid chromatography detection, *J. Agric. Food Chem.* 48(10): 4487–4491. 10.1021/jf000477u.

ten Brink, B., Damink, C., Joosten, H. M. L. J. and Huis in't Veld, J. H. J. (1990). Occurrence and formation of biologically active amines in foods, *Int. J. Food Microbiol.* 11(1): 73–84.

Thiex, N. and Richardson, C. R. (2003). Challenges in measuring moisture content of feeds, *J. Anim. Sci.* 81(12): 3255–3266. 10.2527/2003.81123255x.

Tkaczy, A. and Jedziniak, P. (2021). Mycotoxin biomarkers in pigs – current state of knowledge and analytics, *Toxins* 13(8): 586. 10.3390/toxins13080586.

Tran, T. H. T., Everaert, N. and Bindlle, J. (2018). Review on the effects of potential prebiotics on controlling intestinal enteropathogens *Salmonella* and *Escherichia coli* in pig production, *J. Anim. Physiol. Anim. Nutr. (Berl)* 102(1): 17–32. 10.1111/jpn.12666.

Urriola, P. E., Li, M., Kerr, B. J. and Shurson, G. C. (2014). Evaluation of prediction equations to estimate gross, digestible, and metabolizable energy content of maize dried distillers grains with solubles (DDGS) for swine based on chemical composition, *Anim. Feed Sci. Technol.* 198: 196–202. 10.1016/j.anifeedsci.2014.09.006.

van Dijk, A. J., Everts, H., Nabuurs, M. J. A., Margry, R. J. C. F. and Beynen, A. C. (2001). Growth performance of weaned pigs fed spray-dried animal plasma: a review, *Livest. Prod. Sci.* 68(2–3): 263–274. 10.1016/S0301-6226(00)00229-3.

van Kempen, T. A. T. G., Regmi, P. R., Matte, J. J. and Zijlstra, R. T. (2010). *In vitro* starch digestion kinetics, corrected for estimated gastric emptying, predict portal glucose appearance in pigs, *J. Nutr.* 140(7): 1227–1233. 10.3945/jn.109.120584.

Vanhoutte, I., Audenaert, K. and De Gelder, L. (2016). Biodegradation of mycotoxins: tales from known and unexplored worlds, *Front. Microbiol.* 7: 561. 10.3389/fmicb.2016.00561.

Van Middelaar, C. E., van Zanten, H. H. E. and de Boer, I. J. M. (2019). Future of animal nutrition: the role of life cycle assessment. In: *Poultry and Pig Nutrition – Challenges of the 21st Century* Hendriks, W. H., Verstegen, M. W. A. and Babinsky, L. (Eds). Wageningen Academic Publishers, pp.307–314. 10.3920/978-90-8686-884-1_14.

Velayudhan, D. E., Kim, I. H. and Nyachoti, C. M. (2015). Characterization of dietary energy in swine feed and feed ingredients: a review of recent research results, *Asian Australas. J. Anim. Sci.* 28(1): 1–13. 10.5713/ajas.14.0001R.

Verhelst, R., Schroyen, M., Buys, N. and Niewold, T. (2014). Dietary polyphenols reduce diarrhea in enterotoxigenic Escherichia coli (ETEC) infected post-weaning piglets, *Livest. Sci.* 160: 138–140. 10.1016/j.livsci.2013.11.026.

Vitaglione, P., Napolitano, A. and Fogliano, V. (2008). Cereal dietary fibre: a natural functional ingredient to deliver phenolic compounds to the gut, *Trends Food Sci. Technol.* 19(9): 451–463. 10.1016/j.tifs.2008.02.005.

Wang, J., Qu, Y., Liu, Z. and Zhou, H. (2021). Formation, analytical methods, change tendency, and control strategies of biogenic amines in canned aquatic products: A systematic review, *J. Food Prot.* 84(11): 2020–2036. 10.4315/JFP-21-120.

Wang, L., Hu, Q., Li, P., Lai, C., Li, D., Zang, J. and Ni, S. (2021). Development and validation of equations for predicting the metabolizable energy value of double-low rapeseed cake for growing pigs, *Animals (Basel)* 11(4): 1168. 10.3390/ani11041168.

Wang, L., Yao, D., Urriola, P. E., Hanson, A. R., Saqui-Salces, M., Kerr, B. J., Shurson, G. C. and Chen, C. (2018). Identification of activation of tryptophan-NAD^+ pathway as a prominent metabolic response to thermally oxidized oil through metabolomics-guided biochemical analysis, *J. Nutr. Biochem.* 57: 255–267. 10.1016/j.jnutbio.2018.04.009.

Wang, L., Csallany, A. S., Kerr, B. J., Shurson, G. C. and Chen, C. (2016). Kinetics of forming aldehydes in frying oils and their distribution in French fried revealed by LC-MS-based chemometrics, *J. Agric. Food Chem.* 64(19): 3881–3889. 10.1021/acs.jafc.6b01127.

Wealleans, A. L., Bierinckx, K., Witters, E., di Benedetto, M. and Wiseman, J. (2021). Assessment of the quality, oxidative status and dietary energy value of lipids used in non-ruminant animal nutrition, *J. Sci. Food Agric.* 101(10): 4266–4277. 10.1002/jsfa.11066.

Winkler-Moser, J. K., Hwang, H. S. and Kerr, B. J. (2020). Changes in markers of lipid oxidation and thermal treatment in feed-grade fats and oils, *J. Sci. Food Agric.* 100(8): 3328–3340. 10.1002/jsfa.10364.

Wiseman, J., Powles, J. and Salvador, F. (1998). Comparison between pigs and poultry in the prediction of the dietary energy value of fats, *Anim. Feed Sci. Technol.* 71(1–2): 1–9. 10.1016/S0377-8401(97)00142-9.

Woyengo, T. A., Beltranena, E. and Zijlstra, R. T. (2017). Effect of anti-nutritional factors of oilseed co-products on feed intake of pigs and poultry, *Anim. Feed Sci. Technol.* 233: 76–86. 10.1016/j.anifeedsci.2016.05.006.

Wu, F., Johnston, L. J., Urriola, P. E., Hilbrands, A. M. and Shurson, G. C. (2016a). Evaluation of ME predictions and the impact of feeding maize distillers dried grains with solubles with variable oil content on growth performance, carcass composition, and pork fat quality of growing-finishing pigs, *Anim. Feed Sci. Technol.* 213: 128–141. 10.1016/j.anifeedsci.2016.01.013.

Wu, F., Johnston, L. J., Urriola, P. E., Hilbrands, A. M. and Shurson, G. C. (2016b). Evaluation of NE predictions and the impact of feeding maize distillers dried grains with solubles (DDGS) with variable NE content on growth performance and carcass characteristics of growing-finishing pigs, *Anim. Feed Sci. Technol.* 215: 105–116. 10.1016/j.anifeedsci.2016.02.023.

Wu, G. (2014). Dietary requirements of synthesizable amino acids by animals: a paradigm shift in protein nutrition, *J. Anim. Sci. Biotechnol.* 5(1): 34. Available at: http://www.jasbsci.com/content/S/1/34.

Wu, G. (2010). Functional amino acids in growth, reproduction, and health, *Adv. Nutr.* 1(1): 31–37. 10.3945/an.110.1008.

Xu, R., Kiarie, E. G., Yiannikouris, A., Sun, L. and Karrow, N. A. (2022). Nutritional impact of mycotoxins in food animal production and strategies for mitigation, *J. Anim. Sci. Biotechnol.* 13(1): 69. 10.1186/s40104-022-00714-2.

Yang, P., Zhao, J. B., Zhang, G., Ni, J. J. and Huang, C. F. (2021). Determination and prediction of digestible and metabolizable energy of soybean meal and wheat bran for finishing pigs, *Livest. Sci.* 254: 104741. 10.1016/j.livsci.2021.104741.

Yang, P., Ni, J. J., Zhao, J. B., Zhang, G. and Huang, C. F. (2020). Regression equations of energy values of corn, soybean meal, and wheat bran developed by chemical composition for growing pigs, *Animals (Basel)* 10(9): 1490. 10.3390/ani10091490.

Yang, Z. and Liao, S. F. (2019). Physiological effects of dietary amino acids on gut health and functions of swine, *Front. Vet. Sci.* 6: 169. 10.3389/fvets.2019.00169.

Ye, H., Langendijk, P., Jaworski, N. W., Wu, Y., Bai, Y., Lu, D., Page, G., Kemp, B., Han, D., Soede, N. M. and Wang, J. (2022). Protein digestion kinetics influence maternal protein loss, litter growth, and nitrogen utilization in lactating sows, *Front. Nutr.* 9: 862823. 10.3389/fnut.2022.862823.

Yu, P. and Prates, L. L. (2017). Molecular structure of feeds in relation to nutrient utilization and availability in animals: A novel approach, *Engineering*=730 3(5): 726–730. 10.1016/J.ENG.2017.03.007.

Zeng, Z. K., Shurson, G. C. and Urriola, P. E. (2017). Prediction of the concentration of standardized ileal digestible amino acids and safety margins among sources of distillers dried grains with solubles for growing pigs: A meta-analysis approach, *Anim. Feed Sci. Technol.* 231: 150–159. 10.1016/j.anifeedsci.2017.07.010.

Zentek, J., Buchheit-Renko, S., Ferrara, F., Vahjen, W., van Kessel, A. G. and Pieper, R. (2011). Nutritional and physiological role of medium-chain triglycerides and medium-chain fatty acids in piglets, *Anim. Health Res. Rev.* 12(1): 83–93. 10.1017/S1466252311000089.

Zhang, F. and Adeola, O. (2017). Techniques for evaluating digestibility of energy, amino acids, phosphorus, and calcium in feed ingredients for pigs, *Anim. Nutr.* 3(4): 344–352. 10.1016/j.aninu.2017.06.008.

Zhang, X., Lu, P., Xue, W., Wu, D., Wen, C. and Zhou, Y. (2017). An evaluation of heat on protein oxidation of soy protein isolate or soy protein isolate mixed with soybean oil in vitro and its consequences on redox status of broilers at an early age, *Asian Australas. J. Anim. Sci.* 30(8): 1135–1142. 10.5713/ajas.16.0683.

Zhang, Z., Nie, D., Fan, K., Yang, J., Guo, W. and Meng, J. (2019). A systematic review of plant-conjugated masked mycotoxins: occurrence, toxicology, and metabolism, *Crit. Rev. Food Sci. Nutr.* 60: 1523–1537. 10.1080/10408398.2019.1578944.

Zhou, L. J., Zhang, L. Y., Zhang, E. X., Li., J. T., Yang, W. J. and Wang, Z. Y. (2012). Rapid determination of swine available energy and amino acids in corn distillers dried grains with solubles by near-infrared reflectance spectroscopy, *Anim. Feed Sci. Technol.* 175(3–4): 198–202. 10.1016/j.anifeedsci.2012.06.001.

Zhou, P., Theil, P. K., Wu, D. and Bach Knudsen, K. E. B. (2018). *In vitro* digestion methods to characterize the physiochemical properties of diets varying in dietary fibre source and content, *Anim. Feed Sci. Technol.* 235: 87–96. 10.1016/j.anifeed sci.2017.11.012.

Zijlstra, R. T., Jha, R., Woodward, A. D., Fouhse, J. and van Kempen, T. A. T. G. (2012). Starch and fiber properties affect their kinetics of digestion and thereby digestive physiology in pigs, *J. Anim. Sci.* 90 (Suppl. 4): 49–58. 10.2527/jas.53718.

Chapter 6

Ensuring pig feed safety

Chad Paulk, Jordan Gebhardt, Cassandra Jones, and Jason Woodworth, Kansas State University, USA

1 Introduction

The global swine industry has made substantial gains in herd health by implementing farm biosecurity practices. Many of the primary routes of pathogen entry into the farm (i.e. other pigs, farm employees, visitors, air, etc.) have been minimized in high-health systems, while less research has characterized the risk of feed as a vector of disease transmission. However, several examples of the introduction of diseases via the feed supply chain exist. The introduction of foot and mouth disease (FMD) in both Japan and South Korea has been linked to feedstuffs (Sugiura et al., 2001; Park et al., 2016) and the feeding of silage that was harvested from areas with wild boars infected with Classical Swine Fever Virus (CSFV) has led to illness in naïve pigs (Ribbens et al., 2004). The classical route of transmission of African Swine Fever Virus (ASFV) involves feeding food waste commonly known as 'garbage feeding'. This practice involves feeding wild or domesticated pigs food waste containing contaminated pork products and presents a significant opportunity for pathogen transmission. However, in modern swine production where the use of food waste is relatively limited in scale, there is relatively little information known regarding the stability of viruses in commercial ingredients or finished feed.

http://dx.doi.org/10.19103/AS.2024.0140.10

Understanding the potential risk of feed as a vector of disease transmission became important during the porcine epidemic diarrhea virus (PEDV) outbreak in the United States beginning in 2013. During the outbreak, some epidemiological studies found evidence of PEDV in plasma products; however, no study conclusively identified the source of the outbreak (Pasick et al., 2014), and other investigations led to proposed alternate routes of entry (Russell et al., 2020). Nonetheless, the identification of PEDV RNA in feed ingredients prompted further discussion and research which has continued to expand in recent years.

While not all pathogens are strong candidates for feed-based transmission, it is important to characterize the risk for the feed supply chain to serve as a vehicle for pathogen entry into farms.

The type of feed manufactured is going to determine which biological hazards are of greatest concern. Examples of hazards include pathogenic bacteria, such as *Salmonella* spp. and *Escherichia coli*, and viruses, such as PEDV, PRRSV, ASFV, Senecavirus A (SVV-1), CSFV, Pseudorabies Virus (PRV), and FMD.

2 Pathogens of concern for feed-based transmission

2.1 Prions and parasites

Prior to 2010, the primary biological hazard of concern for U.S. feed mills was transmissible spongiform encephalopathies (TSE) caused by prions, such as bovine spongiform encephalopathy (BSE). While TSE transmission to pigs has been established in experimental conditions, it is not known to occur naturally in swine (Hedman et al., 2016). Regulations established by the Food and Drug Administration (FDA) in 1997 and 2008 have successfully controlled the domestic spread of the disease. Therefore, the United States has negligible BSE risk according to the World Health Organization, and the probability of a swine-based TSE is remote (World Organization for Animal Health, 2018).

Both *Trichinella spiralis* and *Toxoplasma gondii* have historically been linked to feed-based transmission and have a potential impact on human health (Guo et al., 2015; Slifko, Smith, & Rose, 2000). However, biosecurity improvements and indoor all-in/all-out management have largely eradicated the hazards from the domesticated swine herd (Davies et al, 1998). In a study of North Carolina swine farms in 1998, just 1 of 2,175 pigs housed in confinement was seropositive for *Toxoplasma gondii* and 1 of 2,183 pigs was positive for antibodies against *Trichinella spiralis* (Davies et al., 1998). Even then, it is unlikely for illness to result as typical carcass processing and meat preparation is adequate to prevent transmission.

2.2 Bacteria

While the poultry feed and pet food industries have evolved to control Salmonella spp. in the feed supply chain, reports of pathogenic bacteria transmission from feed to swine have been limited. The FDA recognizes that swine feed is rarely in direct contact with immunocompromised people, and therefore, the risk of salmonellosis in humans from swine feed is negligible. Furthermore, the FDA has determined that feed containing most Salmonella serotypes, such as *Salmonella* Enteritidis or *Salmonella* Typhimurium, are not likely to cause swine disease. For these reasons, *Salmonella* Choleraesuis is the only serotype considered by the FDA to be an adulterant in swine feed (Food and Drug Administration, 2013). In its routine surveillance for Salmonella spp. in animal feeds and ingredients, the FDA found that 19.4% of ingredients and 5.6% of complete feeds contained the pathogen (Li et al., 2012). However, no *Salmonella* Choleraesuis was identified, causing the FDA to conclude that it is unlikely for feed to cause salmonellosis in pigs.

Outside of the regulatory environment, there is still concern for other serotypes of *Salmonella* spp. In particular, there is rising concern with an emerging serotype, *Salmonella enterica* serotype 4,[5],12:i:- in swine. This is a potential monophasic variant of *Salmonella* Typhimurium, and is highly resistant to multiple drugs, including ampicillin, streptomycin, sulfonamides, and tetracyclines (Elnekave et al., 2018). The serotype is most commonly linked to pork products, and was responsible for a substantial recall of whole roaster hogs from the state of Washington, but was not been linked to feed (Elnekave et al., 2018). Recently, *S. enterica* serotype 4,[5],12:i:- has again been linked to swine; a recall of pig ears intended to be used as pet treats was linked to illness in at least 93 people in 27 states (CDC, 2019). A 2016–2017 study reported the presence of *S. enterica* serotype 4,[5],12:i:- in 12 United States swine feed mills sampled from feed and mill surfaces over three different seasons (Magossi et al., 2018). Five of 696 feed mill environmental samples (0.72%) and 0 of 39 feed samples were identified to include the pathogen via RT-PCR. Of the positive environmental samples, three were from dust collected on the floor of the bulk ingredient receiving area, one from the bulk ingredient receiving pit, and one from the control room floor. The five positive samples were from three mills located in North Carolina, Kansas, and Indiana, with two positive samples recovered in Fall 2016 and three positive samples recovered in Summer 2017. Thus, this serotype has been found in United States feed mills, but not in a finished feed or on a direct feed-contact surface. Based on the available data, it is concluded that *S. enterica* serotype 4,[5],12:i:-poses very high severity and unlikely probability for occurrence, therefore having a moderate overall risk. Additional research is needed to further assess the root cause and prevalence of *S. enterica* serotype 4,[5],12:i:- in the feed-to-pig-to-pork supply chain due to its significance on human health and presence in the environment of swine feed mills.

There are limited links to feed-based transmission of other Salmonella serotypes resulting in animal or human health concerns. Across all samples from the study above, 8% of feed samples, 13% of bulk ingredient receiving pits, 11% of feed mill floors, and 14% of worker shoes were positive for Salmonella spp. (Magossi et al., 2018). Further investigation determined that the isolates belonged to 15 different serotypes of *S. enterica*: Agona (14), Mbandaka (13), Seftenberg (7), Schwarzengrund (6), Rissen (3), Hartford (2) and Typhimurium (2) Bareilly (1), Braenderup (1), Cubana (1), Javiana (1), Kiambu (1), Poona (1), Soerenga (1), Worthington (1), and I 4,[5],12:i:– (1) (Lomonaco et al., 2018). Notably, *Salmonella* Choleraesuis was not identified. Based on limited data linking other Salmonella serotypes to animal health or human illness, the severity of Salmonella spp. beyond *Salmonella* Choleraesuis and I 4,[5],12:i:- is characterized as very low, while the probability is almost certain. As the swine industry has continued pressure to eliminate sources of antibiotic-resistant bacteria from pork, food safety from feed to fork will continue to be an emerging concern. It will become important to characterize the serotypes of Salmonella more fully and understand their prevalence in feed and feed ingredients and potential transmission to impact animal or human health.

2.3 Viruses

Several examples of virus introduction via the feed supply chain exist. The introduction of FMDV in Japan and South Korea has been linked to feedstuffs (Sugiura et al., 2001; Park et al., 2016). The feeding of silage that was harvested from areas with wild boars infected with CSFV has led to illness in naïve pigs (Ribbens et al., 2004). The U.S. swine industry first recognized the significant role that the feed supply chain can play in pathogen transfer in 2013–2014 with PEDV.

The porcine epidemic diarrhea virus caused substantial mortality and morbidity in the U.S. swine herd. In its 2015 report evaluating the root cause of PEDV, the U.S. Department of Agriculture determined that 'the use of Flexible Intermediate Bulk Containers (aka: FIBC or "tote bags") best fit the criteria established for entry into the United States, rapid and wide spread transmission across the country, and introduction onto individual farms' (United States Department of Agriculture Animal and Plant Health Inspection Service Veterinary Services, 2015). Furthermore, feed or ingredients were linked to outbreaks of PEDV in both the United States and Canada (Aubry et al., 2016; Pasick et al., 2014). Additionally, Senecavirus A, a swine pathogen that causes similar clinical signs as FMDV, is endemic in swine production systems in the

United States and was recently found to be spread via SVV-1-contaminated ingredients and feed in Brazil (Leme et al., 2019).

As foreign animal diseases, such as ASFV, CSFV, and FMD, circulate among several global swine producers, the concern for feed and ingredients as a vector of transboundary disease entry has increased. For example, feeding contaminated feed has been categorized as a risk factor for ASFV transmission (EFSA, 2021; Belyanin, 2013). Belyanin et al. (2013) documented feed being associated with 35% of 284 reported outbreaks in Russia. More recently, reports indicate that 1–2% of tested ingredients from modern Chinese feed mills were ASFV-positive, including corn, soybean meal, rice, wheat, and corn-dried distillers grains with solubles (Dee and Niederwerder, 2019). Recent research has also shown that ASFV can survive in various feed ingredients during transboundary, transatlantic shipping (Dee et al., 2021). Because ASFV can survive in feed during shipping, virus-naïve countries are concerned that a contaminated feed or ingredient will introduce ASFV into the swine population. Regardless of its method of entry, there is concern that infection of pigs may result in contamination of the feed supply chain, and rapid and widespread distribution of the virus similar to field evidence collected in Vietnam (Gebhardt et al., 2022).

3 Feed contamination

As foreign animal diseases, such as ASFV, CSFV, and FMD, circulate among global swine producers, the concern for ingredients as a vector of transboundary disease entry has increased. This poses a tremendous threat to swine industries that have naïve pig populations. Previous examples described in the text, clearly demonstrate FAD contamination risk in animal and non-animal origin feed and ingredients; however, the respective virus must survive transport at sufficient levels to cause animal infection to be of substantial risk. Although it is extremely important to keep FAD out of one's country, it is equally important to reduce the risk of spreading endemic viruses.

For some pathogens such as PEDV which is spread via fecal-oral transmission, there is a risk of contamination through processes such as bulk ingredient receiving. Schumacher et al. (2016) described that if 1 g of feces from an acutely PEDV-infected pig entered an ingredient receiving pit, it could potentially contaminate 500 metric tons of feed, with each gram of feed having a dose capable of causing a pig infection. That is the equivalent of twenty 24-ton feed trucks being contaminated by 1 g of PEDV feces, all of them carrying infectious material to different swine operations, leading to simultaneous infection of multiple herds. This is what was thought to occur in the 2013–2014 incursion of PEDV, when the incidence of the disease dramatically spiked in a manner that had never been previously observed.

Dee et al. (2019) determined the ASFV infection risk with contaminated feed is relatively low for a single exposure (10^4 $TCID_{50}$/g for one meal of 100 g of feed). However, the risk of ASF transmission through feed in an industry setting is substantially higher because it is unlikely that exposure to the ASF agent would be a single exposure event. Feed industry equipment and production processes are designed to efficiently mix low-inclusion products uniformly throughout a batch of feed, and then deliver feed in a way that provides multiple meals with the same consistency to all pigs housed in the same location. This means that instead of a single exposure to a pathogen, it is likely that a contaminated ingredient would be mixed into larger volumes of feed, which would initially dilute the contaminant, but also substantially increase the number of exposures to a population of animals. For example, a single batch of feed delivered to a barn of 1200 pigs, would be consumed in 1.5 days in a series of 20 meals per pig. Instead of a single pig being exposed to a high level of pathogen one time, the exposure concentration is diluted, but the number of exposures increased to 20 meals or exposures per pig (or 24,000 exposures for 1200 pigs) in 36 h. This exponentially increases the probability of infection and decreases the dose likely to cause infection to be 100 $TCID_{50}$/g for twenty 100-g meals of feed. The risk is expanded even further when one considers the potential role a contaminated feed mill may have on subsequent batches of feed. From this data, we can conclude that if ASFV enters the feed supply chain, ASFV infections in pigs are possible, regardless of the ASFV concentration in the feed.

Schumacher et al. (2017) reported that entry of PEDV into a feed mill leads to nearly 100% of surfaces being contaminated, including non-contact surfaces such as walls and floors, and that subsequent batches of virus-free feed do not flush contaminated material out of the conveying system. Subsequently, Gebhardt et al. (2018) reported that material collected from these non-contact surfaces can cause PEDV infection in animals. Similarly, Elijah et al. (2021) determined that the distribution of ASFV into the feed manufacturing environment is widespread and persists even after manufacturing additional feed batches initially free of ASFV. This is like what is observed with PEDV (Schumacher et al., 2017) and indicates that it is extremely important for the feed mills to prevent the entry of ASFV into the mills because once ASFV is in a feed mill, it will remain in its environment for an extended period of time.

Huss et al. (2017) reported that the cleaning and disinfection necessary to sanitize a virus-contaminated feed mill includes complete organic material removal, followed by wet cleaning with multiple sanitizers. The U.S. feed industry is not designed for these types of cleaning and disinfection processes, so the primary focus must be on keeping pathogenic viruses out of feed mills, particularly because feed mills are a central point of cross-traffic among multiple farms or sites. Depending on mill size, it can receive dozens of ingredient delivery vehicles daily. The mill subsequently mixes ingredients together and distributes them to dozens of farms at different locations allowing for the feed supply chain to unknowingly spread the virus to naïve pigs.

The swine industry has developed highly effective biosecurity procedures that prevent people or transport vehicles from serving as fomites for viral transfer (Kim et al., 2017; Allerson et al., 2013). These include protocols for changing shoes, clothing, and/or showering and dynamic biosecurity pyramids for transport vehicles. Many modern swine production systems implement biocontainment practices similar to BSL2 or 3 laboratories but on a much larger scale. However, these same procedures have not been implemented at feed mills and with personnel or feed trucks. Therefore, feed mills and feed delivery vehicles have the potential to become the primary route for pathogen distribution onto farms.

The overall goal of feed mill biosecurity is to prevent contamination of feed with infectious pathogens. In addition to pathogen prevention, if a mill or ingredient does become contaminated it can be almost impossible to control because most feed manufacturing facilities were not hygienically designed. This illustrates the importance of pathogen mitigation in the feed biosecurity protocol. Pathogen mitigation could include strategies such as thermal processing or decontamination via chemical additives. Although these mitigation steps can be included in one's feed mill biosecurity plan, managing a facility to focus on preventing the entry of a pathogen into the feed system is key to reducing the risk of disease spread. To do so, feed mill managers should focus on ingredient sourcing, receiving practices, finished feed separation, human traffic, employee training, trucking, and creating a culture around producing safe feed. To obtain more details than what have been provided in this article please refer to the following resources:

- Developing Biosecurity Practices for Feed & Ingredient Manufacturing, https://www.afia.org/pub/?id=E348BF9F-98ED-09DB-A45D-504737FE7AE2
- Feed Mill Biosecurity Resources, https://www.grains.k-state.edu/research/AnimalFeedandPetFood/feed_science_research_extension/quality_assurance_guidelines_resources/index.html

4 Prevention of feed contamination

4.1 *Ingredient sourcing*

The goal of feed biosecurity is to prevent ingredients or complete feed from coming into contact with a pathogen. This may occur at the ingredient stage, where a combination of geographic location, agricultural practices, and transportation methods may lead the ingredient to be contaminated. Ingredient sourcing is the first key step to prevent biological hazards from entering the feed manufacturing facility via ingredients.

Ingredients from areas experiencing disease outbreaks have a greater potential for being high risk. Supplier identification is important to maintain transparency across the feed supply chain. To accurately identify ingredient risk, knowledge of the ingredient supply chain should extend from the point of ingredient manufacture through transportation to the feed mill, including any intermediaries or blending locations. It is also important to identify biosecurity practices that may be implemented throughout this supply chain. Some of these practices include heat treating the ingredients/complete feed or holding the ingredient outside of the mill long enough that the virus or bacteria becomes inactive.

Defined time and temperatures have been established for quarantining ingredients or complete feed; however, they are based on relatively limited information (SHIC, 2020). Previous research determined that when using a viral isolation model, infectious ASFV particles were still present when meat and bone meal was held at 22°C for 7 days and spray-dried porcine plasma at 4°C for 14 days (Fischer et al., 2021; McOrist et al., 2022). Fischer et al. (2021) noted that ASFV particles failed to cause infection in SDPP from d 14 to 35 when held at 22°C, but this is the only study where ASFV inactivation has been documented in porcine-derived ingredients under quarantine conditions.

For PEDV, the necessary length of quarantine differed based on the porcine-derived ingredient. Blood meal, meat and bone meal, and meat meal held at 25°C maintained detectable levels of infectious viral particles longer than spray-dried porcine plasma in multiple studies (Pujols and Segales, 2014; Trudeau et al., 2017; Gebhardt et al., 2019). Trudeau et al. (2017) observed a similar pattern for porcine delta-coronavirus and transmissible gastroenteritis virus. Porcine epidemic diarrhea virus particles also remained detectable for longer periods of time as the temperature decreased (Pujols and Segales, 2014).

The necessary time and temperature to store ingredients during quarantine is dependent on the pathogen and ingredient of concern. Limited research is available that defines the time and temperatures needed to inactivate viruses, especially ASFV or CSFV, in ingredients or complete feed. Regardless of where an ingredient is being sourced from, each step of the shipping and transportation process from manufacturing to the feed mill should be tracked. Identified biological hazards will depend on the species for which feed is being manufactured. However, the entire feed mill needs to be considered as opposed to species-specific diets. To prevent mill contamination via ingredients, a balance is required for quality, feed safety, and procurement areas of the mill to all work together on ingredient sourcing.

4.2 Receiving and loadout practices

In addition to the risk of ingredients, the movement of trucks and people through receiving poses a significant risk in terms of feed mill biosecurity. The main goal of a feed mill biosecurity plan should be to prevent contaminated material from entering the receiving pit. Previous research suggests that transportation vehicles can play a role in the transmission of bacterial (Fedorka-Cray et al., 1997) and viral pathogens (Dee et al., 2004; Lowe et al., 2014; Elijah et al., 2022; Gebhardt et al., 2022).

Ensuring the underside of the truck is clean before entering the receiving bay also helps reduce the chance of debris entering the pit. Instituting a tire and undercarriage wash prior to trucks entering the mill will help to minimize the chance of contaminants being brought into the receiving bay. In addition, using pit covers while trucks are entering and leaving the receiving bay helps to prevent any debris from falling into the pit and contaminating ingredients. Sweeping spilled ingredients into the pit is another thing to be avoided. Due to the nature of trucks entering and leaving, the ground in the receiving area cannot be considered 'clean' or safe for biosecurity, and sweeping spilled ingredients into the pit is a risk of introducing disease into the mill.

In addition to the vehicles themselves, truck drivers pose another entry point of contamination. Drivers delivering ingredients or picking up finished feed should not enter the mill beyond the receiving bay to prevent tracking potential contaminants to other areas. Implementing a degree of separation between feed delivery drivers and employees in the mill also prevents contamination from trucks that have been on farms from entering the rest of the mill. Having drivers receive paperwork through a window or from the other side of a door at the office can create this separation. Also having drivers observe potential biosecurity risks at the farms they deliver to can help determine better delivery route options. Farms where exhaust fans are directed toward the feed bins or carcass disposal is near the driveway may need to receive feed later in the day to reduce the chance of bringing the disease back to the mill. Having a biosecurity kit in feed delivery trucks helps reduce the risk of contamination. This kit should contain disposable plastic booties for whenever the driver needs to exit their truck, disinfectant to wipe or spray down their cabs, and a method to dispose of these items. In winter months, multiple sets of rubber overshoes could be used instead of plastic booties, but each pair would need to be disinfected after use.

4.3 Zones in the mill

Feed mill managers should focus on ways to separate the mill into specific zones and limit employee crossover between zones. This helps prevent employees from carrying contaminants from their shoes or clothing through the mill. For instance, an employee may help in receiving and carrying contaminated feed dust on their clothes as they walk through the mill to change a die in pelleting. That contamination is no longer limited to the receiving area and now can come in contact with finished feed. Having one zone designated for receiving, another for mixing and processing, and another at loadout helps to prevent this sort of cross-over by preventing employees from moving between areas without going through a decontamination process. This zoning of the mill also needs to cover tools and equipment that might be shared between areas. To have this process be effective, each zone needs its own sets of tools, brooms, ladders, etc. to prevent these items from carrying contamination through the mill as well.

4.4 Feed handling

Preventing finished feed from contacting surfaces that also handle raw ingredients adds another layer of feed security. Disease mitigation measures like heat treatment (pelleting or extrusion) are effective at killing or inactivating bacteria and viruses in the feed but do not prevent recontamination. As a result, sharing equipment between finished feed and raw ingredients should be avoided. When complete separation is not possible, equipment should be thoroughly disinfected before the finished feed is run through it. Dust and screenings are also known to contain higher levels of pathogens than the bulk of the ingredient load. These residues should be discarded instead of being added back to ingredients or feed to prevent the accumulation of pathogens.

4.5 Employee training

Employees need to receive proper training for any biosecurity plan to succeed. Ensuring that they are aware of the mill's biosecurity plan, FSMA regulations, and pathogen mitigation and prevention strategies helps to encourage compliance. Proper disinfecting procedures should not be overlooked. Many disinfectants require a certain amount of time in contact with a surface to kill pathogens. The presence of dust, grain, or other materials can also limit the effectiveness of disinfectants. All of this should be communicated to employees and biosecurity training should be implemented at intervals throughout the year to keep the plan up-to-date and fresh in employee's minds.

5 Mitigation of pathogens in feed

Once biosecurity practices have been reviewed for ingredients, transportation logistics, and personnel, feed suppliers and producers can consider implementing active mitigation steps within the feed biosecurity program. While either prevention or mitigation alone can result in a meaningful reduction in risk, it is the combination of these strategies that result in the greatest risk reduction through the application of multiple 'hurdles' within the biosecurity system. Two primary approaches can be taken to mitigate the hazard of swine viruses in ingredients or feed. These two approaches can be summarized as: (1) point-in-time mitigation strategies and (2) residual effect mitigation strategies.

5.1 Point-in-time mitigation strategies

Point-in-time strategies include approaches such as extended holding time of ingredients, thermal processing, or irradiation. While these strategies can be effective and result in a product that does not contain an infectious virus at the end of treatment, these ingredients are susceptible to re-contamination.

Quarantining ingredients to allow for natural viral degradation may be an effective method of mitigation; however, there is limited information to fully understand the conditions required to mitigate the pathogen. The concept of viral decay can be used to calculate half-life estimates for quarantine time recommendations (Dee et al., 2014). Defined time and temperature recommendations have been established for quarantining ingredients or complete feed (SHIC, 2020).

Thermal processing is another key point-in-time strategy that may be considered in the feed manufacturing process. The most common type of thermal processing in feed mills is the pelleting process. The pelleting process begins by feeding dry mash feed into a conditioning chamber where heat and moisture are applied via steam. As the feed exits the conditioner, it is compressed through a die to form pellets and then cooled before being stored prior to shipment. Different amounts of heat and moisture are required to adequately pellet different types of feeds. Traditionally, it has been determined that conditioned mash moisture should approach 170 g/kg to 180 g/kg and temperature should reach approximately 82.2°C when pelleting corn and soybean meal-based diets. In addition to heat and moisture, it is important to note that the feed experiences pressure and friction while extruding through the pellet die. Although pelleting is the most common form of thermal processing, more extreme types of processing, such as extrusion, expansion, and long-term conditioning using a hygienist do exist but are not commonly found in feed mills.

Pelleting has demonstrated success in reducing the infectivity of PEDV in feed (Cochrane et al., 2017). Cochrane et al. (2017) observed no signs of infection in pigs inoculated with PEDV-contaminated feed conditioned to either 68.3, 79.4, or 90.6°C with retention times ranging between 45 s and 180 s; however, PEDV RNA was still detected in the majority of the pelleted feed samples. When utilizing lower conditioning temperatures, Cochrane et al. (2015) again detected PEDV RNA in the pelleted feed but found no evidence of infection in pigs consuming feed conditioned at or greater than 54.4°C. Likewise, Harrison et al. (2023a) determined that when feed is conditioned at 82°C, viral RNA was still detected in the finished pelleted product but was considered non-infectious for SVV1 and PEDV as no signs of infection were observed in the bioassay.

The success of thermal processing relies on time × temperature combinations and type of pathogen which have not yet been fully established. In the case of ASFV, pelleting is not a plausible mitigation measure because thermal decay curves involving temperatures achieved through a traditional steam conditioner have never been established. Furthermore, quarantine time and thermal processing are both considered point-in-time mitigation measures. Both may, under ideal conditions, lead to viral inactivation. However, neither protects the ingredient nor feed from subsequent downstream cross-contamination that may occur during conveyance, loadout, or transportation.

5.2 Mitigation strategies with residual effects

Residual strategies most commonly involve the addition of feed additives to the ingredient or diet that would be expected to retain feed mitigation properties after the initial time point in which the additive is applied. Research has documented with multiple feed additives that this residual activity does occur (Gebhardt et al., 2020; Niederwerder et al., 2020). There are several scientific publications that evaluate the efficacy of feed additive compounds for reducing the detection of viral genetic material as well as measures of infectivity *in vitro* and *in vivo*. With a wide variety of products available, it is challenging for feed manufacturers and swine producers to effectively compare different commercial feed additive products to select the product best suited for their particular goals. To assist with this process, colleagues at Kansas State University have compiled and continue to update a feed additive document summarizing peer-reviewed literature that compiles relevant information to allow effective decision-making:

- Kansas State University (www.ksuswine.org)

It is important to recognize, however, that as of Spring 2024, there are no products that have been reviewed by regulatory authorities in the United States and granted permission to make any claims of antiviral activity in ingredients or feed. While scientific data of various forms is available for interpretation, currently no claims of efficacy can be made by feed additive suppliers. Nonetheless, many swine producers currently choose to use feed additives for other production purposes, with the potential benefit of having scientific data supporting antiviral activity within swine feed.

6 Feed mill decontamination

As discussed previously, it has been determined that ASFV, SVV-1, and PEDV can be distributed throughout a feed mill environment and a feed mill contaminated with either of these viruses can be detrimental to a production system. This information becomes very pertinent to the swine industry because feed manufacturing facilities are not designed to easily be cleaned and disinfected. If a feed manufacturing facility becomes contaminated with a virus, there are no current practical recommendations for best practices to clean and disinfect these facilities. Wet disinfection has been found to be the most effective feed mill equipment surface decontamination method (Muckey, 2016). However, this is not practical in most current commercial feed production settings. Initial research has been conducted to determine optimal methods for disinfecting traditional feed manufacturing facilities. The three main methods of disinfection that have been investigated to date are (1) flushing procedures; (2) room heat-up; and (3) fumigation.

Research was conducted to assess sequencing batches of PEDV-negative feed following an inoculated batch of feed to determine the effectiveness of reducing the risk of viral transmission (Schumacher et al., 2016). While this may be a practical mitigation technique for feed mills to implement, there remains a significant quantity of viral particles on feed-contact surfaces including dust production and distribution throughout the facility (Schumacher et al, 2017). This dust may pose a risk for contamination of subsequent diets. To address these challenges, Gebhardt et al. (2018) proposed including chemical mitigants such as formaldehyde or medium chain fatty acids (MCFA) with a flush step during the feed manufacturing process. Rice hulls were selected as the carrier for this chemical flush because of the relatively low cost and high degree of abrasiveness, which may help facilitate the removal of viral contamination on equipment surfaces. The authors determined that dust generated during the 10% MCFA flush and subsequent batch of feed was not infectious. However, it is important to consider that in commercial mills dust would accumulate over additional batches of feed, and the potential of recontamination needs to be considered.

Harrison et al. (2023b) conducted a similar experiment evaluating the distribution of SVV1 inoculated feed in the environment and feed of a feed mill before and after the use of flushing procedures with chemical mitigants. Flush batches between the purposefully inoculated batch and subsequent virus-free batches of feed were able to reduce viral RNA regardless of treatment type. Chemical mitigants reduced SVV-1 RNA more quickly than the non-chemical ground corn flush; however, viral RNA was still present in feed and within the environment after the final batch. The flush treatments were able to reduce SVV1 infectivity of the subsequent feed and dust samples when given to pigs in a bioassay. The authors concluded that flushing feed manufacturing systems following a pathogen introduction can reduce the initial contamination but did not eliminate the risk of recontamination in both the feed and the environment. When determining flushing procedures to use, one should consider how to increase the surface area that the chemical mitigant is able to come in contact with.

Harrison et al. (2023a) conducted a second experiment evaluating the use of ClO_2 fumigation or portable heaters to decontaminate feed manufacturing equipment. Five decontamination treatments were included in the study and were tested on five separate pieces of feed manufacturing equipment (mixer, bucket elevator, feed cleaner, drag conveyor, and distributor with a downspout. Decontamination treatments followed the inoculation batch and were (1) complete feed mill decontamination - removal of organic matter with heated pressure washing, disinfection with 1% Virkon, disinfection with 5% household bleach, environmental heat held at 60°C for 48 h; (2) ClO_2 application; (3) organic matter removal + ClO_2 application; (4) heat for 48 h; and (5) organic matter removal + heat. Except for the complete feed mill decontamination treatment, there was no statistical difference between the inoculation and post-treatment steps regardless of location. The complete facility decontamination (removal of organic matter with heated pressure washing, disinfection with 1% Virkon, disinfection with 5% household bleach, environmental heat held at 60°C for 48 h) was the only decontamination treatment where SVV1 RNA was non-detectable after completion of all steps. The removal of organic material, chlorine dioxide, and heat treatments reduced the detectable quantity of RNA, but viral particles were still detectable across mill surfaces. Although SVV1 RNA was detected via PCR, replication of SVV1 particles was not found in any rooms evaluating the decontamination treatments or the removal of organic manner. The lack of viral replication in pigs provides evidence that the decontamination treatments were able to inactivate SVV1 particles even though RNA was still detectable via PCR.

7 Conclusion

Much has been learned through producer experiences and controlled research that has led to a fundamental shift in the way the swine feed manufacturing industry approaches biosecurity. Ensuring pig feed safety is not something that has immediate effects within the mill but comes with added efforts to accomplish it successfully. It can be easy to overlook during day-to-day operations. An effective biosecurity plan requires the identification and evaluation of hazards, as well as procedures to control biological hazards within a particular facility. These control procedures can include the prevention of entry of the pathogen during ingredient receiving, monitoring and exclusion of foot traffic by high-risk individuals, prevention of cross-contamination, and proactive mitigation using thermal processing or chemical additives. Biosecurity as a whole continues to gain importance in the global animal health industry with focus being placed on disease prevention - and feed biosecurity will continue to be an important component of a comprehensive biosecurity program.

8 Where to look for further information

8.1 Additional resources

- www.ksu.swine.edu
- https://www.afia.org/pub/?id=E348BF9F-98ED-09DB-A45D-504737FE7AE2
- https://www.asi.k-state.edu/research/feedsafetyresources/

9 References

Allerson, M. W., Cardona, C. J. and Torremorell, M. (2013). Indirect transmission of influenza A virus between pig populations under two different biosecurity settings, *Plos One* 8(6), e67293. doi:10.1371/journal.pone.0067293.

Aubry, P., Thompson, J., Pasma, T., Furness, M. C. and Tataryn, J. (2016). Weight of the evidence linking feed to an outbreak of porcine epidemic diarrhea in Canadian swine herds, *JSHAP* 25(2), 69–72.

Belyanin, S., Vasilev, A., Kolbasov, D., Tsybanov, S., Balyshev, V. and Kurinnov, V. (2013). Virulence of African swine fever isolates, *Vet. Kubani* 5, 9–10.

CDC (2019). Outbreak of multi-drug resistant *Salmonella* infections linked to contact with pig ear dog treats, *Centers for Disease Control and Prevention Investigation Notice*. Available at: https://www.cdc.gov/salmonella/pet-treats-07-19/index.html.

Cochrane, R., Dritz, S., Woodworth, J., Huss, A. R., Stark, C. R., Hesse, R. A., Zhang, J., Tokach, M. D.., Bai, J. and Jones, C. K. (2015). Evaluating Chemical Mitigation of Porcine Epidemic Diarrhea Virus (PEDV) in Swine Feed and Ingredients. *KS ag. Exp. Stat., Res. Rep.* 1(7). doi:10.4148/2378-5977.1110.

Cochrane, R. A., Schumacher, L. L., Dritz, S. S., Woodworth, J. C., Huss, A. R., Stark, C. R., DeRouchey, J. M., Tokach, M. D., Goodband, R. D., Bia, J., Chen, Q., Zhang, J., Gauger, P. C., Derscheid, R. J., Magstadt, D. R., Main, R. G. and Jones, C. K. (2017). Effect of pelleting on survival of porcine epidemic diarrhea virus-contaminated feed, *J. Anim. Sci.* 95(3). doi:10.2527/jas.2016.0961.

Davies, P. R., Morrow, W. E., Deen, J., Gamble, H. R. and Patton, S. (1998). Seroprevalence of *Toxoplasma gondii* and *Trichinella spiralis* in finishing swine raised in different production systems in North Carolina, USA, *Prev. Vet. Med.* 36(1), 67–76. doi:10.1016/S0167-5877(98)00072-5.

Dee, S., Bauermann, F. V., Niederwerder, M. C., Singrey, A., Clement, T., de Lima, M., Long, C., Patterson, G., Sheahan, M. A., Stoian, A. M. M., Petrovan, V., Jones, C. K., De Jong, J., Ji, J., Spronk, G. D., Minion, L., Christopher-Hennings, J., Zimmerman, J. J., Rowland, R. R. R., Nelson, E., Sundberg, P. and Diel, D. G. (2018). Survival of viral pathogens in animal feed ingredients under transboundary shipping models, *Plos One* 13(3), e0194509. doi:10.1371/journal.pone.0194509.

Dee, S., Clement, T., Schelkopf, A., Nerem, J., Knudsen, D., Christopher-Hennings, J. and Nelson, E. (2014). An evaluation of contaminated complete feed as a vehicle for porcine epidemic diarrhea virus infection of naive pigs following consumption via natural feeding behavior: proof of concept, *BMC Vet. Res.* 10, 176. doi:10.1186/s12917-014-0176-9.

Dee, S., Deen, J., Otake, S. and Pijoan, C. (2004). An experimental model to evaluate the role of transport vehicles as a source of transmission of porcine reproductive and respiratory syndrome virus to susceptible pigs, *Can. J. Vet. Res.* 68, 208–214.

Dee, S. and Niederwerder, M. (2019). The foreign animal disease risk of feed. In: *Proceedings of the AASV Seminar* (vol. 12). American Association of Swine Veterinarians, Perry, IA.

Dee, S., Shah, A., Jones, C., Singrey, A., Hanson, D., Edler, R., Spronk, G., Niederwerder, M. and Nelson, E. (2021). Evidence of viral survival in representative volumes of feed and feed ingredients during long-distance commercial transport across the continental United States, *Trans. Emer. Dis.* 69, 149–156. doi:10.1111/tbed.14057Elijah.

Elijah, C. G., Harrison, O., Blomme, A., Woodworth, J., Jones, C., Paulk, C. and Gebhardt, J. (2022). Understanding the role of feed manufacturing and delivery within a series of porcine Deltacoronavirus investigations, *J. Swine Health Prod.* 30(1), 17–23. doi:10.54846/jshap/1250.

Elijah, C. G., Trujillo, J. D., Jones, C. K., Gaudreault, N. N., Stark, C. R., Cool, K. R., Paulk, C. B., Kwon, T., Woodworth, J. C., Morozov, I., Gallardo, C., Gebhardt, J. T. and Richt, J. A. (2021). Evaluating the distribution of African swine fever virus within a feed mill environment following manufacture of inoculated feed, *Plos One* 16(8), e0256138. doi:10.1371/journal.pone.0256138.

Elnekave, E., Hong, S., Mather, A. E., Boxrud, D., Taylor, A. J., Lappi, V., Johnson, T. J., Vannucci, F., Davies, P., Hedberg, C., Perez, A. and Alvarez, J. (2018). *Salmonella enterica* serotype 4,[5],12:i:-in swine in the United States Midwest: an emerging multidrug-resistant clade, *Clin. Infect. Dis.* 66(6), 877–885. doi:10.1093/cid/cix909.

EFSA Panel on Animal Health and Welfare (AHAW), Nielsen, S. S., Alvarez, J., Bicout, D. J., Calistri, P., Canali, E., Drewe, J. A., Garin-Bastuji, B., Gonzales Rojas, J. L., Gortázar Schmidt, C., Herskin, M., Miranda Chueca, M. Á., Michel, V., Padalino, B., Pasquali, P., Sihvonen, L. H., Spoolder, H., Stahl, K., Velarde, A., Viltrop, A., Winckler, C., Boklund, A., Botner, A., Gervelmeyer, A., Mosbach-Schulz, O. and Roberts, H. C. (2021). Ability of different matrices to transmit African swine fever virus. Available at: https://efsa.onlinelibrary.wiley.com/doi/epdf/10.2903/j.efsa.2021.6558 Accessed 4 December 2023.

Fedorka-Cray, P., Hogg, A., Gray, J., Thomas, L. A. and Ferris, K. (1997). Feed and feed trucks as sources of Salmonella contamination in swine, *J. Swine Health Prod.* 5(5), 189–193.

Fischer, M., Pikalo, J., Beer, M. and Blome, S. (2021). Stability of African swine fever virus on spiked spray-dried porcine plasma, *Transbound. Emerg. Dis.* 68(5), 2806-2811. doi:10.1111/tbed.14192.

Food and Drug Administration (2013). Compliance policy guide, *Salmonella in Food for Animals*. Available at: https://www.fda.gov/downloads/iceci/compliancemanuals/compliancepolicyguidancemanual/ucm361105.pdf Sec.690.800.

Gebhardt, J. T., Cochrane, R. A., Woodworth, J. C., Jones, C. K., Niederwerder, M. C., Muckey, M. B., Stark, C. R., Tokach, M. D., DeRouchey, J. M., Goodband, R. D., Bai, J., Gauger, P. C., Chen, Q., Zhang, J., Main, R. G. and Dritz, S. S. (2018). Evaluation of the effects of flushing feed manufacturing equipment with chemically treated rice hulls on porcine epidemic diarrhea virus cross-contamination during feed manufacturing, *J. Anim. Sci.* 96(10), 4149–4158. doi:10.1093/jas/sky295.

Gebhardt, J., Dritz, S., Elijah, C., Jones, C. K., Paulk, C. B. and Woodworth, J. C. (2022). Sampling and detection of African swine fever virus within a feed manufacturing and swine production system, *Trans. Emerg. Dis.* 6(9), 103–114. doi:10.1111/tbed.14335.

Gebhardt, J. T., Thomson, K. A., Woodworth, J. C., Dritz, S. S., Tokach, M. D., DeRouchey, J. M., Goodband, R. D., Jones, C. K., Cochrane, R. A., Niederwerder, M. C., Fernando, S., Abbas, W. and Burkey, T. E. (2020). Effect of dietary medium-chain fatty acids on nursery pig growth performance, fecal microbial composition, and mitigation properties against porcine epidemic diarrhea virus following storage, *J. Anim. Sci.* 98(1), 1–11. doi:10.1093/jas/skz358.

Gebhardt, J. T., Woodworth, J. C., Jones, C. K., Tokach, M. D., Gauger, P. C., Main, R. G., Zhang, J., Chen, Q., DeRouchey, J. M., Goodband, R. D., Stark, C. R., Bergstrom, J. R., Bai, J. and Dritz, S. S. (2019). Determining the impact of commercial feed additives as potential porcine epidemic diarrhea virus mitigation strategies as determined by polymerase chain reaction analysis and bioassay, *Transl. Anim. Sci.* 3(1), 93–102. doi:10.1093/tas/txy100.

Guo, M., Dubey, J. P., Hill, D., Buchanan, R. L., Gamble, H. R., Jones, J. L. and Pradhan, A. K. (2015). Prevalence and risk factors for *Toxoplasma gondii* infection in meat animals and meat products destined for human consumption, *J. Food Prot.* 78(2), 457–476. doi:10.4315/0362-028X.JFP-14-328.

Harrison, O. L., Otott, H. K., Bai, J., Hamill, V. A., Singrey, A., Gauger, P. C., Almeida, M. N., Woodworth, J. C., Stark, C. R., Pogranichniy, R. M., Jones, C. K., Gebhardt, J. T. and Paulk, C. B. (2023a). Understanding the reduction of Porcine Epidemic Diarrhea Virus, Porcine reproductive and respiratory syndrome virus, and Seneca Valley Virus 1 RNA in inoculated feed and the environment following thermal processing, *KS Ag. Exp. St. Res. Rep.* 9(7), 7. doi:10.4148/2378-5977.8539.

Harrison, O. L., Otott, H. K., Bai, J., Hamill, V. A., Singrey, A., Gauger, P. C., Almeida, M. N., Woodworth, J. C., Stark, C. R., Pogranichniy, R. M., Jones, C. K., Gebhardt, J. T. and Paulk, C. B. (2023b). Evaluating the distribution of porcine epidemic diarrhea virus, porcine reproductive and respiratory syndrome virus, and Seneca valley virus 1 inoculated feed after the use of physical or chemical mitigants to flush a feed manufacturing facility, *KS Ag. Exp. St. Res. Rep.* 9(7), 7. doi:10.4148/2378-5977.8538.

Hedman, C., Bolea, R., Marin, B., Cobrière, F., Filali, H., Vazquez, F., Pitarch, J. L., Vargas, A., Acín, C., Moreno, B., Pumarola, M., Andreoletti, O. and Badiola, J. J. (2016). Transmission of sheep-bovine spongiform encephalopathy to pigs, *Vet. Res.* 47, 14. doi:10.1186/s13567-015-0295-8.

Huss, A. R., Schumacher, L. L., Cochrane, R. A., Poulsen, E., Bai, J., Woodworth, J. C., Dritz, S. S., Stark, C. R. and Jones, C. K. (2017). Elimination of porcine epidemic diarrhea virus in an animal feed manufacturing facility, *Plos One* 12(1), e0169612. doi:10.1371/journal.pone.0169612.

Kim, Y., Yang, M., Goyal, S. M., Cheeran, M. C. J. and Torremorell, M. (2017). Evaluation of biosecurity measures to prevent indirect transmission of porcine epidemic diarrhea virus, *BMC Vet. Res.* 13(1), 89. doi:10.1186/s12917-017-1017-4.

Leme, R. A., Miyabe, F. M., Dall Agnol, A. M., Alfieri, A. F. and Alfieri, A. A. (2019). Seneca valley virus RNA detection in pig feed and feed ingredients in Brazil, *Transbound. Emerg. Dis.* 66(4), 1449–1453. doi:10.1111/tbed.13215.

Li, X., Bethune, L. A., Jia, Y., Lovell, R. A., Proescholdt, T. A., Benz, S. A., Schell, T. C., Kaplan, G. and McChesney, D. G. (2012). Surveillance of *Salmonella* prevalence in animal feeds and characterization of the *Salmonella* isolates by serotyping and antimicrobial susceptibility, *Foodborne Pathog. Dis.* 9(8), 692–698. doi:10.1089/fpd.2011.1083.

Lomonaco, S., Magossi, G., Leon, M., Miller, D., Kastanis, G. J., Tallent, S. M., Allard, M. W., Brown, E. W. and Trinetta, V. (2018). Draft genome sequences of 57 *Salmonella enterica* strains from selected U.S. swine feed mills, *Microbiol. Resour. Announc.* 7(17). doi:10.1128/MRA.01191-18.

Lowe, J., Gauger, P., Harmon, K., Zhang, J., Connor, J., Yeske, P., Loula, T., Levis, I., Dufresne, L. and Main, R. (2014). Role of transportation in spread of porcine epidemic diarrhea virus infection, United States, *Emerg. Infect. Dis.* 20(5), 872–874. doi:10.3201/eid2005.131628.

Magossi, G., Cernicchiaro, N., Dritz, S., Houser, T., Woodworth, J., Jones, C. and Trinetta, V. (2019). Evaluation of *Salmonella* presence in selected United States feed mills, *Micro. Open* 8(5), e00711. doi:10.1002/mbo3.711.

McOrist, S., Scott, P. C., Jendza, J., Paynter, D., Certoma, A., Izzard, L. and Williams, D. T. (2022). Analysis of acidified feed components containing African swine fever virus, *Res. Vet. Sci.* 152, 248–260. doi:10.1016/j.rvsc.2022.08.014.

Muckey, M. B., Jones, C. K., Woodworth, J. C., Paulk, C. B., Dritz, S. S. and Gebhardt, J. T. (2021). Using environmental sampling to evaluate the effectiveness of decontamination methods to reduce detection of porcine epidemic diarrhea virus RNA on feed manufacturing surfaces. *Trans. Anim. Sci.* 5, txab121. doi:10.1093/tas/txab121.

Niederwerder, M., Dee, S., Diel, D., Stoian, A. M. M., Constance, L. A., Olcha, M., Petrovan, V., Patterson, G., Cino-Ozuna, A. G., & Rowland, R. R. R. (2020). Mitigating the risk of African swine fever virus in feed with anti-viral chemical additives, *Transbound. Emerg. Dis.* 00, 1–10. doi:10.1111/tbed.13699.

Park, J. H., Tark, D., Lee, K. N., Lee, S. Y., Ko, M. K., Lee, H. S., Kim, S. M., Ko, Y. J., Seo, M. G., Chun, J. E., Lee, M. H. and Kim, B. (2016). Novel foot-and-mouth disease virus in Korea, July-August 2014, *Clin. Exp. Vaccin. Res.* 5(1), 83–87. doi:10.7774/cevr.2016.5.1.83.

Pasick, J., Berhane, Y., Ojkic, D., Maxie, G., Embury-Hyatt, C., Swekla, K., Handel, K., Fairles, J. and Alexandersen, S. (2014). Investigation into the role of potentially contaminated feed as a source of the first-detected outbreaks of porcine epidemic diarrhea in Canada, *Transbound. Emerg. Dis.* 61(5), 397–410. doi:10.1111/tbed.12269.

Pujols, J. and Segales, J. (2014). Survivability of porcine epidemic diarrhea virus (PEDV) in bovine plasma submitted to spray drying processing and held at different time by temperature storage conditions, *Vet. Microbiol.* 174(3–4), 427–432. doi:10.1016/j.vetmic.2014.10.021.

Ribbens, S., Dewulf, J.,Koenen, F., Laevens, H., Mintiens, K. and de Kruif, A. (2004). An experimental infection (II) to investigate the importance of indirect classical swine fever virus transmission by excretions and secretions of infected weaner pigs, *J. Vet. Med.* 51(10), 438–442. doi:10.1111/j.1439-0450.2004.00783.x.

Russell, L. E., Polo, J. and Meeker, D. (2020). The Canadian 2014 porcine epidemic diarrhoea virus outbreak: important risk factors that were not considered in the epidemiological investigation could change the conclusions, *Transbound. Emerg. Dis.* 67(3), 1101–1112. doi:10.1111/tbed.13496.

Schumacher, L. L., Huss, A. R., Cochrane, R. A., Stark, C. R., Woodworth, J. C., Bai, J., Poulsen, E. G., Chen, Q., Main, R. G., Zhang, J., Gauger, P. C., Ramirez, A., Derscheid, R. J., Magstadt, D. M., Dritz, S. S. and Jones, C. K. (2017). Characterizing the rapid spread of porcine epidemic diarrhea virus (PEDV) through an animal food manufacturing facility, *Plos One* 12(11), e0187309. doi:10.1371/journal.pone.0187309.

Schumacher, L. L., Woodworth, J. C., Jones, C. K., Chen, Q., Zhang, J., Gauger, P. C., Stark, C. R., Main, R. G., Hesse, R. A., Tokach, M. D. and Dritz, S. S. (2016). Evaluation of the minimum infectious dose of porcine epidemic diarrhea virus in virus-inoculated feed, *Amer. J. Vet. Res.* 77(10), 1108–1113. doi:10.2460/ajvr.77.10.1108.

Swine Health and Information Center (SHIC) (2020). Holding time calculations for feed ingredients to mitigate virus transmision. Available at: https://www.swinehealth.org/wp-content/uploads/2020/02/Holding-Time-Calculations-for-Feed-Ingredients-to-Mitigate-Virus-Transmission-Print-02.04.20.pdf.

Slifko, T. R., Smith, H. V. and Rose, J. B. (2000). Emerging parasite zoonoses associated with water and food, *Int. J. Parasitol.* 30(12–13), 1379–1393. doi:10.1016/S0020-7519(00)00128-4.

Sugiura, K., Ogura, H., Ito, K., Ishikawa, K., Hoshino, K. and Sakamoto, K. (2001). Eradication of foot and mouth disease in Japan, *Revue Scientifique et Technique de l'OIE* 20(3), 701–713. doi:10.20506/rst.20.3.1301.

Trudeau, M. P., Verma, H., Sampedro, F., Urriola, P. E., Shurson, G. C. and Goyal, S. M. (2017). Environmental persistence of porcine coronaviruses in feed and feed ingredients', *Plos One* 12(5), e0178094. doi:10.1371/journal.pone.0178094.

United States Department of Agriculture Animal and Plant Health Inspection Service Veterinary Services. *Swine Enteric Coronavirus Introduction to the United States: Root Cause Investigation Report*. (2015). Available at: https://www.aphis.usda.gov/animal_health/animal_dis_spec/swine/downloads/secd_final_report.pdf Accessed 4 December 2023.

World Organisation for Animal Health (2018). Recognition of the bovine spongiform encephalopathy risk status of members resolution no. 26. Available at: http://www.oie.int/fileadmin/Home/eng/Animal_Health_in_the_World/docs/pdf/Resolutions/2018/A26_RESO_2018_BSE.pdf.

Part 3

The role of feed additives in optimising pig nutrition

Chapter 7

Understanding and optimizing the use of amino acids in pig nutrition

Sung Woo Kim[1], *Hyunjun Choi, Yesid Garavito Duarte, and Alexa R. Gormley, North Carolina State University, USA*

1 Introduction

Protein is one of the fundamental components of the pig body. On a dry matter basis, protein makes up the largest proportion of the pig's body at birth. This proportion remains high even though protein accretion slows down as the pig grows whilst fat accretion accelerates. While fat can be stored in the cell as triacylglycerides, protein cannot be stored in the body. The amount of protein in a pig's body is the result of the net balance between protein synthesis and protein degradation. Proteins turn over quickly, taking as short as a few seconds and as long as many days depending on the types of protein in the body. A pig can gain body protein when the rate of protein synthesis exceeds the rate of protein degradation or can lose body protein in an opposite case.

Proteins are provided in feeds through protein supplements and supplemental amino acids. Of course, other feedstuffs such as cereal grains contribute dietary protein as well. Dietary protein is digested and broken down into free amino acids and small peptides (di- and tri-peptide) that can be

[1] Correspondence: sungwoo_kim@ncsu.edu.

http://dx.doi.org/10.19103/AS.2024.0140.11

absorbed into the body and eventually free amino acids will be transported to target tissues and cells for utilization. Any excess dietary amino acids beyond what pigs can utilize won't increase body protein but rather will be catabolized to glucose or ketone bodies as energy sources releasing nitrogen. It's not necessary to provide excess dietary protein, but it is important to focus on the amino acid content in feed to ensure their balance and that the amount included meets the requirement. It is an important task for pig nutritionists to pay attention to the amino acid balance and requirements, as it is also directly related to the sustainability of animal agriculture.

In addition to the importance of considering the balance and amount of amino acids in pig diets, increased attention has been given to select amino acids possessing unique functions in the body. These are grouped as 'functional amino acids' and some of these concepts have been implemented in pig nutrition.

Increased availability and affordability of supplemental amino acids allow pig nutritionists to proactively apply the concept of ideal protein, low protein formulation, and precision feeding in pig production. Currently more than ten amino acids are commercially available as feed-grade supplemental amino acids, including seven essential amino acids. With these advances and advantages, more precise and refined nutritional programs and interventions can be used, improving the efficiency of pig production for greater profitability and environmental sustainability than ever before.

In this chapter, principles of understanding amino acid needs and estimating requirements, practical aspects of meeting the amino acid requirements, use of supplemental amino acids, and the concept and application of functional amino acids are included. It is advised that research data cited in this chapter may not cover the entire spectrum of literature but rather directly focused on the subject matters.

2 Protein and amino acid metabolism

Proteins are made of 20 amino acids, specifically, a sequence of 20 amino acids. These amino acids are linked by peptide bonds to form a polypeptide and the interactions among these amino acids within a polypeptide or between polypeptides further give structure to a protein. The sequence of these amino acids makes each protein uniquely different in their composition, properties, structure, and function.

Proteins can be generally grouped by structure protein (collagen, elastins, actin, myosin, keratin, etc.) in an animal body, storage proteins (glycinin and β-conglycinin in soybeans, zein in corn, kafirin in sorghum, gluten in wheat, etc.) in plant tissues, and proteins with specific functions (enzymes, peptide hormones, antibodies, globulins, hemoglobin, myoglobin, lipoprotein, albumin, etc.). Storage proteins in feedstuffs are the major source of amino acids for animals. Depending on the type of feedstuff, certain storage proteins are highly digestible, whereas most others from plant feedstuffs have limitations, especially when anti-nutritional compounds are included. Proper processing can reduce anti-nutritional compounds in these feedstuffs.

Once ingested, proteins are denatured to a linear form of polypeptides by hydrochloric acid and partly hydrolyzed to smaller peptides by pepsin in the stomach. These peptides reach the small intestine where they are further hydrolyzed to tri-peptides, di-peptides, and free amino acids (single amino acids) by multiple proteases from the pancreas and the small intestine. These proteases include trypsin, chymotrypsin, elastase, carboxypeptidase A and B, amino-peptidase, and di-peptidase. Undigested proteins or partly digested peptides remain and move to the large intestine, whereas most tri-peptides, di-peptides, and amino acids are absorbed through peptide transporters (PEPT) and amino acid transporters (Systems) on the plasma membrane of the enterocytes lining villi in the small intestine.

Tri-peptides and di-peptides are further hydrolyzed into free amino acids and enter blood circulation with other amino acids. Blood carries amino acids throughout the body and delivers them to cells. It's well understood and investigated that a significant portion of absorbed amino acids are utilized in the intestine (20–50% of absorbed amino acids) with the remaining amino acids carried to other tissues, including the liver and skeletal muscle.

Once amino acids are delivered to the target cells of various body tissues, they are either used for protein synthesis or catabolized after deamination to provide carbon skeletons for use in building other amino acids, fatty acids, or glucose. Protein synthesis is a highly coordinated process initiated by the activation of several initiation factors (IFs), transcription of messenger RNA (mRNA), and translation of mRNA to make a polypeptide. Each step of the translation of the genetic information yields specific amino acids depending on the type of codon, and these amino acids are connected by peptide bonds, forming a polypeptide. A newly constructed polypeptide is further processed to form a protein, including a secondary and a tertiary structure constructed with α-helix, β-pleated sheet, disulfide bonds, and hydrogen bonds. Depending on the type of protein, it can be secreted from a cell or become cellular protein.

From the process of protein digestion, absorption and transport of amino acids, and protein synthesis, an animal gains protein in the body. The newly made protein is predetermined for its use which can be a variety of things, including body maintenance, tissue repair and replacement, tissue growth, and production of conceptus and milk.

When preparing a feed for pigs, it is important to know how much of each amino acid will be needed by a pig and how much of each amino acid can be provided from the feed. Considering that protein is made of a sequence of amino acids, it is the amino acid that pigs need, not protein. There is no such term for 'protein requirement' in pig nutrition, but the important goal is to meet the amino acid requirement.

Among the 20 amino acids making proteins, 9 of them need to be provided from feeds because these amino acids cannot be synthesized or sufficiently synthesized in an animal cell. Therefore, these amino acids are called 'essential amino acids' or 'indispensable amino acids' and include lysine, methionine, threonine, tryptophan, valine, leucine, isoleucine, histidine, and phenylalanine. Among these essential amino acids, lysine, methionine, threonine, and tryptophan are major 'limiting amino acids' that are typically deficient in feeds used to meet the pig's requirements. Valine, leucine, and isoleucine are called 'branched chain amino acids' because their side chains possess a branched hydrocarbon chain. Valine and isoleucine can be deficient in feeds occasionally, whereas leucine is typically sufficient in feeds. Histidine and phenylalanine are typically sufficient in feeds exceeding requirements for pigs. Most essential amino acids (with the exception of phenylalanine) are commercially available as feed grade that can be supplemented in pig feeds.

3 Principles of amino acid needs

Pigs need amino acids to sustain life and to grow. Amino acids are needed for body maintenance including enzymes, peptide hormones, transporters, receptors, carriers, mucins, immunoglobulins, as well as proteins for the replacement of damaged tissue. Amino acids are also used for growth, such as the needs for hyperplasia and hypertrophy of muscle fiber and for the production of fetal tissues, conceptus, mammary parenchymal tissue, and milk protein.

3.1 Maintenance

Amino acids are used to synthesize proteins that are essential in sustaining life and body maintenance. These include the use of amino acids for the synthesis of enzymes and peptide hormones, as well as for the repair of tissues. Amino acids are also used to synthesize cytokines, immunoglobulins, and other signaling molecules that function in immune reactions as a part of body maintenance. Each of these proteins and peptides have a unique amino acid sequence and profile contributing to the amino acid needs for body maintenance. The compositional difference of amino acids in these proteins is shown in Table 1.

Enzymes are biological catalysts that are needed for biochemical reactions in the body. Enzymes are therefore critically important and needed in the

Table 1 Amino acid composition in various proteins essential in body maintenance

	Albumin[1]		Mucin[2]		Insulin[3]		Pepsin[4]		Trypsin[5]	
	mg/g	AA:Lys	mg/g	AA:Lys	mg/g	AA:Lys	mg/g	AA:Lys	mg/g	AA:Lys
EAA										
Lys	100.0	1.00	34.7	10.0	18.8	1.00	3.1	1.00	44.6	1.00
Met	8.0	0.08	-	-	-	-	12.2	4.00	8.9	0.20
Thr	43.0	0.43	148.4	42.8	37.3	1.98	82.6	27.00	44.6	1.00
Trp	-	-	-	-	-	-	15.3	5.00	17.9	0.40
Val	59.0	0.59	39.2	11.3	70.6	3.75	67.3	22.00	71.4	1.60
Leu	93.0	0.93	-	-	118.2	6.28	82.6	27.00	71.4	1.60
Ile	34.0	0.34	83.5	24.1	31.2	1.66	76.5	25.00	71.4	1.60
His	20.0	0.20	34.7	10.0	38.6	2.05	3.1	1.00	17.9	0.40
Phe	50.0	0.50	-	-	60.2	3.20	42.8	14.00	17.9	0.40
NEAA									Ala	83.0
Ala	83.0	0.83	66.8	19.3	40.4	2.15	48.9	16.00	67.0	1.50
Gly	29.0	0.29	76.4	22.0	79.2	4.21	107.0	35.00	111.6	2.50
Ser	38.0	0.38	152.9	44.1	52.2	2.77	131.5	43.00	116.1	2.60
Tyr	33.0	0.33	-	-	74.5	3.96	45.9	15.00	35.7	0.80
Glu	-	-	131.0	37.8	137.5	7.30	39.8	13.00	80.4	1.80
Gln	169.0	1.69	-	-	-	-	39.8	13.00	-	-
Asp	-	-	71.3	20.6	58.8	3.13	91.7	30.00	107.1	2.40
Asn	84.0	0.84	-	-	-	-	36.7	12.00	67.0	1.50
Arg	45.0	0.45	39.8	11.5	19.0	1.01	06.1	2.00	17.9	0.40

(*Continued*)

Table 1 (*Continued*)

	Albumin[1]		Mucin[2]		Insulin[3]		Pepsin[4]		Trypsin[5]	
	mg/g	AA:Lys	mg/g	AA:Lys	mg/g	AA:Lys	mg/g	AA:Lys	mg/g	AA:Lys
Pro	55.0	0.55	131.7	38.0	19.2	1.02	45.9	15.00	44.6	1.00
Cys	58.0	0.58	-	-	110.8	5.89	18.3	6.00	53.6	1.20
Total	1000		1000		1000		1000		1000	

[a]Carlsson et al. (1977)
[b]Fogg et al. (1996)
[c]Chance et al. (1968)
[d]Tang et al. (1973)
[e]Walker and Keil (1973)

process of feed digestion, facilitation of nutrient metabolism, maintenance of cellular activities, excretion of waste and toxins, etc. Enzymes are proteins composed of a sequence of amino acids. Amino acids are also used in the synthesis of peptide hormones, such as insulin, glucagon, and growth hormone, with roles in the regulation of nutrient metabolism and nutrient partitioning. Immune cells secrete various proteins and peptides including cytokines and immunoglobulins which are needed for immune reactions and body protection.

Digestive secretions contribute to a significant amount of proteins used for body maintenance. These include many digestive enzymes and mucins. Once secreted for their functions in feed digestion, enzymes are not recovered but passed to the large intestine and excreted, unless used by bacteria in the large intestine. Digestive secretions, therefore, contribute to the loss of amino acids from the body. As the turnover rate of intestinal epithelium is fast, there is continuous proliferation and differentiation of intestinal stem cells to epithelial cells in crypts that migrate to the tip of villi as they functionally mature. Those cells in the tip of villi will, then, slough off as new cells replace the old cells. Amino acids are continuously used in this process. Sloughed cells from villi contribute to the loss of amino acids from the body. The loss of amino acids from digestive secretions and sloughed cells are collectively termed 'endogenous amino acid loss'. The amount of endogenous amino acid loss is influenced by various factors, including protein levels, anti-nutritional compounds, dry matter intake (DMI), and body weight. Increased protein levels and anti-nutritional compounds in the diet increase endogenous amino acid loss, whereas increased DMI and body weight decrease endogenous amino acid loss. Table 2 shows the amounts of endogenous amino acid loss as influenced by various factors.

3.2 Growth

Amino acids are used to support the growth of animals that is largely due to the increase in the muscle tissue. Major types of muscle tissue include skeletal muscle, smooth muscle, and cardiac muscle. Skeletal muscles are largely composed of muscle fibers (skeletal muscle cells) that are attached to bone by tendons. The growth of skeletal muscles is a combination of the hyperplasia and hypertrophy of muscle fibers. The hyperplasia of muscle fibers occurs during fetal development and is completed prior to birth, whereas the hypertrophy of muscle fibers occurs throughout the lifetime. The growth of muscles is an outcome of increased protein synthesis and/or decreased protein degradation.

Table 2 Endogenous amino acid loss (g/kg DMI) by various factors in pigs

	Protein,[a] %			Arabinoxylan,[b] %			DMI,[c] kg/d			Body weight,[d] kg	
	4.6	8.6	17.5	0.0	4.0	7.5	0.7	1.5	2.2	25.9	93.2
EAA											
Lys	0.26	0.46	0.52	0.28	0.40	0.60	1.39	0.96	0.71	0.33	0.20
Met	0.09	0.16	0.27	-	-	-	0.41	0.28	0.21	0.10	0.07
Thr	0.38	0.81	1.10	0.80	0.92	1.27	1.53	0.92	0.68	0.51	0.28
Trp	-	-	-	-	-	-	0.36	0.23	0.16	-	-
Val	0.31	0.61	0.76	0.59	0.65	0.92	1.48	0.99	0.70	0.41	0.24
Leu	0.38	0.69	0.80	0.60	0.66	0.99	1.72	1.19	0.85	0.53	0.25
Ile	0.24	0.47	0.54	0.35	0.43	0.65	1.20	0.83	0.60	0.30	0.17
His	0.14	0.26	0.38	0.27	0.32	0.50	0.53	0.34	0.26	0.15	0.10
Phe	0.20	0.36	0.41	0.26	0.29	0.43	1.05	0.74	0.51	0.26	0.15
NEAA											
Ala	0.30	0.58	0.67	0.48	0.58	0.89	2.05	1.40	1.03	-	-
Gly	0.48	0.75	0.70	1.47	1.60	2.05	4.30	2.54	1.97	-	-
Ser	0.50	1.04	1.29	0.66	0.79	1.18	2.05	1.21	0.91	-	-
Tyr	0.16	0.32	0.35	0.27	0.30	0.48	0.92	0.66	0.46	-	-
Glu	0.82	1.98	3.00	1.86	1.82	3.27	3.31	2.38	1.67	-	-
Asp	0.50	1.08	1.51	0.90	0.99	1.53	2.57	1.87	1.35	-	-
Arg	0.23	0.42	0.46	0.30	0.36	0.55	1.53	0.87	0.77	0.26	0.15
Pro	0.46	1.01	1.25	0.83	0.92	1.73	10.54	7.36	6.88	-	-
Cys	0.13	0.22	0.25	-	-	-	0.53	0.34	0.24	-	-

[a] Hodgkinson et al. (2000)
[b] Morel et al. (2003)
[c] Moter and Stein (2004)
[d] Presto et al. (2010)

Table 3 Amino acid ratio to lysine in various structural proteins

	Longissimus dorsi muscle[a]	Collagen[b]	Elastin[c]	Keratin[d]
EAA				
Lys	1.00	1.00	1.00	1.00
Met	0.31	0.10	0.45	0.20
Thr	0.52	0.65	1.36	1.71
Trp	-	-	-	0.34
Val	0.60	0.00	11.18	1.66
Leu	0.91	1.05	5.55	2.26
Ile	0.56	0.35	2.09	1.03
His	0.54	1.90	0.55	0.31
Phe	0.46	0.50	3.18	0.80
NEAA				
Ala	0.61	5.55	19.00	1.23
Gly	0.53	17.65	24.82	1.23
Ser	0.46	1.35	1.18	2.46
Tyr	0.42	1.35	1.91	1.00
Glu	1.77	1.05	3.36	4.74
Asp	1.04	0.35	1.45	2.17
Arg	0.72	3.95	0.91	2.66
Pro	0.43	6.40	10.73	2.51
OH-Pro	-	6.65	2.09	0.00
Cys	0.13	0.15	-	6.29

[a]Conde-Aguilera et al. (2014)
[b]Ahn et al. (2021)
[c]Grant (1965)
[d]Moran et al. (1967)

Synthesis of proteins in muscles would use specific amino acid profile depending on the types of proteins. Table 3 shows amino acid composition in various structural proteins.

3.3 Production

Amino acids are used to synthesize proteins for the production of fetal tissues, uterine, placenta, mammary parenchymal tissues, and milk proteins depending on specific physiological stages of pigs. Quantitative growth of the fetus is fairly slow until 70 days of conception and the major increase in fetal tissues occurs from day 70 (McPherson et al., 2004; Ji et al., 2005). Protein gain in the fetus increases almost 18 folds after 70 days of conception (Kim et al., 2009, 2010). Similarly, quantitative growth of mammary parenchymal tissue mainly occurs

Table 4 Amino acid accretion and its relative ratio to lysine in fetal and mammary parenchymal tissues of sows during gestation

	Fetal tissue		Mammary parenchymal tissue		Fetal tissue		Mammary parenchymal tissue	
	Gain, g/day				Ratio to Lys			
	Day 0 to 70	Day 70 to farrowing	Day 0 to 70	Day 70 to farrowing	Day 0 to 70	Day 70 to farrowing	Day 0 to 70	Day 70 to farrowing
Protein	0.25	4.63	0.14	3.41				
Lys	0.019	0.283	0.011	0.256	1.00	1.00	1.00	1.00
Thr	0.010	0.162	0.006	0.145	0.53	0.57	0.55	0.57
Met	0.006	0.092	0.003	0.068	0.32	0.33	0.27	0.27
Trp	0.003	0.056	0.002	0.040	0.16	0.20	0.18	0.16
Val	0.013	0.211	0.008	0.194	0.68	0.75	0.73	0.76
Leu	0.020	0.332	0.012	0.286	1.05	1.17	1.09	1.12
Ile	0.009	0.142	0.006	0.141	0.47	0.50	0.55	0.55

Source: Adapted from: Mateo et al. (2007, 2008); Lin et al. (2009), Kim et al. (2009); Kim (2010).

after day 70 of pregnancy (Ji et al., 2006). Protein gain in mammary parenchymal tissue increases almost 24 folds after day 70 of pregnancy (Kim et al., 2009, 2010). Compositional changes among amino acids for protein gain in fetus and in mammary parenchymal tissues throughout pregnancy are shown in Table 4.

Secretions from the mammary gland are a critical source of nutrients for newborn pigs. During the first few days following parturition, mammary secretion is called colostrum as it gradually changes to mature milk. Colostrum is high in dry matter mainly due to high protein content. Dry matter and protein contents reduce as the mammary secretion changes to mature milk. Colostrum is also characterized by its high content of immunoglobulins providing

Table 5 Protein content and amino acid composition in colostrum and milk

	Colostrum[a]		Milk[b]	
DM, g/kg	243 ± 8		166 ± 4	
Fat, g/kg DM	219 ± 21		416 ± 51	
Protein, g/kg DM	497 ± 23		286 ± 2	
Immunoglobulin G, g/L	8.96 ± 0.15		4.83 ± 0.13	
Amino acid (EAA)	g/kg DM	Ratio to Lys	g/kg DM	Ratio to Lys
Lys	65.9 ± 2.6	1.00	25.2 ± 0.7	1.00
Met	10.5 ± 0.6	0.16	6.4 ± 0.1	0.25
Thr	26.7 ± 1.6	0.41	13.8 ± 0.3	0.55
Trp	6.6 ± 0.1	0.10	4.1 ± 0.1	0.16
Val	32.5 ± 2.0	0.49	15.2 ± 0.4	0.60
Leu	44.6 ± 2.4	0.68	28.0 ± 0.8	1.11
Ile	18.6 ± 1.3	0.28	13.9 ± 0.4	0.55
His	11.7 ± 0.6	0.18	5.7 ± 0.1	0.23
Phe	20.3 ± 1.5	0.31	12.1 ± 0.3	0.48
NEAA				
Ala	20.3 ± 0.9	0.31	11.6 ± 0.3	0.46
Gly	19.6 ± 0.9	0.30	6.8 ± 0.2	0.27
Ser	27.4 ± 1.2	0.42	14.7 ± 0.3	0.58
Tyr	18.5 ± 1.0	0.28	11.8 ± 0.3	0.47
Glu	78.0 ± 5.0	1.18	58.7 ± 1.4	2.33
Asp	38.1 ± 3.1	0.58	32.4 ± 0.8	1.29
Arg	21.2 ± 0.9	0.32	9.0 ± 0.2	0.36
Pro	56.3 ± 0.9	0.85	34.9 ± 0.7	1.38
Cys	7.3 ± 0.1	0.11	4.5 ± 0.1	0.18

[a]Colostrum was collected within 12 h post partum. Source: Adopted from: Lin et al., 2009 and Shen et al., 2011.
[b]Milk was collected on days 17–21 of lactation. Source: Adopted from: Mateo et al., 2008, Mateo et al., 2009; and Shen et al., 2011.

passive immunity to newborn pigs. Lin et al. (2007) showed that about 25% of immunoglobulins in colostrum could be delivered to suckling piglets during the first 3 days after birth, without being destroyed. High immunoglobulin contents in colostrum affects its amino acid profile, making colostrum uniquely different from mature milk (Table 5).

4 Estimation of amino acid requirements

There have been tremendous research efforts to evaluate and determine amino acid requirements for pigs. Utilizing available research data, several recommendations of amino acid requirements have been published and used in formulating feeds for pigs. Examples of recommendations include those from the ARC (1981) from the UK, the NRC (2012, as well as its previous editions) from the USA, and the Brazilian Table (2019, as well as its previous editions) from Brazil. These recommendations have widely been accepted and referenced. The NRC (2012) has been cited over 7000 times by other peer-reviewed papers since its publication in 2012 with high citation and impact on pig nutrition and production.

Amino acid requirements have traditionally been estimated based on the growth performance of pigs fed feed with various level of a targeting amino acid (empirical approach), whereas some recent estimations have been based on a mechanistic approach considering the use and needs of amino acids for various components including maintenance, lean growth, and production. Both empirical and mechanistic approaches should be used in estimating accurate amino acid requirements.

4.1 Empirical approach

Amino acid requirements have been estimated empirically. The requirements can be estimated by titrating a target amino acid in diets fed to pigs to determine the specific level at which the growth or growth efficiency is optimized and reaches a plateau. This specific level can be considered 'requirement' of this amino acid for the optimal growth of pigs. It is typical to have at least four diets (or commonly five to seven diets) with varying levels of a target amino acid to be fed to pigs for the duration of a target body weight. Different statistical approaches can be used to determine a requirement level, but a broken-line analysis providing a breakpoint has extensively been used. The requirements of amino acids are measured for each amino acid using this empirical approach.

Traditionally, amino acid requirements have been estimated by measuring growth of pigs fed diets with titrated levels of a targeted amino acid, including threonine (Jayaraman et al., 2015), tryptophan (Clizer et al., 2023), valine (Barea et al., 2009; Soumeh et al., 2015), isoleucine (Htoo et al., 2014), and histidine (Izquierdo et al., 1988; Cemin et al., 2018). The recent availability of feed-grade histidine and improvement of the genetic potential for lean gain in pigs has prompted the re-investigation of the requirement of histidine or the ideal ratio of histidine to lysine. Considering the unique functional roles of histidine, Cheng et al. (2023) evaluated histidine requirements for nursery pigs from 7 to 11 kg body weight, based on the availability of His-containing proteins (carnosine, hemoglobin, and pancreatic trypsinogen), showing that the ideal SID His to Lys ratio is between 0.35 and 0.41.

There can be limitations of solely relying on the empirical approach as the estimation from the empirical approach is rather fixed and does not consider any possible interactions or external factors. To overcome these limitations, the mechanistic approach has been considered to estimate amino acid requirements.

4.2 Mechanistic approach

Amino acid requirements can be estimated based on an understanding of the individual factors and components influencing the requirements. Alteration of each factor and component comprising the requirement would influence the requirement. In growing pigs, for example, amino acid needs for body maintenance and amino acid needs for muscle growth are the two major components comprising the amino acid requirement. For pregnant pigs, amino acid requirements can be built based on amino acid needs for body maintenance, fetal growth, uterine and placental growth, mammary gland development, and maternal growth, in case of young sows. For lactating sows, amino acid requirements can be built based on amino acid needs for body maintenance, mammary gland growth, milk production, and maternal growth. Again, in case of young sows, however, there can be some contribution of amino acids from regression of the uterus and from maternal tissue mobilization. Each component can be altered depending on growth stage, physiological status, health status, environment (such as ambient temperature), housing condition, etc. A mechanistic approach provides a way to estimate amino acid requirement considering the dynamic changes of these factors and components. Kim et al. (2019) estimated the amino acid requirements for pregnant pigs and lactating pigs using a mechanistic approach. Amino acid needs for fetal growth (McPherson et al., 2004), mammary gland growth (Kim et al., 1999a,b,c; Ji et al., 2006), milk production, maternal tissue gain or mobilization (Kim and Easter, 2001; Kim et al., 2001; Ji et al., 2005b), and maternal maintenance (Ji et al., 2005a) were considered together to estimate the amino acid requirements of sows.

5 Availability and use of supplemental amino acids

All 20 amino acids can be synthesized chemically or from microbial fermentation. These are used in medical, pharmaceutical, health, and food industries. Production efficiency or fermentation yields vary depending on the target of the amino acid. At least nine amino acids, however, are commercially available as feed grade and they are often supplemented in feed to meet the needs and requirements of target animals. Commercially available feed-grade amino acids include lysine, methionine, threonine, tryptophan, valine, isoleucine, histidine, arginine, and glutamine. The number of feed-grade amino acids will increase depending on demand, even though each of these will go through a strict registration process by the controlling agencies of each country or regions.

Feed-grade amino acids are produced by microbial fermentation in a large scale typically with a range of 100–500 kilo liters. Microbial fermentation produces only L-type amino acids, the type of amino acids that can be used by animal cells (compared with D-type amino acids). Depending on the chemical properties, acid (such as hydrochloric acid or sulfuric acid) can be added to neutralize select amino acids. This includes L-lysine HCl, L-arginine HCl, L-histidine HCl, and L-lysine sulfuric acid. Typical microorganisms used in the production of amino acids are *Corynebacterium glutamicum* and *Escherichia coli*, although *Corynebacterium glutamicum* is most typically used due to the acceptance and safety, regardless of lower yield compared with *Escherichia coli*.

Once amino acids are secreted from the microbial cells, the whole fermentation broth is filtered to separate the microbial cells from amino acids, and further processing increases the purity of the amino acids. Feed-grade amino acids traditionally have fairly high purity at near 98.5%, excluding acids added to stabilize the precipitation of basic amino acids. However, this purification adds to the production cost significantly. More recently, new generation feed-grade amino acids include impurities such as microbial cells and other metabolites due to skipping the purification process to reduce the production cost and product price. This new generation of feed-grade amino acids have 60–80% purity and are widely used by pig nutritionists globally. Microbial cells included in feed-grade amino acids are mostly *C. glutamicum*, a gram-positive bacteria that belongs to the phylum Actinomycetota. *C. glutamicum* included in feed-grade amino acids has recently been shown to have postbiotic effects, reducing populations of potentially harmful bacteria in the mucosal layer of the small intestine in pigs, in turn reducing intestinal inflammation and oxidative damages.

In the case of methionine, uniquely, there are two production process including chemical synthesis and microbial fermentation resulting in three types of products: DL-methionine, L-methionine, and DL-methionine hydroxy analog (DL-MHA or DL-2-hydroxy-(4-methylthio)butanoic acid). As methionine is a sulfur-containing amino acid, the fermentation process including sulfur has reduced production yield significantly, and therefore, chemical synthesis has been used traditionally to produce DL-methionine and DL-MHA. There have been many animal studies conducted indicating that DL-methionine and DL-MHA are equally bioavailable compared with L-methionine, mainly in poultry, and also in pigs. Some recent animal studies, however, provide evidence that L-methionine has higher bioefficacy than DL-methionine, especially for the use in the intestine, because D-methionine needs to be converted to L-methionine in the liver prior to be available to animal cells. Table 6 shows the summary of

Table 6 Nutritional efficacy of L-methionine, DL-methionine, and DL-methionine hydroxyl analog

Comparison	RBA[a]	Response criteria	Body weight	Reference
DL-MHA to DL-Met	0.68	Weight gain	12–22 kg	Opapeju et al. (2012)
DL-MHA to DL-Met	0.63	Gain-to-feed ratio	12–22 kg	Opapeju et al. (2012)
DL-MHA to DL-Met	0.66	AA oxidation	13–21 kg	Shoveller et al. (2010)
DL-MHA to DL-Met	0.63	N retention	15–18 kg	Wang et al. (2020)
DL-MHA to DL-Met	0.68	N retention	16–21 kg	Wang et al. (2020)
DL-MHA to DL-Met	0.71	N retention	19–22 kg	Opapeju et al. (2012)
DL-MHA to L-Met	0.80	Gain-to-feed ratio	1.9–5.2 kg	Zhang et al. (2019)
DL-MHA to L-Met	0.77	Villus height	1.9–5.2 kg	Zhang et al. (2019)
L-Met to DL-Met	1.06	Gain-to-feed ratio	1.9–5.2 kg	Zhang et al. (2019)
L-Met to DL-Met	1.08	Villus height	1.9–5.2 kg	Zhang et al. (2019)
L-Met to DL-Met	1.44	Weight gain	7–14 kg	Shen et al. (2014)
L-Met to DL-Met	1.23	Gain-to-feed ratio	7–14 kg	Shen et al. (2014)
L-Met to DL-Met	1.67	Glutathione	7–14 kg	Shen et al. (2014)
L-Met to DL-Met	1.06	Malondialdehyde	7–14 kg	Shen et al. (2014)
L-Met to DL-Met	1.14	Protein carbonyl	7–14 kg	Shen et al. (2014)
L-Met to DL-Met	1.15	Villus height	7–14 kg	Shen et al. (2014)
L-Met to DL-Met	1.43	N retention	12–16 kg	Zhou et al. (2021)
L-Met to DL-Met	1.01	N retention	15–16 kg	Zhou et al. (2021)
L-Met to DL-Met	1.02	Plasma urea N	21–48 kg	Yang et al. (2019)

[a]Relative bioavailability (coefficient).

existing literature evaluating and comparing nutritional efficacy among three methionine sources.

Feed-grade amino acids have been called crystalline amino acids, synthetic amino acids, or supplemental amino acids, depending on the literature. However, in this chapter, a term 'supplemental amino acids' will be used because synthetic and crystalline do not fit for the current purpose because (1) amino acids are mostly produced by microbial fermentation (disqualifying a term 'synthetic') and (2) new generation feed-grade amino acids are not crystalized (including impurities; disqualifying a term 'crystalline').

Supplemental amino acids are extensively used in formulating feeds for poultry and pigs as they offset limitations of the use of protein supplements deficient in specific amino acids (limiting amino acids). Therefore, the use of protein supplements can be reduced when supplemental amino acids are used to meet the requirements of limiting amino acids resulting in the benefits of reducing feed costs, reducing crude protein in feeds, and enhancing intestinal health by reducing ammonia-producing bacteria in the intestine. Nutritionally, supplemental amino acids can provide all needed amino acids, almost eliminating the use of protein supplements in the feed. However, there have been limitations associated with formulating feeds extensively with supplemental amino acids until certain levels. This topic is further discussed in the later section 'Low-protein formulation.'

Use of supplemental amino acids has increased gradually and the introduction of a new feed-grade supplemental amino acid then increases

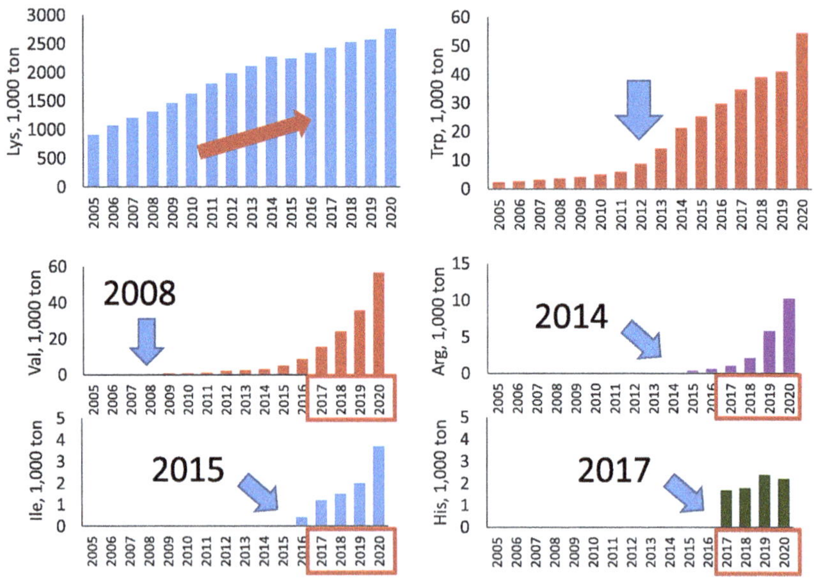

Figure 1 The global use of supplemental amino acids.

the needs of existing supplemental amino acids in formulating pig feeds. The use of L-lysine, as an example, has increased by 300% (three-fold) from 0.9 million ton in 2005 to 2.8 million ton in 2022 (Figure 1). The use of L-tryptophan was insignificant until 2011 but rapidly increased by 1100% (11-fold) from 5000 ton in 2011 to 55 000 ton in 2022 (Figure 1). The use of L-valine and L-isoleucine has been limited but it increases rapidly (Figure 1). With increased use of supplemental amino acids, the production costs reduce with increased affordability and availability. There are only a limited number of manufacturers of supplemental amino acids in the world, but their contribution to animal agriculture and environmental sustainability should be acknowledged.

6 Concept and application of functional amino acids

When certain amino acids are in excess at the time of protein synthesis, these excess amino acids are catabolized through oxidation. This process is initiated with deamination removing an amino group from an amino acid. After deamination, what remains is mainly composed of carbon and hydrogen, called a carbon skeleton. Depending on the type of amino acid (which is related to the structure of side chain), the carbon skeleton is utilized differently, with lysine and leucine being metabolized to ketone bodies, whereas others are metabolized to glucose or glucose metabolites. In the case of five amino acids, including tryptophan, threonine, tyrosine, phenylalanine, and isoleucine, their carbon skeletons can be metabolized to both ketone bodies and glucose metabolites. Amino groups that are removed from amino acids will be taken up by the liver and converted to urea via the urea cycle. Newly synthesized urea will be taken by the kidney and will be excreted as a component of urine. Therefore, excessive amino acids do not increase protein synthesis, but rather are catabolized and used to generate energy with increased nitrogen excretion.

There are, however, exceptions. Select amino acids, when they are in excess, are also metabolized to other bioactive compounds with specific functions in the body. These amino acids are called 'functional amino acids' and include arginine, tryptophan, methionine, phenylalanine, glutamine, glutamic acid, etc. This term 'functional amino acids' was first coined and used in 2007 (Kim et al., 2007; Wu and Kim, 2007) and is widely used in pig nutrition (Le Floc'h et al., 2018; Chalvon-Demersay et al., 2021; Rodrigues et al., 2021).

6.1 Arginine

Arginine is typically sufficient in feeds and can also be synthesized in the animal body primarily in the kidney. Arginine will be used in protein synthesis along with others of the 19 amino acids, but it can also be used as a precursor for other bioactive compounds including creatine, ornithine, polyamines, agmatine, and nitric oxide (NO). The major role of creatine is in energy metabolism by transferring phosphorus to adenosine diphosphate (ADP) for the conversion to adenosine triphosphate (ATP) in muscle and nerve tissue. Ornithine can be used in the synthesis of glutamate, glutamine, and polyamine. Polyamines include spermine and spermidine, possessing triple amine groups in their structure. Polyamines provide amine groups used in the proliferation and differentiation of cells and shown to help angiogenesis. Agmatine functions in the inhibition of NO synthesis and polyamine synthesis. Nitric oxide is generated when arginine is converted to citrulline by NO synthase. When NO is generated in endothelial cells lining a blood vessel, NO will activate guanylate cyclase increasing cyclic guanosine 3′,5′-monophosphate (cGMP) concentrations in the cell signaling relaxation of vascular smooth muscle causing vasodilation.

Considering the functional benefits in angiogenesis and vasodilation, arginine has been supplemented in feeds for breeding pigs, showing a reduced number of stillborns (Mateo et al., 2007) and increased milk production in early lactation (Mateo et al., 2008). Supplemental arginine to milk replacer fed to nursing piglets increased their growth with elevated levels of plasma insulin and growth hormone (Kim and Wu, 2004).

6.2 Methionine

Methionine is one of the key limiting amino acids typically deficient in feeds for pigs. In order to support growth and lean gain of pigs, methionine often has to be supplemented. In addition to its importance in protein synthesis, methionine is used as a precursor of other bioactive compounds used in the body including betaine, choline, creatine, cysteine, glutathione, polyamine, and taurine. Methionine contributes to the synthesis of betaine and choline by providing multiple methyl groups. Betaine and choline are methyl donors in the body critically important in 1-carbon metabolism used in the synthesis of creatine, DNA, polyamines, and phospholipids with essential roles in reproduction and growth. Choline is also used to synthesize betaine and acetylcholine. Together with arginine, methionine is used to synthesize creatine that is needed for energy metabolism in muscle and nerve tissue. Methionine (thorough cysteine) together with glutamate and glycine are used in glutathione synthesis. Glutathione functions as an effective antioxidant in the cell. Taurine is made from methionine through cysteine. Taurine is an important component of bile acid and is also one of the effective antioxidants in the cell.

6.3 Tryptophan

In typical pig feeds, tryptophan is one of the major limiting amino acids together with lysine, methionine, and threonine. Feeds with distillers' dried grain with solubles (DDGS) should be carefully checked for tryptophan deficiency as corn protein is low in tryptophan. Supplemental tryptophan is most likely used in feeds to meet the tryptophan requirement to support the optimal growth of pigs. Tryptophan, however, is also used to provide important bioactive compounds needed in the normal function of the body. Tryptophan is used as a precursor for anthranilic acid, tryptamine, serotonin, melatonin, and niacin. Anthranilic acid is converted from tryptophan via kynurenine by indoleamine 2,3-dioxygenase (IDO), functioning in the suppression of T-cell activation and autoimmune reactions. Antigen-presenting cells are known to express IDO using tryptophan to produce anthranilic acid that is known to be important in maintaining pregnancy. Tryptamine has multiple effects on the body including vasoconstriction, increasing blood pressure, and neuromodulation. Tryptamine is synthesized from tryptophan by 5-OH-Trp decarboxylase. Serotonin is one of the major neurotransmitters synthesized from tryptophan by 5-OH-Trp decarboxylase causing depression of CNS activity and functioning as a sedative. Serotonin can be further metabolized to melatonin, especially in the pineal gland, but also in other body tissues. Melatonin has a central role in circadian rhythms. Through the kynurenine pathway, some tryptophan can also be converted to niacin, vitamin B3, with the role in energy metabolism as a part of nicotinamide adenine dinucleotide (NAD) and nicotinamide adenine dinucleotide phosphate (NADP). However, the efficiency of niacin synthesis from tryptophan is very low (1:60) and thus, tryptophan is not an effective precursor of niacin.

Tryptophan has been supplemented at high levels beyond its nutritional requirement in order to utilize its functional role in affecting the behavior and physiological response of pigs under stress conditions. Supplementation of L-tryptophan up to 0.6% increased the concentration of serotonin in the hypothalamus of nursery pigs reducing the concentration of cortisol and oxidative damage products, possibility affecting the growth performance of nursery pigs under social mixing stress (Shen et al., 2012a, 2015). When the amount of large neutral amino acids in the diet was reduced, the utilization of supplemental L-tryptophan by the hypothalamus was enhanced as they share to receptor to cross the blood-brain barrier (Shen et al., 2012b). When pigs were consuming a diet contaminated with deoxynivalenol, the catabolism of tryptophan in the hypothalamus was increased, reducing the concentration of serotonin (Shen et al., 2021), implying that the needs of tryptophan could be increased with the presence of mycotoxins.

6.4 Phenylalanine

Irreversible hydroxylation, occurring in the liver and kidneys, transforms phenylalanine into tyrosine (Fitzpatrick, 2012). Tyrosine acts as a precursor for various neurotransmitters, including melanin, catecholamines, and thyroid hormones (Meyers, 2000). Melanin, a pigment responsible for skin, hair, and eye color, is derived from phenylalanine and plays a role in protecting the skin from ultraviolet radiation (Sarangarajan and Apte, 2006). The synthesis of thyroid hormone, crucial for basal metabolism and growth, is influenced by the supply of phenylalanine (Li et al., 2012). This process involves the release of triiodothyronine and thyroxine.

6.5 Glutamine

Glutamine is the most abundant amino acid in the animal body and is considered one of non-essential amino acids. However, glutamine can become essential due to increased utilization by immune cells when the immune system is activated (Haynes et al., 2009; Wu et al., 2014; Duan et al., 2016). Glutamine serves as a major energy substrate for highly proliferating cells, especially in enterocytes and immunologically challenged lymphocytes (Curi et al., 2005). Glutamine acts as a precursor for the synthesis of purine and pyrimidine nucleotides, which are essential for cell proliferation (Gao et al., 2022). As a precursor of glutamate, glutamine plays a significant role in the synthesis of glutathione, demonstrating antioxidant properties in the intestinal cells (Haynes et al., 2009; Stipanuk et al., 2009). Glutamine has also been shown to increase protein synthesis in skeletal muscle and small intestinal mucosa through the activation of the mTOR signaling pathway, preventing intestinal atrophy in pigs (Zhu et al., 2015; Qin et al., 2018). Glutamine increases the production of NO in the cells (Yin et al., 2015), a process highly related to gaseous signaling molecules in the body (Li et al., 2009).

7 Practical aspects of feed formulation meeting amino acid needs

7.1 Ideal protein

To sustain a life and to support the growth and production, the animal body is required to make various proteins. All these proteins have their own unique amino acid sequences and compositions and thus supporting their synthesis requires all amino acids to be sufficiently present without any deficiency, i.e. limiting amino acids. If the amount of each amino acid needed for an animal is known, a feed can be formulated to meet the need, and this is the concept of ideal protein in formulating an animal feed.

A feed with the ideal protein provides ideally balanced amino acids to pigs maximizing the efficiency of protein synthesis and minimizing the loss of unused amino acids. The consideration of ideal protein in feed formulation can benefit pig production by reducing feed cost, reducing use of protein supplements in feeds, reducing nitrogen excretion from pig production, and reducing intestinal challenges especially for young pigs. In order to achieve ideal protein, multiple feedstuffs are used to adjust the amino acid profile and supplemental amino acids can provide specific limiting amino acids in the feed. The availability of an increased number of supplemental feed-grade amino acids allows pig nutritionists to formulate feeds with more precise ideal protein than before.

The ideal protein concept is introduced and applied in pig nutrition by ARC (1981) and further developed by Wang and Fuller (1989) and Chung and Baker (1992). For growing pigs, ideal protein is based on amino acid needs for the body maintenance and the needs for lean growth. Therefore, ideal protein can be affected by the stage of growth with different levels of amino acids needed for body maintenance and lean growth. In the case of breeding pigs, however, ideal protein is influenced by multiple factors including body maintenance, maternal growth (if any), fetal and conceptus tissue growth, mammary gland growth, milk protein production, and the contribution of amino acids from maternal tissue mobilization (if any). Ideal protein for pregnant pigs is influenced by the stage of pregnancy due to the growth rate of the fetus, conceptus, and mammary glands. Ideal protein for lactating pigs is influenced by the age of sows (or gilts) and their ability to eat (i.e. voluntary feed intake). The amounts and ratios of amino acids needed for milk production and mammary gland growth are different from the amounts and ratios of amino acids that can be contributed from maternal tissue mobilization. Thus, the amino acid contents and profiles in the feeds should be adjusted to balance these differences. Amino acid needs for milk production and mammary gland growth has metabolic priority to maternal tissues, and thus, any deficits in amino acid needs would be provided by the mobilization of maternal tissues if the dietary source is not sufficient. Application of the ideal protein should be based on the stage of pregnancy and the age of sows considering the dynamic changes of ideal protein during gestation and lactation.

7.2 Low-protein formulation

Consideration of ideal protein concept in feed formulation and the availability of supplemental amino acids allow the pig nutritionists to reduce the use of protein supplements resulting in reduced crude protein in pig feeds. Most protein supplements from plant sources contain more non-essential amino acids than what pigs need, whereas protein supplements from animal sources contain high levels of essential amino acids, however, their uses are limited due to high cost and regulations. The use of animal protein compared with the use of plant protein can reduce crude protein in the feed while meeting the requirements of essential amino acids if the feed cost and regulations are not a constraint.

Supplemental feed-grade amino acids provide opportunities to reduce the use of protein supplements both from plant and animal sources. Until recently, commercially available supplemental feed-grade amino acids include lysine, methionine, threonine, tryptophan, valine, isoleucine, histidine, arginine, and glutamine, and they are available in various forms (free or salt, and L- or DL-) as previously mentioned in 'Availability and use of supplemental amino acids.'

In pig feeds typically used in commercial production, the use of L-lysine HCl and L-threonine allows for the reduction of 1.4% crude protein which can be equal to the reduction of 8.5% soybean meal, as an example (Table 7). When DL- or L-methionine is used, crude protein can be further reduced by 2.7% with a reduction of soybean meal by 7.5%. When L-tryptophan, L-valine, and L-isoleucine are used in addition to the previous three amino acids, 2.3% crude protein can be further reduced. In modern pig production, L-lysine HCl, DL(L)-methionine, L-threonine, L-tryptophan, and L-valine are most likely used whereas L-isoleucine can also be used depending on the types of feedstuffs used.

Based on the reviews with multiple meta-analysis of data from existing literature, it has been shown that up to 0.55% and 0.31% L-Lys HCl can be used with other supplemental amino acid to reduce the use of protein supplements without negatively affecting the growth performance of nursery pigs and growing-finishing pigs, respectively (Rocha et al., 2022). When crude protein

Table 7 Examples of the use of supplemental amino acids and reduction of crude protein in pig feeds

%	SBM	+ Lys, Thr	+ Met	+ Trp, Val, Ile	+ His, Phe
Soybean meal	38.5	30.0	22.5	15.5	8.9
L-Lys HCl		0.10	0.32	0.54	0.75
L-Thr		0.12	0.22	0.32	0.41
DL-Met or L-Met			0.07	0.13	0.19
L-Trp				0.04	0.08
L-Val				0.10	0.21
L-Ile				0.04	0.12
L-His					0.07
L-Phe					0.12
Calculated composition, %					
Crude protein	23.1	19.9	17.2	14.9	12.7
SID Lys	0.98	0.98	0.98	0.98	0.98

was further reduced with an increased use of supplemental amino acids, growth and/or feed efficiency of pigs were reduced. It is speculated that extensive reduction of the use of protein supplements in feeds limits nitrogen utilization for the production of some non-essential amino acids and/or limits the supply of certain bioactive compounds included in protein supplements. However, these observations are mostly based on animal feeding studies with empirical approaches and predictions, whereas the exact mode of actions are still to be evaluated and validated.

7.3 Precision feeding

As previously discussed, the application of the ideal protein concept can maximize the efficiency of protein synthesis whilst minimizing the loss of unused amino acids, which has economic, physiological, and environmental implications. Considering modern, intensive farming practices, phase-feeding was once thought to be the best approach to minimizing the feeding of excess amino acids when meeting the ever-changing amino acid needs of pigs, especially in growing pigs. However, with recent developments in technology, the pig producers have been looking for an opportunity to maximize performance and minimize waste by achieving true ideal protein feeding for each individual pig.

The development of precision feeding has opportunities to improve efficiency in all areas of pig production. For growing pigs, rapid growth and development lead to daily changes in the amino acid requirements. It was thought that if a prediction model for the optimal amino acid composition and amount throughout the grow-finish phase could be established, the model could then be applied to an individual feeding system. The first instance of a precision feeding system developed for pigs was a prototype model-based feeding system that evaluated individual pigs daily to make weekly dietary adjustments to the provided feed based on the average needs of all the pigs in the pen (Parsons et al., 2007). The development of a real-time prediction model for the amino acid needs of growing pigs was later established utilizing both empirical and mechanistic approaches to estimate the optimal daily amino acid requirements of individual animals, using body weight and daily feed intake to estimate protein deposition and net energy intake, and thus the optimal standardized ileal digestible (SID) lysine was estimated in relation to net energy intake (Hauschild et al., 2012). The rest of the amino acids included in the model were estimated in relation to SID lysine. This and other similar prediction models have been employed with success in grow-finish pigs, generally having minimal impact on performance whilst decreasing feed costs and waste outputs (Andretta et al., 2014; Andretta et al., 2016; Cloutier et al., 2015; Andretta et al., 2018; Pomar et al., 2014).

Precision feeding also has applications in the feeding of gilts and sows. It is understood that the nutritional needs of gilts and sows vary greatly throughout development, pregnancy, and lactation and are also influenced by the stage of pregnancy and the parity (or age). To mitigate the changing nutritional needs throughout gestation, it is typical to provide two gestation diets, and some pig producers will also choose to provide a more nutrient-dense diet for their gestating gilts to account for the simultaneous maternal gain that occurs during the first pregnancy. By implementing precision feeding, nutritionists can supply the exact amino acids needed to support the current nutritional needs of each individual animal to improve fetal growth, maternal lactation and recovery, and subsequent reproductive performance whilst simultaneously minimizing waste. Research suggests that the precision provision of amino acids during gestation can improve offspring performance, namely in young sows, suggesting that the mass supply of a single gestation diet may be insufficient to support the nutritional needs for both maternal and fetal growth (Hansen et al., 2021). In addition to performance, precision feeding leads to decreased feeding costs and nutrient excretion in sows, similar to grow-finish pigs (Gaillard and Dourmad, 2022; Gauthier at al., 2022). Based on advancing technology, improved prediction models, and promising research, the use of precision feeding across the pig production proves to be a potential solution to maximizing productivity while minimizing cost as well as negative physiologic and environmental outcomes associated with feeding an excess of amino acids.

8 Conclusion

Amino acids are critically important nutrients for the survival and growth of pigs. Essential amino acids should be provided to pigs sufficiently to meet their requirements for body maintenance, growth, and/or production. Non-essential amino acids are mostly sufficient in typical pig feeds unless the formulation limits the use of protein supplements. The availability of supplemental amino acids allows the nutritionists to formulate pig feeds to match ideal protein with reduced use of protein supplements. However, extensive uses of supplemental amino acids may reduce feed efficiency of pigs. This restriction is possibly due to deficiency of nitrogen sources and/or limited supply of bioactive compounds from reduced use of protein supplements. Understanding the ways of effectively reducing dietary protein with supplemental amino acids warrants further research. Precise feed formulation meeting amino acid requirement would improve growth efficiency, reduce production cost, and enhance environmental sustainability of pig production. However, certain amino acids would have unique biological functions when they are provided sufficiently that is greater than nutritional requirements for protein synthesis.

These are called functional amino acids and their functional benefits have been extensively studied recently. The availability of an increased number of feed-grade supplemental amino acids, increased affordability, and advances in control and process technology provide exciting opportunities for pig production allowing low protein formulation and precision feeding.

9 Where to look for further information

For the complete overview of protein and amino acids as they relate to swine nutrition, see the *Nutrient Requirements of Swine: Eleventh Revised Edition* (see the References for full details).

Universities with an extension program focused in swine will provide locally relevant information regarding nutrient recommendations. As an example, the North Carolina State University Swine Extension Program offers information about nutrition on their website: https://swine.ces.ncsu.edu/.

For further information regarding the role of amino acids in reproducing swine, please see the following works by the author:

- Kim, S. W. and Easter, R. A. (2003). Amino acid utilization for reproduction in sows. In (ed. J. P. F. D'Mello) *Amino Acids in Animal Nutrition*. CABI Publishing, pp. 203–222.
- Kim, S. W. and Wu, G. (2009). Regulatory role for amino acids in mammary gland growth and milk synthesis. *Amino Acids* 37:89–95. doi: 10.1007/s00726-008-0151-5.

Information regarding optimal amino acid usage in swine diets is made available to the swine producer or lay audience through a variety of websites and can be found using a search of keywords. Examples of such websites include:

- https://porkcheckoff.org/pork-production-management/swine-health/swine-nutrition/.
- https://porkgateway.org/resources/category/swine-nutrition/ .
- https://www.pigprogress.net/health-nutrition/nutrition/.
- https://www.nationalhogfarmer.com/livestock-management/hog-nutrition.

10 References

Ahn, H., Gong, D. J., Lee, H. H., Seo, J. Y., Song, K. M., Eom, S. J. and Yeo, S. Y. (2021). Mechanical properties of porcine and fish skin-based collagen and conjugated collagen fibers, *Polymers* 13(13):2151. doi: 10.3390/polym13132151.

Andretta, I., Pomar, C., Rivest, J., Pomar, J., Lovatto, P. A. and Neto, J. R. (2014). The impact of feeding growing-finishing pigs with daily tailored diets using precision feeding techniques on animal performance, nutrient utilization, and body and carcass composition, *J. Anim. Sci.* 92(9):3925–3936. doi: 10.2527/jas.2014-7643.

Andretta, I., Pomar, C., Kipper, M., Hauschild, L. and Rivest, J. (2016). Feeding behavior of growing-finishing pigs reared under precision feeding strategies, *J. Anim. Sci.* 94(7):3042–3050. doi: 10.2527/jas.2016-0392.

Andretta, I., Hauschild, L., Kipper, M., Pires, P. G. S. and Pomar, C. (2018). Environmental impacts of precision feeding programs applied in pig production, *Animal* 12(9):1990–1998. doi: 10.1017/S1751731117003159.

ARC. (1981). *The Nutrient Requirements of Pigs*. Slough: Commonwealth Agricultural Bureaux.

Barea, R., Brossard, L., Le Floc'h, N., Primot, Y., Melchior, D. and van Milgen, J. (2009). The standardized ileal digestible valine-to-lysine requirement ratio is at least seventy percent in postweaned piglets, *J. Anim. Sci.* 87:935–947. doi: 10.2527/jas.2008-1006.

Carlsson, R. N. K., Ingvarsson, B. I. and Karlsson, B. W. (1977). Isolation and characterization of albumin from porcine serum, colostrum and urine, *Int. J. Biochem.* 8(4):285–294. doi: 10.1016/0020-711X(77)90134-3.

Cemin, H. S., Vier, C. M., Tokach, M. D., Dritz, S. S., Touchette, K. J., Woodworth, J. C., DeRouchey, J. M. and Goodband, R. D. (2018). Effects of standardized ileal digestible histidine to lysine ratio on growth performance of 7- to 11-kg nursery pigs, *J. Anim. Sci.* 96(11):4713–4722. doi: 10.1093/jas/sky319.

Chalvon-Demersay, T., Luise, D., Le Floc'h, N., Tesseraud, S., Lambert, W., Bosi, P., Trevisi, P., Beaumont, M. and Corrent, E. (2021). Functional amino acids in pigs and chickens: Implication for Gut health, *Front. Vet. Sci.* 8:663727. doi: 10.3389/fvets.2021.663727.

Chance, R. E., Ellis, R. M. and Bromer, W. W. (1968). Porcine proinsulin: Characterization and amino acid sequence, *Science* 161:165–167. doi: 10.1126/science.161.3837.165.

Chen, Y., Li, D., Dai, Z., Piao, X., Wu, Z., Wang, B., Zhu, Y. and Zeng, Z. (2014). L-methionine supplementation maintains the integrity and barrier function of the small-intestinal mucosa in post-weaning piglets, *Amino Acids* 46(4):1131–1142. doi: 10.1007/s00726-014-1675-5.

Cheng, Y. C., Lee, H. L., Hwang, Y. and Kim, S. W. (2023). The effects of SID His to Lys ratio on growth performance, intestinal health, and mobilization of histidine-containing proteins in pigs at 7 to 11 kg body weight, *J. Anim. Sci.* 100:skac396. doi: 10.1093/jas/skac396.

Chung, T. K. and Baker, D. H. (1992). Ideal amino acid pattern for 10-kilogram pigs, *J. Anim. Sci.* 70(10):3102–3111. doi: 10.2527/1992.70103102x.

Clizer, D. A., Tostenson, B. J., Frederick, B., Cline, P. M. and Samuel, R. S. (2023). Performance response of increasing the standardized ileal digestible tryptophan:lysine ratio in diets containing 40% dried distiller grains with solubles, *J. Anim. Sci.* 101:skad264. doi: 10.1093/jas/skad264.

Cloutier, L., Pomar, C., Létourneau Montminy, M. P., Bernier, J. F. and Pomar, J. (2015). Evaluation of a method estimating real-time individual lysine requirements in two lines of growing-finishing pigs, *Animal* 9(4):561–568. doi: 10.1017/S1751731114003073.

Conde-Aguilera, J. A., Cobo-Ortega, C., Mercier, Y., Tesseraud, S. and van Milgen, J. (2014). The amino acid composition of tissue protein is affected by the total sulfur amino acid supply in growing pigs, *Animal* 8(3):401–409. doi: 10.1017/S1751731113002425.

Curi, R., Lagranha, C. J., Doi, S. Q., Sellitti, D. F., Procópio, J., Pithon-Curi, T. C., Corless, M. and Newsholme, P. (2005). Molecular mechanisms of glutamine action, *J. Cell. Physiol.* 204(2):392–401. doi: 10.1002/jcp.20339.

Duan, J., Yin, J., Ren, W., Liu, T., Cui, Z., Huang, X., Wu, L., Kim, S. W., Liu, G., Wu, X., Wu, G., Li, T. and Yin, Y. (2016). Dietary supplementation with L-glutamate and L-aspartate alleviates oxidative stress in weaned piglets challenged with hydrogen peroxide, *Amino Acids* 48(1):53–64. doi: 10.1007/s00726-015-2065-3.

Fitzpatrick, P. F. (2012). Allosteric regulation of phenylalanine hydroxylase, *Arch. Biochem. Biophys.* 519(2):194–201. doi: 10.1016/j.abb.2011.09.012.

Fogg, F. J. J., Huyyon, D. A., Jumel, K., Pearson, J. P., Harding, S. E. and Allen, A. (1996). Characterization of pig colonic mucins, *Biochem. J.* 316(3):937–942. doi: 10.1042/bj3160937.

Gao, L. M., Liu, G. Y., Wang, H. L., Wassie, T. and Wu, X. (2022). Maternal pyrimidine nucleoside supplementation regulates fatty acid, amino acid and glucose metabolism of neonatal piglets, *Anim. Nutr.* 11:309–321. doi: 10.1016/j.aninu.2022.07.011.

Grant, R. A. (1965). Variations in the amino acid composition of Aortic elastin from different species, *Br. J. Exp. Pathol.* 47(2):163–167. PMID: 5943817; PMCID: PMC2094569.

Hauschild, L., Lovatto, P. A., Pomar, J. and Pomar, C. (2012). Development of sustainable precision farming systems for swine: Estimating real-time individual amino acid requirements in growing-finishing pigs, *J. Anim. Sci.* 90(7):2255–2263. doi: 10.2527/jas.2011-4252.

Haynes, T. E., Li, P., Li, X., Shimotori, K., Sato, H., Flynn, N. E., Wang, J., Knabe, D. A. and Wu, G. (2009). L-glutamine or L-alanyl-L-glutamine prevents oxidant-or endotoxin-induced death of neonatal enterocytes, *Amino Acids* 37(1):131–142. doi: 10.1007/s00726-009-0243-x.

Hodgkinson, S. M., Moughan, P. J., Reynolds, G. W. and James, K. A. C. (2000). The effect of dietary peptide concentration on endogenous ileal amino acid loss in the growing pig, *Br. J. Nutr.* 83(4):421–430. PMID: 10858700.

Htoo, J. K., Zhu, C. L., Huber, L., de Lange, C. F., Quant, A. D., Kerr, B. J., Cromwell, G. L. and Lindemann, M. D. (2014). Determining the optimal isoleucine:lysine ratio for ten- to twenty-two-kilogram and twenty-four- to thirty-nine-kilogram pigs fed diets containing nonexcess levels of leucine, *J. Anim. Sci.* 92(8):3482–3490. doi: 10.2527/jas.2013-6934.

Izquierdo, O. A., Wedekind, K. J. and Baker, D. H. (1988). Histidine requirement of the young pig, *J. Anim. Sci.* 66(11):2886–2892. doi: 10.2527/jas1988.66112886x.

Jayaraman, B., Htoo, J. K. and Nyachoti, C. M. (2015). Effects of dietary threonine:lysine ratioes and sanitary conditions on performance, plasma urea nitrogen, plasma-free threonine and lysine of weaned pigs, *Anim. Nutr.* 1(4):283–288. doi: 10.1016/j.aninu.2015.09.003.

Ji, F., Wu, G., Blanton, J. R. Jr. and Kim, S. W. (2005a). Changes in weight and composition in various tissues of pregnant gilts and their nutritional implications, *J. Anim. Sci.* 83(2):366–375. doi: 10.2527/2005.832366x.

Ji, F., Wu, G., Blanton, J. R. and Kim, S. W. (2005b). Weight and compositional changes in pregnant gilts and its implication to nutrition, *J. Anim. Sci.* 83(2):366–375. doi: 10.2527/2005.832366x.

Ji, F., Hurley, W. L. and Kim, S. W. (2006). Characterization of mammary gland development in pregnant gilts, *J. Anim. Sci.* 84(3):579–587. doi: 10.2527/2006.843579x.

Kim, S. W. (2010). Recent advances in sow nutrition, *Rev. Bras. Zootecn.* 39:303–310. doi: 10.1590/S1516-35982010001300033.

Kim, S. W. and Easter, R. A. (2001). Nutrient mobilization from body tissues as influenced by litter size in lactating sows, *J. Anim. Sci.* 79(8):2179–2186. doi: 10.2527/2001.7982179x.

Kim, S. W. and Wu, G. (2004). Dietary arginine supplementation enhances the growth of milk-fed young piglets, *J. Nutr.* 134(3):625–630. doi: 10.1093/jn/134.3.625.

Kim, S. W., Hurley, W. L., Han, I. K. and Easter, R. A. (1999a). Changes in tissue composition associated with mammary gland growth during lactation in the sow, *J. Anim. Sci.* 77(9):2510–2516. doi: 10.2527/1999.7792510x.

Kim, S. W., Hurley, W. L., Han, I. K., Stein, H. H. and Easter, R. A. (1999b). Effect of nutrient intake on mammary gland growth in lactating sows, *J. Anim. Sci.* 77(12):3304–3315. doi: 10.2527/1999.77123304x.

Kim, S. W., Osaka, I., Hurley, W. L. and Easter, R. A. (1999c). Mammary gland growth as affected by litter size in lactating sows: Impact on lysine requirement, *J. Anim. Sci.* 77(12):3316–3321. doi: 10.2527/1999.77123316x.

Kim, S. W., Mateo, R. D., Yin, Y. -L. and Wu, G. (2007). [AJAS Outstanding Research Award] Functional amino acids and fatty acids for enhancing production performance of sows and piglets, *Asian-Australas. J. Anim. Sci.* 20(2):295–306. doi: 10.5713/ajas.2007.295.

Kim, S. W., Hurley, W. L., Wu, G. and Ji, F. (2009). [Invited: Early career achievement award] Ideal amino acid balance for sows during gestation and lactation, *J. Anim. Sci.* 87:E123–E132. doi: 10.2527/jas.2008-1452.

Kim, S. W., Baker, D. H. and Easter, R. A. (2014). Dynamic ideal protein and limiting amino acids for lactating sows: The impact of amino acid mobilization, *J. Anim. Sci.* 79(9):2356–2366. doi: 10.2527/2001.7992356x.

Le Floc'h, N., Wessels, A., Corrent, E., Wu, G. and Bosi, P. (2018). The relevance of functional amino acids to support the health of growing pigs, *Anim. Feed Sci. Technol.* 245:104–116. doi: 10.1016/j.anifeedsci.2018.09.007.

Li, Q., Mair, C., Schedle, K., Hammerl, S., Schodl, K. and Windisch, W. (2012). Effect of iodine source and dose on growth and iodine content in tissue and plasma thyroid hormones in fattening pigs, *Eur. J. Nutr.* 51(6):685–691. doi: 10.1007/s00394-011-0247-7.

Li, X., Bazer, F. W., Gao, H., Jobgen, W., Johnson, G. A., Li, P., McKnight, J. R., Satterfield, M. C., Spencer, T. E. and Wu, G. (2009). Amino acids and gaseous signaling, *Amino Acids* 37(1):65–78. doi: 10.1007/s00726-009-0264-5.

Li, Y., Zhang, H., Chen, Y. P., Ying, Z. X., Su, W. P., Zhang, L. L. and Wang, T. (2017). Effects of dietary l-methionine supplementation on the growth performance, carcass characteristics, meat quality, and muscular antioxidant capacity and myogenic gene expression in low birth weight pigs, *J. Anim. Sci.* 95(9):3972–3983. doi: 10.2527/jas2017.1652.

Lin, C., Mahan, D. C., Wu, G. and Kim, S. W. (2009). Protein digestibility of colostrums by neonatal pigs, *Livest. Sci.* 121(2–3):182–186. doi: 10.1016/j.livsci.2008.06.006.

Mateo, R. D., Wu, G., Moon, H. K., Carroll, J. A. and Kim, S. W. (2008). Effects of dietary arginine supplementation during gestation and lactation on the performance of lactating primiparous sows and nursing piglets, *J. Anim. Sci.* 86(4):827–835. doi: 10.2527/jas.2007-0371.

Mateo, R. D., Carroll, J. A., Hyun, Y., Smith, S. and Kim, S. W. (2009). Effect of dietary supplementation of omega-3 fatty acids and high protein on reproductive outcome of primiparous sows for two parities, *J. Anim. Sci.* 87:948–959. doi: 10.1016/j.theriogenology.2017.05.023.

McPherson, R. L., Ji, F., Wu, G., Blanton, J. R. Jr. and Kim, S. W. (2004). Fetal growth and compositional changes of fetal tissues in the pigs, *J. Anim. Sci.* 82(9):2534–2540. doi: 10.2527/2004.8292534x.

Meyers, S. (2000). Use of neurotransmitter precursors for treatment of depression, *Altern. Med. Rev.* 5(1):64–71. PMID: 10696120.

Morales, A., Sánchez, V., Pérez, B., Camacho, R. L., Arce, N., Avelar, E., González-Vega, J. C., Htoo, J. K. and Cervantes, M. (2022). Effect of DL-methionine supplementation above requirement on performance; intestinal morphology, antioxidant activity, and gene expression; and serum concentration of amino acids in heat stressed pigs, *J. Anim. Sci.* 101:skac379. doi: 10.1093/jas/skac379.

Moran, E. T., Summers, J. D. and Slinger, S. J. (1967). Keratins as sources of protein for the growing chick: 2. Hog hair, a valuable source of protein with appropriate processing and amino acid balance, *Poult. Sci.* 46(2):456–465. doi: 10.3382/ps.0460456.

Morel, P. C. H., Padilla, R. M. and Ravindran, G. (2003). Effect of non-starch polysaccharides on mucin secretion and endogenous amino acid losses in pigs, *Asian-Australas. J. Anim. Sci.* 16(9):1332–1338. doi: 10.5713/ajas.2003.1332.

Moter, V. and Stein, H. H. (2004). Effect of feed intake of endogenous losses and amino acid and energy digestibility by growing pigs, *J. Anim. Sci.* 82(12):3518–3525. doi: 10.2527/2004.82123518x.

NRC. (2012). *Nutrient Requirements of Swine*. Eleventh Revised Edition, 12th ed. Washington, DC: The National Academies Press.

Opapeju, F. O., Htoo, J. K., Dapoza, C. and Nyachoti, C. M. (2012). Bioavailability of methionine hydroxy analog-calcium salt relative to DL-methionine to support nitrogen retention and growth in starter pigs, *Animal* 6(11):1750–1756. doi: 10.1017/S1751731112000869.

Parsons, D. J., Green, D. M., Schofield, C. P. and Whittemore, C. T. (2007). Real-time control of pig growth through an integrated management system, *Biosyst. Eng.* 96(2):257–266. doi: 10.1016/j.biosystemseng.2006.10.013.

Pomar, C., Pomar, J., Dubeau, F., Joannopoulos, E. and Dussault, J. P. (2014). The impact of daily multiphase feeding on animal performance, body composition, nitrogen and phosphorus excretions, and feed costs in growing-finishing pigs, *Animal* 8(5):704–713. doi: 10.1017/S1751731114000408.

Presto, M. H., Lyberg, K. and Lindberg, J. E. (2010). Effect of body weight on ileal endogenous nitrogen and amino acid loss in PVTC-cannulated pigs, *Livest. Sci.* 134(1–3):18–20. doi: 10.1016/j.livsci.2010.06.084.

Qin, Q., Xu, X., Wang, X., Wu, H., Zhu, H., Hou, Y., Dai, B., Liu, X. and Liu, Y. (2018). Glutamate alleviates intestinal injury, maintains mTOR and suppresses TLR4 and NOD signaling pathways in weanling pigs challenged with lipopolysaccharide, *Sci. Rep.* 8(1):15124. doi: 10.1038/s41598-018-33345-7.

Rocha, G. C., Duarte, M. E. and Kim, S. W. (2022). Advances, implications, and limitations of low-crude-protein diets in pig production, Animals (Basel) 12(24):3478. doi: 10.3390/ani12243478.

Rodrigues, L. A., Wellington, M. O., González-Vega, J. C., Htoo, J. K., Van Kessel, A. G. and Columbus, D. A. (2021). Functional amino acid supplementation, regardless of dietary protein content, improves growth performance and immune status of weaned pigs challenged with Salmonella Typhimurium, *J. Anim. Sci.* 99(2):skaa365. doi: 10.1093/jas/skaa365.

Sarangarajan, R. and Apte, S. P. (2006). The polymerization of melanin: A poorly understood phenomenon with egregious biological implications, Melanoma Res. 16(1):3–10. doi: 10.1097/01.cmr.0000195699.35143.df.

Shen, Y. B., Voilque, G., Odle, J. and Kim, S. W. (2012a). Effects of elevating tryptophan intake on growth and hypothalamic serotonin secretion and stress hormone secretion in nursery pigs, *J. Anim. Sci.* 90:2264–2275. doi: 10.2527/jas.2011-4203.

Shen, Y. B., Voilque, G., Odle, J. and Kim, S. W. (2012b). Dietary L-tryptophan supplementation with reduced large neutral amino acids enhances feed efficiency and reduces stress hormone secretion of nursery pigs under social-mixing stress, *J. Nutr.* 142(8):1540–1546. doi: 10.3945/jn.112.163824.

Shen, Y. B., Weaver, A. C. and Kim, S. W. (2014). Effect of feed grade L-methionine on growth performance and gut health in nursery pigs compared with conventional DL-methionine, *J. Anim. Sci.* 92(12):5530–5539. doi: 10.2527/jas.2014-7830.

Shen, Y. B., Coffey, M. T. and Kim, S. W. (2015). Effects of short term supplementation of L-tryptophan and reducing large neutral amino acid along with L-tryptophan supplementation on growth and stress response in pigs, *Anim. Feed Sci. Technol.* 207:245–252. doi: 10.1016/j.anifeedsci.2015.06.020.

Shen, Y. B., Weaver, A. C. and Kim, S. W. (2021). Physiological effects of deoxynivalenol from naturally contaminated corn on cerebral tryptophan metabolism, behavioral response, gastrointestinal immune status and health in pigs following a pair-feeding model, *Toxins* 13(6):393. doi: 10.3390/toxins13060393.

Shoveller, A. K., Moehn, S., Rademacher, M., Htoo, J. K. and Ball, R. O. (2010). Methionine-hydroxy analogue was found to be significantly less bioavailable compared to DL-methionine for protein deposition in growing pigs, *Animal* 4(1):61–66. doi: 10.1017/S1751731109990917.

Soumeh, E. A., van Milgen, J., Sloth, N. M., Corrent, E., Poulsen, H. D. and Nørgaard, J. V. (2015). Requirement of standardized ileal digestible valine to lysine ratio for 8- to 14-kg pigs, *Animal* 9(8):1312–1318. doi: 10.1017/S1751731115000695.

Stipanuk, M. H., Ueki, I., Dominy, J. E., Simmons, C. R. and Hirschberger, L. L. (2009). Cysteine dioxygenase: A robust system for regulation of cellular cysteine levels, *Amino Acids* 37(1):55–63. doi: 10.1007/s00726-008-0202-y.

Su, W., Zhang, H., Ying, Z., Li, Y., Zhou, L., Wang, F., Zhang, L. and Wang, T. (2018). Effects of dietary l-methionine supplementation on intestinal integrity and oxidative status in intrauterine growth-retarded weanling piglets, *Eur. J. Nutr.* 57(8):2735-2745. doi: 10.1007/s00394-017-1539-3.

Tang, J., Sepulveda, P., Marciniszyn, J. Jr., Chen, K. C., Huang, W. Y., Tao, N., Liu, D. and Lanier, J. P. (1973). Amino-acid sequence of porcine pepsin, *Proc. Natl Acad. Sci. U. S. A.* 70(12):3437–3439. doi: 10.1073/pnas.70.12.3437.

Walker, J. E. and Keil, B. (1973). Purification and characterization of different active form of pork trypsin, *Eur. J. Biochem.* 32(3):486–491. doi: 10.1111/j.1432-1033.1973.tb02632.x.

Wang, T. C. and Fuller, M. F. (1989). The optimum dietary amino acid pattern for growing pigs: 1. Experiments by amino acid deletion, *Br. J. Nutr.* 62(1):77–89. doi: 10.1079/BJN19890009.

Wang, M. Q., Huyen, T. T., Lee, J. W., Ramos, S. H., Htoo, J. K., Kinh, V. and Lindemann, M. D. (2020). Bioavailability of the calcium salt of DL-methionine hydroxy analog compared with DL-methionine for nitrogen retention and the preference of nursery pigs for diets based on the 2 forms of methionine, *J. Anim. Sci.* 98(12):skaa349. doi: 10.1093/jas/skaa349.

Wu, G. and Kim, S. W. (2007). Functional amino acids in animal production. In Pond, W. G. and Bell, A. W. (Eds.), *Encyclopedia of Animal Science*. Taylor & Francis Publication. ISBN: 0824754964 and 0824754956.

Wu, M., Xiao, H., Ren, W., Yin, J., Hu, J., Duan, J., Liu, G., Tan, B., Xiong, X., Oso, A. O., Adeola, O., Yao, K., Yin, Y. and Li, T. (2014). An NMR-based metabolomic approach to investigate the effects of supplementation with glutamic acid in piglets challenged with deoxynivalenol, *PLoS One* 9(12):e113687. doi: 10.1371/journal.pone.0113687.

Yang, Z., Hasan, M. S., Htoo, J. K., Burnett, D. D., Feugang, J. M., Crenshaw, M. A. and Liao, S. F. (2019). Effects of dietary supplementation of l-methionine vs. DL-methionine on performance, plasma concentrations of free amino acids and other metabolites, and myogenesis gene expression in young growing pigs, *Transl. Anim. Sci.* 3(1):329–339. doi: 10.1093/tas/txy109.

Yin, J., Liu, M., Ren, W., Duan, J., Yang, G., Zhao, Y., Fang, R., Chen, L., Li, T. and Yin, Y. (2015). Effects of dietary supplementation with glutamate and aspartate on diquat-induced oxidative stress in piglets, *PLoS One* 10(4):e0122893. doi: 10.1371/journal.pone.0122893.

Zhang, H., Li, Y., Chen, Y., Ying, Z., Su, W., Zhang, T., Dong, Y., Htoo, J. K., Zhang, L. and Wang, T. (2019). Effects of dietary methionine supplementation on growth performance, intestinal morphology, antioxidant capacity and immune function in intra-uterine growth-retarded suckling piglets, *J. Anim. Physiol. Anim. Nutr.* 103(3):868–881. doi: 10.1111/jpn.13084.

Zhou, H., Yuan, Z., Chen, D., Wang, H., Shu, Y., Gao, J., Htoo, J. K. and Yu, B. (2021). Bioavailability of the DL-methionine and the calcium salt of DL-methionine hydroxy analog compared with l-methionine for nitrogen retention in starter pigs, *J. Anim. Sci.* 99(6):skab151. doi: 10.1093/jas/skab151.

Zhu, Y., Lin, G., Dai, Z., Zhou, T., Li, T., Yuan, T., Wu, Z., Wu, G. and Wang, J. (2015). Lglutamine deprivation induces autophagy and alters the mTOR and MAPK signaling pathways in porcine intestinal epithelial cells, *Amino Acids* 47(10):2185–2197. doi: 10.1007/s00726-014-1785-0.

Chapter 8

Understanding and optimizing the use of prebiotics in pig nutrition

Knud Erik Bach Knudsen, Élisabeth Chassé, Niloofar Razmgah, Mette Skou Hedemann and Nuria Canibe, Aarhus University, Denmark

1 Introduction

The diverse collection of microorganisms (termed 'the microbiota') in the gastrointestinal (GI) tract of pigs performs numerous roles. It is involved in the digestion and fermentation of carbohydrates and other nutrients, production of vitamins, maintenance of normal functionality of the intestinal villi, regulation of the immune response and protection from pathogenic bacteria (Buffie and Pamer, 2013; Kamada et al., 2013). The gut microbiota is a complex dynamic ecosystem whose composition shifts over time and along the entire GI tract (Kim et al., 2011; Kim and Isaacson, 2015; Zhao et al., 2015).

The colonization of the GI tract is initiated at birth, with the mother's milk creating a milk-oriented microbiome characterized by a high density of lactic acid bacteria. This changes gradually over the suckling period (Isaacson and Kim, 2012). The most drastic change occurs in connection with the weaning process when the piglets are shifted abruptly from a liquid milk-based diet to a solid plant-based diet rich in carbohydrates and proteins. In the immediate post-weaning period, a low and variable feed intake and unstable environment

http://dx.doi.org/10.19103/AS.2024.0140.12

give rise to a highly unstable microbial community. It typically takes 5–10 days for the intestinal bacterial community to adapt to a new feeding situation and to stabilize the concentration of short-chain fatty acids (SCFAs) in the colonic content (Janczyk et al., 2007; Pieper et al., 2008).

Several studies conducted during the weaning transition have reported a decrease in bacteria belonging to the *Lactobacillus* group and a reduction in microbial diversity, while *Clostridium* spp., *Prevotella* spp. or facultative anaerobes such as Proteobacteriaceae, including *E. coli*, tend to increase (Gresse et al., 2017). While members of the lactobacilli family are the dominating bacterial groups in post-weaned piglets, there is a gradual increase in Firmicutes and a decrease in Bacteroidetes as the pigs get older (Kim et al., 2011).

Overall, the GI microbiota is influenced by several factors including genetics, host physiology (age, disease, stress, etc.) and environmental factors such as diet and medication. Diet is a key environmental factor influencing the composition and metabolic function of the GI microbiota. In this context, consumption of specific diets and ingredients such as dietary fiber, prebiotics and probiotics can modulate the microbiota.

In this chapter, we discuss the use of prebiotics in the modulation of the GI microbiota and the consequences this may have for pig health. The prebiotics we discuss are limited to carbohydrate-derived prebiotics, with the exception of human milk oligosaccharides and the potential polyphenols prebiotic group (Gibson et al., 2017). Other groups, proteins, protein hydrolysates, peptides, amino acids and nucleotides, which are also claimed to have prebiotic effects (Kiernan et al., 2023), are not covered.

2 Dietary fiber and prebiotics

Conventional diets for pigs are based on carbohydrate- and protein-rich feedstuffs. Carbohydrate-rich feedstuffs are typically derived from cereals and by-products, which differ from region to region – maize and associated products in the USA, South America and Asia and wheat, barley, rye and oats in Europe, Canada and Oceania. A similar diversity exists for protein-rich feedstuffs, with soybean meal dominant in the US and Asia, while a diversity of others such as rape seed meal or cake, peas, fava beans, sunflower meal or cake, lupins, etc. are used elsewhere.

It is well known that the main factor regulating microbiota activity is the content and composition of the amount of dietary fiber (DF) (Bach Knudsen et al., 2023). DF is defined as the sum of non-digestible oligosaccharides (NDOs), resistant starch (RS), non-starch polysaccharides (NSPs) and lignin (Cummings et al., 2009; Phillips and Cui, 2011). NSP is quantitatively the most important compound, followed by RS and/or NDO, depending on diet composition (Bach Knudsen and Lærke, 2018; Bach Knudsen et al., 2023). The complex NSP fraction is the main component of the cell walls in which cereals provide an array of arabinoxylans and mixed β(1,3;1,4)-glucan, while protein-rich feedstuffs provide an array of pectins, and both feedstuff types provide cellulose. The RS in conventional feedstuffs is usually a result of incomplete digestion of starch and derives mostly from cereals, except when peas and fava beans are major dietary ingredients (Bach Knudsen et al., 2023). NDO derives primarily from protein-rich feedstuffs in the form of raffinose oligosaccharides, but fructans from cereals also contribute to the NDO content in compound feeds (Bach Knudsen and Lærke, 2018).

Almost 30 years ago, a class of compounds, later termed 'prebiotics', were recognized for their specific ability to manipulate host microbiota to the benefit of the host (Gibson and Roberfroid, 1995; Gibson et al., 2017). These studies were originally primarily performed on human subjects but were later also carried out on production and companion animals (Gibson et al., 2017). At that time, fructans (fructo-oligosaccharides (FOS) and inulin) and galactans (galacto-oligosaccharides or GOS) fitted that category, with their effects acting through enrichment of *Lactobacillus* spp. and/or *Bifidobacterium* spp. (Roberfroid et al., 2010b; Gibson et al., 2017).

Although dietary oligosaccharides had already been used to benefit human health, particularly in Asia, it was not until 1995 that the prebiotic concept was defined as a 'non-digestible food ingredient that beneficially affects the host by selectively stimulating the growth and/or activity of one or a limited number of bacteria already resident in the colon' (Gibson and Roberfroid, 1995). The prebiotic concept was originally built on the probiotic concept defined in 2001 and reaffirmed in 2014 (FAO & WHO, 2006; Hill et al., 2014). However, while probiotics use live microorganisms, prebiotics are non-viable substances that serve as nutrients for beneficial indigenous (resident) microorganisms harbored by the host and probiotic strains.

The definition of prebiotic has been modified on several occasions (for an overview see Gibson et al. (2017)) and most recently updated by the International Scientific Association for Probiotics and Prebiotics, which defines prebiotics as 'a substrate that is selectively utilized by host microorganisms conferring a health benefit' (Gibson et al., 2017). The recent definition is broader than the original as it expands the concept of prebiotics to include non-carbohydrate substances, application to body sites other than the GI tract and diverse categories other than food (Fig. 1). The expansion in the prebiotic concept is partly due to advances in tools for microbiome research, which have improved our knowledge of the composition of microbiota. In addition, advances in new technologies have also made it clear that a broader range of

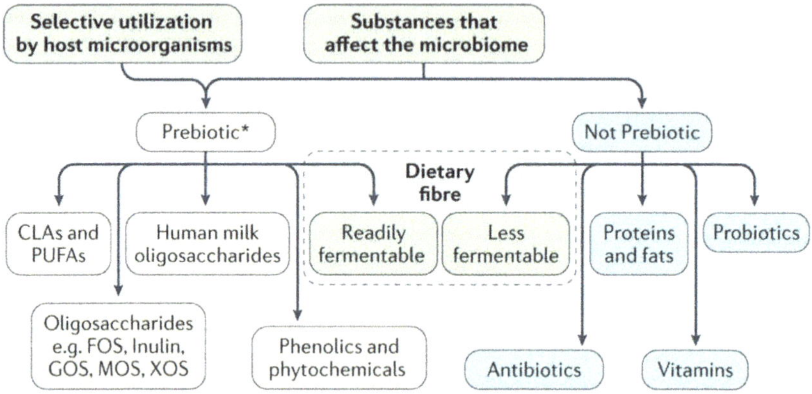

Figure 1 Distinguishing what is considered a prebiotic with the International Association for Probiotics and Prebiotics consensus statement on the definition and scope of prebiotics. *The figure shows candidate as well as accepted prebiotics, in that levels of evidence currently vary, with FOS and GOS being the most researched prebiotics. CLA, conjugated linoleic acid; PUFA, polysaturated fatty acids; FOS, fructo-oligosaccharides; GOS, galacto-oligosaccharides; MOS, mannanoligosaccharides; XOS, xylo-oligosaccharides. Source: Adapted from: Gibson et al. (2017).

bacteria than *Lactobacillus* spp. and/or *Bifidobacterium* spp. are affected by prebiotics (Pandey et al., 2023; Zhou et al., 2023).

3 Types of prebiotics: oligosaccharides, polysaccharides and non-carbohydrate prebiotics

3.1 Oligosaccharides

The best-known prebiotics are FOS, inulin-type fructans and GOS, as evidenced by numerous studies across species (Roberfroid et al., 2010b; Gibson et al., 2017). Inulin-type fructans are commonly isolated from chicory root for commercial use (Kelly, 2009) and comprise a heterogenous group of oligomers and polymers consisting of linear chains of β(2-1)-linked fructose monomers. The chains vary in degree of polymerization (DP) and can have either a glucose (GF series) or fructose (F series) moiety at the reducing end (Kelly, 2009). Inulin-type fructans are predominantly the inulin-type GF series with DP 2-60 but can also be further hydrolyzed to obtain FOS with DP 2-10 (Singh and Singh, 2010) or separated physically to obtain long-chain inulin (DP 10-60) (lcFOS).

GOS are produced commercially by fungal, yeast or bacterial β-galactosidases, which catalyze the hydrolysis and transgalactosylation of lactose (Torres et al., 2010). This process results in the production of a highly complex mixture of galactose chains with a terminal reducing glucose or galactose unit, varying in type of glycosidic linkage (β(1-3), β(1-4), β(1-6), β(1-2) and DP (Torres et al., 2010)) and with chains that can be either linear or branched with a disubstituted terminal reducing glucose or galactose unit (Coulier et al., 2009). Non-reducing GOS containing a β(1-1)-linkage between the acceptor's terminal glucose residue and the enzymatically attached galactose residue have also been reported (Fransen et al., 1998; Coulier et al., 2009).

Beyond FOS, inulin-type fructans and GOS, several other oligosaccharides have potential prebiotic effects. Pectin oligosaccharides (POS) have promising prebiotic effects (Onumpai et al., 2011; Gomez et al., 2014) and can be obtained by hydrolysis of pectin, mostly from associated products such as sugar beet pulp and citrus peels. Pectin has high structural complexity (Caffall and Mohnen, 2009), which leads to the formation of a broad range of oligosaccharides upon hydrolysis. These include arabinogalacto-oligosaccharides, galacto-oligosaccharides and arabino-oligosaccharides and acidic oligosaccharides such as oligo-galacturonides and rhamnogalacturonan-oligosaccharides. These oligosaccharides can be further modified with methyl- or acetyl ester groups (Babbar et al., 2016).

Hemicelluloses can also be a source of oligosaccharides when exposed to hydrolysis with acids or enzymes (Falck et al., 2014; Rios-Rios et al., 2021; Yan et al., 2022). Upon hydrolysis of xylan from straw and cornstalks, xylo-oligosaccharides (XOS) are produced that are comprised of linear chains of β(1-4)-linked xylose units ranging from DP 2-10 (Moniz et al., 2016; Rios-Rios et al., 2021). Another source of hemicellulose-derived oligosaccharides is arabinoxylan-oligosaccharides (AXOS), which can be produced by the degradation of arabinoxylan from a product such as wheat bran, resulting in linear chains of β(1-4)-linked xylan units substituted with α-L-arabinofuranose units via a α(1-2) and/or α(1-3)-linkage (Falck et al., 2014).

Other plant-based non-digestible carbohydrates of interest include isomalto/malto-polysaccharides (IMMP) and isomalto-oligosaccharide (IMO) preparations, which are enzymatically produced from potato starch (Hu et al., 2013; Leemhuis et al., 2014). IMMP are polysaccharides comprised of mainly α(1-4)- and α(1-6)-linked glucose moieties, whereas an IMO preparation is a highly complex mixture of α-gluco-oligosaccharides varying in terms of glycosidic linkage and size. Mannose oligosaccharides (MOS) can be produced by enzymatic conversion of mannan-rich plant-waste biomass into oligosaccharides (Hlalukana et al., 2021) or a glucomannan complex deriving from the cell walls of yeast (Agazzi et al., 2020).

3.2 Polysaccharides

Some polysaccharides are claimed to have prebiotic effects such as β-glucan, particularly from yeast, algae and fungi, RS and pectin. The composition of the β-glucan found in yeast, algae and fungi is diverse, with the bonds present in bacteria being predominantly β(1-3) linkages, in yeast, algae (laminarin) and fungi β(1-3) with β(1-6) branches, which can be compared with the β(1-3) and β(1-4) linkages present in cereals (Wood et al., 2003; Izydorczyk and Dexter, 2008). There are also differences in the ratio of the bonds and branches between the different β-glucan sources as seen for yeast and laminarin (Manners et al., 1973). The β-glucans found in fungi are insoluble and large in size with a molecular weight of 100 000 to 500 000 Da (Giavasis, 2014).

Resistant starch has a similar structure to digestible starch with α(1-4) bonds found in amylose and α(1-4) and α(1-6) bonds found in amylopectin. The delimitation between digestible and RS is that RS represents a fraction of starch that is not digested up to the end of the small intestine (Asp, 1992). Pectin represents a heterogeneous group of polysaccharides comprised of as many as 17 different monosaccharides (Caffall and Mohnen, 2009; Zdunek et al., 2021). The main structural elements include homogalacturonan, rhamnogalacturonan types I and II, glucuronoxylan and arabinogalactans types I and II, which vary depending on their source (Caffall and Mohnen, 2009; Zdunek et al., 2021).

3.3 *Non-carbohydrate prebiotics*

Plant polyphenols constitute a diverse class of compounds found in most plant tissues and are present at various concentrations in plants used for animals (Tomás-Barberán et al., 2021). They are often termed 'phytochemicals' and there is accumulating evidence that some phenolic compounds can meet the criteria of prebiotics - although more studies in target hosts are required (Gibson et al., 2017).

4 Modes of actions of prebiotics

The ability to selectively stimulate specific groups of microorganisms is the way in which the effects of prebiotics on the microbiota differ from those of most cell wall components such as cellulose, β-glucan, arabinoxylan and pectin (Gibson et al., 2017). Due to the complex nature of the cell wall polysaccharides, a mix of glycosidic hydrolases (GHs), polysaccharide lyases (PLs) and carbohydrate esterases (CEs) is required to hydrolyse the bonds in the cell walls (Flint et al., 2008; Flint et al., 2012a). A common feature of DF is therefore the ability to stimulate microbiota activity in a broad sense, in contrast to prebiotics, which are metabolized by a limited number of microorganisms and with broader effects on the GI environment (Fig. 2).

The best and most documented prebiotics are fructans and galactans (Rastall and Gibson, 2015) but other oligosaccharides appear to have similar effects such as XOS and plant-based MOS. A common feature of these oligosaccharides is that they are preferentially metabolized by bifidobacteria and lactobacilli (Roberfroid et al., 2010). This phenomenon is linked to the

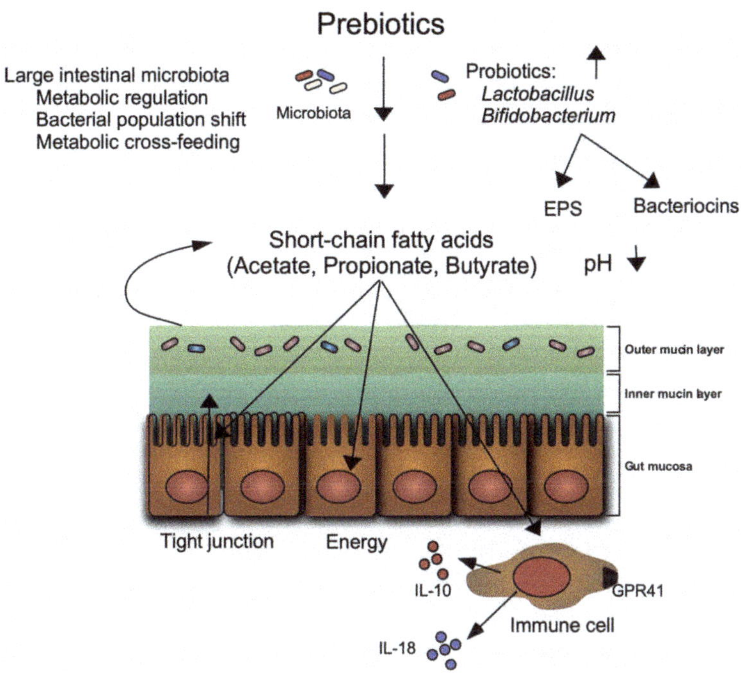

Figure 2. Fermentation of prebiotics and potential effects on the microbiota and environment in the gastrointestinal tract.

structure of the oligosaccharides as the linkage bonds in FOS and GOS can be readily degraded by β-fructanosidase and β-galactosidase enzymes, respectively, which are prevalent in bifidobacteria (Rastall, 2010; Rastall and Gibson, 2015). This genus also seems to preferentially metabolize the chain length typically found in FOS and GOS polymerization in the range of 4–10 (Rastall, 2010; Rastall and Gibson, 2015). Another feature of importance is that the microbes have the appropriate transport machinery to capture and deliver these substances into the microbial cytoplasm, which is essential for the selectivity of prebiotics at the target sites (Goh and Klaenhammer, 2015).

The production of organic acids, lactic acids and SCFA is a key element in the physiological effects of prebiotics (Roberfroid et al., 2010). The main SCFAs - the results of several bacterial metabolic pathways - are acetate, propionate and n-butyrate (Flint, 2004). These SCFAs are crucial for intestinal health and the different SCFAs can modulate various aspects of metabolic activity, including colonocyte function, gut homeostasis, energy gain, immune function, blood lipids and appetite (Flint et al., 2012b; Bach Knudsen et al., 2018). The high production of organic acids facilitated by prebiotics can outperform potential pathogens primarily due to the abundance of genes within lactobacillus and bifidobacterium that encode enzymes capable of degrading carbohydrates, along with carbohydrate uptake proteins, and fermentation pathways tailored for prebiotics (Rastall, 2010; Goh and Klaenhammer, 2015; Rastall and Gibson, 2015). Rapid generation of organic acids will reduce the pH of the digesta, which is unfavorable for the growth of pathogenic bacteria.

However, beyond organic acids, lactobacillus and bifidobacterium may also produce other compounds such as bacteriocins and exopolysaccharides (EPS), which can give these probiotic bacteria a competitive edge compared to other microorganisms in the GI environment. In the natural environment, bacteria-producing bacteriocins are able to compete against other bacteria for nutrients. These bacteria can either belong to the same species (narrow spectrum) or across genera (broad spectrum) but are immune to their own bacteriocin(s) (Cotter et al., 2005). Bacteriocin production takes place most frequently during the late exponential or early stationary phases of growth. It is influenced by cell–cell communication sensing and stress signaling and is regarded as a probiotic trait contributing to the suppression of intestinal pathogens (Martinez et al., 2013a). However, despite several reports of *Lactobacillus*- and *Bifidobacterium*-associated bacteriocins, there are relatively few in-depth studies on the interaction between probiotics and intestinal pathogens (Martinez et al., 2013a; Martinez et al., 2013b), particularly *in vivo*, with a complex interaction between the different nutrient residues, digesta dynamics and the possibility that the produced bacteriocins can cross the gut–blood barrier (Dicks et al., 2018).

There is increasing evidence that bacterial EPS are likely to be effector-molecules involved in beneficial health effects of some bacteria, i.e., probiotics, relative to pathogens. EPS form part of the microbial envelope - a dense layer of polysaccharides that protects the cell from its environment. These polysaccharides may be directly linked to the bacterial membrane (e.g. capsular polysaccharides) or loosely attached as a slime layer, as is the case for EPS. EPS are high-molecular-weight polymers that consist of a single type of polysaccharides or several types of polysaccharides of monosaccharides and non-carbohydrate substituents such as acetate, pyruvate, succinate and phosphate. EPS is produced by all bacteria in the gut, but the beneficial health effect depends on the chemical composition of the EPS, which is bacterial strain- and species-dependent (Oerlemans et al., 2021). A recent review has summarized some of the beneficial effects of EPS in human health. EPS are found to act as polymers that directly influence bacterial interactions with their host, such as bacterial adhesion to the GI epithelium, their influence on immune responses and the intestinal microbiota (Oerlemans et al., 2021).

The intestinal lining has a range of versatile functions that help to regulate nutrient and water absorption while acting as a barrier to prevent microbes and infectious agents from coming into contact with the gut. The harmful effects elicited by pathogenic microbes occur after they attach to enterocytes of the intestinal wall and colonize the environment. Mucin secretion at the brush-border membrane limits the number of bacteria that reach the epithelial layer and influences the metabolic activity of bacteria. Increased mucin production helps to improve gut barrier function, as pathogenic microbes cannot penetrate the dense mucus layer (Hansson, 2012). There are indications that prebiotics can help protect the epithelium, either by increasing mucin secretion, particularly of sulfated mucin, or by enhancing pathogen degradation and reducing pathogenic microbe attachment to colonic epithelial cells (Spring et al., 2015; Rajani et al., 2016). Moreover, soluble mannosyl receptors secreted in mucin competitively bind to type-1 fimbriae of pathogenic microbes such as *Salmonella*, *E. coli* and other gram-negative bacteria. Increased mucin production in the gut caused by MOS, for example, helps clear pathogens by preventing microbial attachment to the intestinal wall (Chacher et al., 2017).

The mucosal immune system of the GI tract is the largest immune sites in the body, containing almost 80% of the body's immune cells. Its main function is to defend the body against potential pathogens that can enter the body across vulnerable surface epithelia (Bailey et al., 2005). An important function of the immune system is to discriminate between pathogen-associated and 'harmless' antigens, expressing active responses against the former and tolerance to the latter. Control of the immune response involves local generation of IgA through interactions with dendritic cells or through a broad spectrum of pathogen-recognition receptors present in a wide variety of immune cells, which play a crucial role in the regulation and initiation of immune response (Bailey et al., 2005). Examples of pathogen-recognition receptors that can recognize microbial metabolism products are toll-like receptors, lectin receptors and nucleotide oligomerization domain-like receptors. Toll-like receptors play a major role in pathogen recognition and initiation of inflammatory and immune responses as the stimulation of these receptors by microbial products leads to the activation of signaling pathways that result in the induction of antimicrobial genes and inflammatory cytokines (Janeway and Medzhitov, 2002). Bacterial-associated molecules that trigger the immune system can, among other things, be exopolysaccharides, which exhibit variations in composition depending on whether they are produced by probiotic bacteria, e.g. bifidobacteria and lactobacillus, or pathogens such as *Salmonella* and *E. coli* (Oerlemans et al., 2021). Another example is GOS, which influences the microbiome composition and bile acid (BA) pool and thereby alleviates LPS-induced immune imbalance, as seen in the small intestine of a porcine model (Tian et al., 2023). In the cited study, it was suggested that GOS reduced the production of pro-inflammatory cytokines and inhibited NF-κB activation via microbe-BA-dependent signalling in LPS-challenged piglets (Tian et al., 2023).

5 Prebiotics in diets for pigs: oligosaccharides and inulin

Several studies have reported the effects of prebiotics on the GI microbiota of piglets and older pigs, such as an increased proportion of *Lactobacillus* spp. (Patterson et al., 2010; Jiao et al., 2014), *Bifidobacterium* spp. (Mølbak et al., 2007; Patterson et al., 2010) and a decreased amount of potentially harmful groups such as *Clostridium* (Patterson et al., 2010; Jiao et al., 2014) and Enterobacteriaceae (Castillo et al., 2008; Patterson et al., 2010) (Table 1).

Maize starch in a maize-soybean meal basal diet (BD) was replaced by either 4% short-chained inulin (P95, DPn: 2–7, average of 4), 4% long-chain inulin (HP, DP: 10–60, average of 25) or a 50:50 mix of P95 and HP (Synergy1) and fed to 32 piglets. Inulin was present in digesta from jejunum and ileum but not from caecum (Fig. 3) (Yasuda et al., 2009). The relative abundance of *Bifidobacterium* spp. and *Lactobacillus* spp. phylotypes in mucosa of all segments of the large intestine was higher for all inulin-containing diets, whereas in the ileum it was only higher when feeding P95 and Synergy1 and in jejunum when feeding P95 (Fig. 4A). (Patterson et al., 2010). A similar pattern was found in the luminal content of the GI tract, albeit not to the same extent as in the mucosa (Fig. 4B) (Patterson et al., 2010) and with marked decreases in less favorable populations such as *Clostridium* spp., *Streptococcus* spp./*Prevotella* spp. and Enterobacteriaceae/*Pseudomonas* spp./*Bacillus* spp. There was also a downregulation in the expression of the tumor necrosis factor, a cytokine involved in systemic inflammation, presumably due to the enhancement of the gram-positive *Bifidobacterium* spp. and *Lactobacillus* spp. not containing lipopolysaccharides at the expense of gram-negative bacteria (Yasuda et al., 2009; Patterson et al., 2010).

A fructan-rich diet (7.8% fructan of DM) from dried chicory roots (a mix of long- and short-chained fructans) was also found to stimulate bifidobacteria, compared to a control diet (1.0% fructans of DM) rich in resistant carbohydrates (Mølbak et al., 2007). The absolute number of *Bifidobacterium thermacidophilum* was higher in the colon of pigs fed the fructan-rich diet. There was also a stimulatory effect of the fructan-rich diet on *Megasphaera elsdenii*, most likely due to cross-feeding by lactate from *B. thermacidophilum* (Mølbak et al., 2007). Through *in situ* hybridizations, it was found that *B. thermacidophilum* and *M. elsdenii* colonized both the lumen and the mucous layer of the epithelium (Mølbak et al., 2007). When combined with the lower luminal pH in the colon (Thomsen et al., 2007), these conditions are likely responsible for protection against experimentally induced infections with *Brachyspira hyodysenteria* causing swine dysentery found in two independent studies performed on two continents (Thomsen et al., 2007; Hansen et al., 2010). However, it is not possible to conclude whether the inhibitory effect of the fructan-rich diet

Table 1 Overview over selected studies presenting effects of including carbohydrate and non-carbohydrate prebiotics on measured of GI tract health parameters

Prebiotic	Animal	Inclusion rate	Effects on GI microbiota	Other effects	Reference
Oligosaccharides					
FOS and TOS	Weaned pigs	1% and 4%	No effects on lactobacilli and bifidobacteria but reduced enterococci in ileum. No effects in feces	Decreased pH at ileum but increased pH in feces	(Houdijk et al., 1999)
FOS and TOS	Weaned pigs	4%	No effects on bifidobacteria in the gastrointestinal tract	FOS and TOS increased the yeast population stronger for TOS than FOS	(Mikkelsen and Jensen, 2004)
FOS and Inulin	Weaned pigs	4%	Increased *Bifidobacterium* spp. and *Lactobacillus* spp. and decease in clostridia and members of Enterobacteriaceae		(Patterson et al., 2010)
Raftifeed (inulin)	Weaned pigs	5% and 15%	Increased *Lactobacillus*:coliform ratio	Pigs on the high dose had fewer cases of diarrhoea and shed fewer ETEC than those on the low dose	(Wellock et al., 2008)
Fructans from chicory roots	Growing pigs	1% and 7.8%	Increased *Bifidobacterium thermacidophilum* and *Megasphaera elsdenii*	Prevention of swine dysentery caused by *Brachyspira hyodysenteriae*	(Mølbak et al., 2007; Thomsen et al., 2007)
XOS	Weaned pigs	1%	Increased abundance of *Lactobacillus* and *Bifidobacterium* in small intestine	Reduced incidence of diarrhoea, increased organic acid production in small intestine	(Sun et al., 2023)
GOS	Weaned pigs	0.05-0.2%	Increased *Lactobacillus* and *Bifidobacteria* and decreased *E. coli* in feces	Decreased in serum concentration of pro-inflammatory cytokines and increase in anti-inflammatory cytokines. Improved growth performance	(Xing et al., 2020)

(*Continued*)

Table 1 (*Continued*)

Prebiotic	Animal	Inclusion rate	Effects on GI microbiota	Other effects	Reference
Polysaccharides					
Raw potato starch	Nursery pigs	14%	No influence on colonic lactobacillus counts but decreased *E. coli* counts	No effect on growth performance	(Bhandari et al., 2009)
High amylose maize starch	Weaned pigs	HAM 63% amylose	Increased caecal and colonic *Bifidobacterium* spp. and decreased *Clostridium* clusters IV and XIVa	Reduced feed intake and growth	(Fouhse et al., 2015)
High amylose maize starch	Weaned pigs	HAM starch 85% amylose	Increased colonic and fecal numbers of lactobacilli and bifidobacteria	Higher concentration and pool of individual and total SCFA and lower pH in colon	(Bird et al., 2007)
High amylose maize starch	Growing pigs	HAM 63% amylose	Selective promotion of *Bifidobacterium* spp. and lactobacillus	Increased fecal SCFA and butyrate content	(Regmi et al., 2011)

FOS, fructo-oligosaccharides; TOS, trans-galacto-oligosaccharides; XOS, xylo-oligosaccharides; GOS, galacto-oligosaccharides; HAM, high-amylose maize.

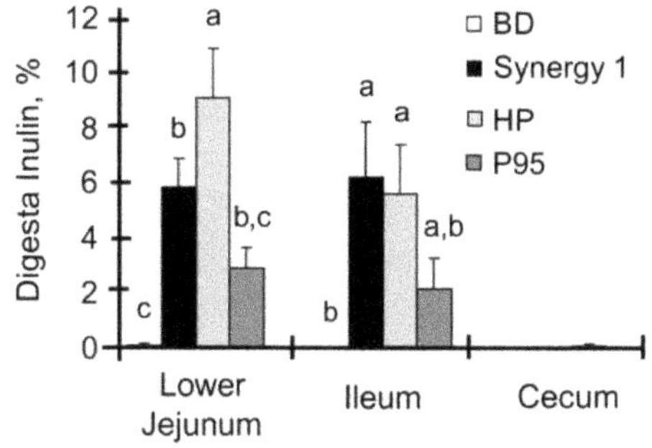

Figure 3 Effects of dietary supplemental inulin type on inulin concentrations of digesta in various gastrointestinal tract segments of pigs. The values (%) are the concentration of inulin detected in the digesta collected at a given site and do not reflect the proportion of inulin recovered from the diet. Values are means ±6 SE, $n = 8$. Means without a common letter differ, $P < 0.05$. Source: Adapted from: Yasuda et al. (2009).

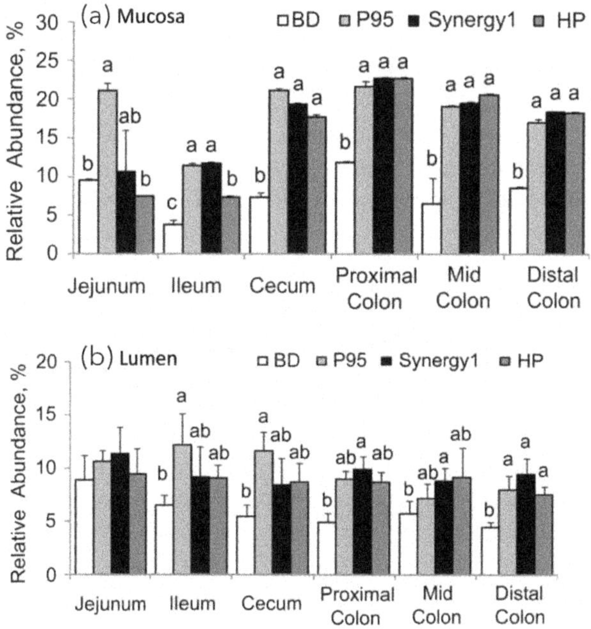

Figure 4 Effects of different types of supplemental inulin on relative amounts of phylotypes corresponding to the sum of *Bifidobacterium* and *Lactobacillus* spp. in the mucosa (A) and the lumen (B) contents of selected intestinal segments in pigs. Values are means ±6 SE, $n = 4$. Within a segment, means without a common superscript letter differ, $P < 0.05$ (A) or tended to differ, $P = 0.05–0.07$. Source: Adapted from: Patterson et al. (2010).

toward *B. hyodysenteria* was due to species-specific antagonistic activity from *B. thermacidophilum* subsp. *porcinum* or from a change in the intestinal environment due to fructan fermentation (Mølbak et al., 2007).

Weaned piglets fed diets containing insoluble NSP from Solka-floc (cellulose) or Raftifeed (long-chained inulin) at two differing levels (50 vs 150 g/kg) were experimentally challenged with enterotoxigenic *E. coli* (ETEC) (Wellock et al., 2008). The inclusion of Raftifeed (Raftifeed diets are labelled low or high in soluble NSP) significantly decreased the occurrence of diarrhea and improved gut health, as indicated by a significantly lower caecal digesta pH and increased *Lactobacillus*:coliform ratio, when compared with the Solka-floc diets on day 14 post-weaning. The NSP source had no effect on ETEC shedding or pig performance. Pigs fed on the high inclusion level had fewer cases of diarrhea and shed less ETEC than those on the low inclusion level and had increased colonic *Lactobacillus*:coliform ratio and SCFA concentration but decreased performance. It was concluded that the inclusion of Raftifeed may have a beneficial effect on gut health and protect against post-weaning colibacillosis.

A meta-analysis based on 33 published papers assessed the effects of dietary inulin supplementation at doses from 0.1% to 25.8% and with median and mean inulin levels of 0.1–2% and 3–4%, respectively, on GI fermentation, digestibility and growth in pigs (Metzler-Zebeli et al., 2017). Very few of the investigated fermentation response variables were found to be influenced by dietary inulin, while there was a strong negative relationship between dietary inulin and gastric pH in weaned pigs, colonic enterobacteria and fecal lactobacilli throughout all production phases. Observed negative relationships between inulin and colonic bifidobacteria and fecal enterobacteria and *E. coli* were of minor physiological relevance. The meta-analysis concluded that dietary inulin across studies had a stimulatory effect on gastric acid secretion, whereas the findings on fecal lactobacilli and bifidobacteria were not in line with expectations. Due to limited information provided in the original studies, such as dietary fructans or fibers, low numbers of observations and low inulin levels, it was concluded that the relationships should be regarded as trends.

Other studies with oligofructose (Shim et al., 2005) or FOS and TOS (Houdijk et al., 2002; Mikkelsen et al., 2003; Mikkelsen and Jensen, 2004), however, have shown limited effects of prebiotics on the *Bifidobacterial* population in the GI tract of piglets. Moreover, a fructan-rich diet (11.6% fructan of DM) compared to a control diet (0.8% fructan of DM) did not protect against post-weaning diarrhea when experimentally challenged with *E. coli* (Hedemann and Bach Knudsen, 2010) but resulted in an improved feed conversion ratio overall.

Xylo-oligosaccharides (XOS), or arabinoxylan (AX), added at a level of 1% to a corn–soybean meal basal diet (CON), significantly reduced the diarrhea incidence of weaned piglets compared with the CON group. XOS and AX also improved the antioxidant capacity, immune function and populations of beneficial bacteria. In addition, microbial fermentation of XOS with a lower polymerization and molecular mass than AX produced more organic acids and increased the abundance of *Lactobacillus* and *Bifidobacterium* in the upper gut of weaned pigs (Sun et al., 2023).

The effects of GOS added to a maize-soybean meal basal diet at doses of 0%, 0.05%, 0.1%, 0.15% and 0.2% GOS were studied in weaned piglets. The parameters included growth performance, fecal microbiota, immune response and antioxidant capability (Xing et al., 2020). The results indicated that GOS supplementation improved growth performance, increased the number of *Lactobacillus* and *Bifidobacteria*, and decreased the number of *E. coli* in feces. A decrease in serum concentration of pro-inflammatory cytokines, but an increase in anti-inflammatory cytokines in a linear or quadratic dose-dependent manner, were also found. Moreover, GOS supplementation promoted the activities of antioxidant enzymes. A study in which mannan oligosaccharides and fructan were added to a negative control diet at levels of 0.1% alone or as combinations (1:1) concluded that mannan oligosaccharides had a beneficial effect on growth performance, nutrient digestibility and decreased diarrhea score in weaning piglets (Zhao et al., 2012).

6 Prebiotics in diets for pigs: polysaccharides and non-carbohydrate prebiotics

Studies performed with β-glucan have indicated a wide range of anti-inflammatory, antioxidant and prebiotic properties, particularly with β-glucan from yeast, algae, bacteria, fungi and to a minor extent from cereals (Lam and Chi-Keung Cheung, 2013; Bouwhuis et al., 2017; Luo et al., 2019; Dowley et al., 2022). In experiments with pigs, β-glucan supplementation has been found to improve pig performance brought about by changes in gut microbial composition, gut morphology and barrier function and improved immune status (Bouwhuis et al., 2017; Luo et al., 2019).

Two other types of polysaccharides that have been investigated for their prebiotic properties include RS (Metzler-Zebeli et al., 2019; Tan et al., 2021) and pectins (Wiese, 2019; Xu et al., 2023). Studies with RS on nursery and growing pigs were shown to increase colonic and fecal lactobacilli and bifidobacteria (Bird et al., 2007; Bhandari et al., 2009; Regmi et al., 2011; Fouhse et al., 2015; Metzler-Zebeli et al., 2019), increase luminal SCFA and decrease pH in digesta and feces of nursery and growing pigs (Bird et al., 2007; Fouhse et al., 2015). However, the studies indicated that relatively high levels (10–15%) of RS are required to lower pH in the proximal, mid-, distal colon and feces by 0.2, 0.6, 0.4 and 0.6 units, respectively. It was also found that a 0.5 log unit increase in lactobacilli and bifidobacteria abundance required an RS level of 10% (Metzler-Zebeli et al., 2019).

Pectin is present in the cell wall of feedstuffs such as beet pulp, citrus pulp and soybean hulls, but citrus and apple residues from food production are also used in pig diets (Drochner et al., 2004). Pectin is known to influence digestion and fermentation as well as formation of fermentation products at all sites of the pig GI tract (Drochner et al., 2004), with possible effects on microbiota and other health parameters in pigs as reviewed by Wiese (2019) and Kiernan et al. (2023). The inclusion of pectin in pig diets can influence the composition of the microbiota and production and composition of SCFA; however, at present there is a lack of understanding of the structure-to-function relationship of pectin and thereby the most appropriate source for inclusion in pig diets.

Among non-carbohydrate prebiotics, polyphenols are deemed to have 'prebiotic-like properties' as they possess microbiota-modulating abilities when included in the diet (Xu et al., 2020). A major part of the polyphenol (>90–95%) is not absorbed from the small intestine and therefore reaches the colon where extensive biotransformation by the colonic microbiota occurs (Clifford, 2004). Polyphenol supplementation has been associated with increases in the abundance of beneficial *Lactobacillus*, *Bifidobacteria* and *Prevotella* and decreases in the amount of harmful *Streptococcus* and *Clostridium* (Fiesel et al., 2014; Jang et al., 2016; Xu et al., 2020).

It has also been found that feeding polyphenol-rich plant products to weaned pigs reduces the amount of harmful bacteria, including *Streptococcus* and *Clostridium*, without affecting the beneficial bacteria, *Lactobacillus* and *Bifidobacterium* (Fiesel et al., 2014). However, supplementation can also lead to an increase in the pH of the feces and a reduction in the concentration of SCFA (Fiesel et al., 2014). The latter effect may be a result of a decrease in *Bacteroidetes*, as polyphenol supplementation can decrease *Bacteroidetes* in the colonic digesta of weaned pigs (Wu et al., 2020). The reduction in SCFA concentration and increase in the pH of the feces noted in Fiesel et al., (2014) indicates a reduction in bacterial fermentation in the GI tract. This is not a desirable effect as SCFA plays an essential role in the regulation of metabolism, the immune system and cell proliferation in the GI tract. The increase in pH is not desirable either as a lower pH in the intestine can help limit the growth of pathogenic bacteria.

7 Conclusion

When the concept of prebiotics was defined in 1995 (Gibson and Roberfroid, 1995), NDOs were the only compound class considered. Gradually the concept has been expanded and, with the most recent update, non-carbohydrate compounds have also been included. While the new technologies available for characterizing the microbiota have made it possible to take a wider view on what stimulates a healthy microbiota, it is becoming increasingly difficult to draw generalized conclusions.

For the oligosaccharides, predominantly FOS, GOS and inulin, it is possible to build a molecule-function relationship as these carbohydrates are preferentially metabolized by Bifidobacterium and Lactobacilli, giving rise to rapid organic acid production, which can influence pH in the lumen and thereby outperform the growth of pathogens potentially brought forward by co-production of bacteriocins and EPS. However, although there appears to be a molecule-function relationship for NDO, it is far from the case that supplementing pig diets with prebiotics in the form of oligofructose, FOS, GOS and inulin (or fructans) will stimulate health-promoting microorganisms over enterobacteria. A reason for this is that the basal diet contains decent amounts of NDO from both protein-rich feedstuffs and cereals, making the additional amount of NDO too small to have major changes in the substrate composition for fermentation in the large intestine and thereby microbiota composition. For human subjects, a significant increase in Bifidobacteria is typically seen with a dose of 10–15 g/day of NDO, which can be compared to DF in the background diet, which provides around 18–20 g/day of fermentable carbohydrates (Van Loo et al., 1999; Roberfroid et al., 2010). For pigs, DF in the background diet is typically much higher (16–18% of DF) and the included dose of NDO lower (in several cases around 4%), which together gives a relatively lower NDO/DF ratio. The degradation of NDO in the stomach and small intestine of pigs is higher than in human subjects (40–50% compared to ~10%) (Bach Knudsen and Hessov, 1995; Bach Knudsen et al., 2023). Since the mode of action of NDO is through modulation of the microbiota, it also raises the question of whether the significant effects found in some studies (Xing et al., 2020; Sun et al., 2023; Zhou et al., 2023) at very low inclusion levels (<1%) are nothing more than an occasional change finding. Experiments with RS have also shown modulations of the microbiota and the pH, and at levels of 10–15% of RS there appear to be consistent effects. The inclusion level of RS is probably too high to be of any practical relevance, but it can be noted that the RS/DF ratio in the pig experiments is in the same order as the NDO/DF ratio in the human experiments. Although the use of β-glucan from yeast, algae, fungi and bacteria and pectin from plants may represent interesting prebiotic perspectives, knowledge of these compounds is still in its infancy and more studies are required to obtain a better molecular/structure-function relationship. The same is the case with polyphenols, which also warrant further investigation to develop a molecule-function relationship.

8 References

Agazzi, A., Perricone, V., Zorini, F. O., Sandrini, S., Mariani, E., Jiang, X. R., Ferrari, A., Crestani, M., Nguyen, T. X., Bontempo, V., Domeneghini, C. and Savoini, G. 2020. Dietary mannan oligosaccharides modulate gut inflammatory response and improve duodenal villi height in post-weaning piglets improving feed efficiency. *Animals-Basel* 10(8). https://doi.org/10.3390/ani10081283

Asp, N. G. 1992. Resistant starch - proceedings from the 2nd plenary meeting of Euresta - European flair concerted action 11 on physiological implications of the consumption of resistant starch in man. *Eur J Clin Nutr* 46, S1.

Babbar, N., Dejonghe, W., Gatti, M., Sforza, S. and Elst, K. 2016. Pectic oligosaccharides from agricultural by-products: production, characterization and health benefits. *Crit Rev Biotechnol* 36(4), 594–606. https://doi.org/10.3109/07388551.2014.996732

Bach Knudsen, K. E. and Hessov, I. 1995. Recovery of inulin from Jerusalem-Artichoke (Helianthus-Tuberosus L) in the small-intestine of man. *Brit J Nutr* 74(1), 101–113. https://doi.org/10.1079/Bjn19950110

Bach Knudsen, K. E. and Lærke, H. N. 2018. Carbohydrates and lignin in the feed – from sugars to complex composed fibres. In: Moughan, P. and Hendriks, W., editors, *Feed Evaluation Science*. Wageningen Academic Publisher, Wageningen. p. 109–140.

Bach Knudsen, K. E., Lærke, H. N., Hedemann, M. S., Nielsen, T. S., Ingerslev, A. K., Nielsen, D. S. G., Theil, P. K., Purup, S., Hald, S., Schioldan, A. G., Marco, M. L., Gregersen, S. and Hermansen, K. 2018. Impact of diet-modulated butyrate production on intestinal barrier function and inflammation. *Nutrients* 10(10). https://doi.org/10.3390/nu10101499

Bach Knudsen, K. E., Lærke, H. N. and Jørgensen, H. 2023. Carbohydrates and carbohydrate utilization in Swine. In: Chiba, L. I., editor, *Sustainable Swine Nutrition*. John Wiley & Sons Ltd. p. 151–187.

Bailey, M., Haverson, K., Inman, C., Harris, C., Jones, P., Corfield, G., Miller, B. and Stokes, C. 2005. The development of the mucosal immune system pre- and post-weaning: balancing regulatory and effector function. *Proc Nut Soc* 64(4), 451–457.

Bhandari, S. K., Nyachoti, C. M. and Krause, D. O. 2009. Raw potato starch in weaned pig diets and its influence on postweaning scours and the molecular microbial ecology of the digestive tract. *J Anim Sci* 87(3), 984–993. https://doi.org/10.2527/jas.2007-0747

Bird, A. R., Vuaran, M., Brown, I. and Topping, D. L. 2007. Two high-amylose maize starches with different amounts of resistant starch vary in their effects on fermentation, tissue and digesta mass accretion, and bacterial populations in the large bowel of pigs. *Brit J Nut* 97(1), 134–144. https://doi.org/10.1017/S0007114507250433

Bouwhuis, M. A., Sweeney, T., Mukhopadhya, A., McDonnell, M. J. and O'Doherty, J. V. 2017. Maternal laminarin supplementation decreases Salmonella Typhimurium shedding and improves intestinal health in piglets following an experimental challenge with S.Typhimurium post-weaning. *Anim Feed Sci Tech* 223, 156–168. https://doi.org/10.1016/j.anifeedsci.2016.11.007

Buffie, C. G. and Pamer, E. G. 2013. Microbiota-mediated colonization resistance against intestinal pathogens. *Nat Rev Immunol* 13(11), 790–801. https://doi.org/10.1038/nri3535

Caffall, K. H. and Mohnen, D. 2009. The structure, function, and biosynthesis of plant cell wall pectic polysaccharides. *Carbohyd Res* 344(14), 1879–1900. https://doi.org/10.1016/j.carres.2009.05.021

Castillo, M., Martín-Orúe, S. M., Taylor-Pickard, J. A., Peérez, J. F. and Gasa, J. 2008. Use of mannan-oligosaccharides and zinc chelate as growth promoters and diarrhea preventative in weaning pigs: effects on microbiota and gut function. *J Anim Sci* 86, 94–101.

Chacher, M. F. A., Kamran, Z., Ahsan, U., Ahmad, S., Koutoulis, K. C., Din, H. G. Q. U. and Cengiz, Ö. 2017. Use of mannan oligosaccharide in broiler diets: an overview of underlying mechanisms. *World Poultry Sci J* 73(4), 831–844. https://doi.org/10.1017/S0043933917000757

Clifford, M. N. 2004. Diet-derived Phenols in plasma and tissues and their implications for health. *Planta Med* 70(12), 1103–1114. https://doi.org/10.1055/s-2004-835835

Cotter, P. D., Hill, C. and Ross, R. P. 2005. Bacteriocins: developing innate immunity for food. *Nat Rev Microbiol* 3(10), 777–788. https://doi.org/10.1038/nrmicro1273

Coulier, L., Timmermans, J., Bas, R., Van Den Dool, R., Haaksman, I., Klarenbeek, B., Slaghek, T. and Van Dongen, W. 2009. In-depth characterization of prebiotic galacto-oligosaccharides by a combination of analytical techniques. *J Agric Food Chem* 57(18), 8488–8495. https://doi.org/10.1021/jf902549e

Cummings, J. H., Mann, J. I., Nishida, C. and Vorster, H. H. 2009. Dietary fibre: an agreed definition. *Lancet* 373(9661), 365–366. https://doi.org/10.1016/S0140-6736(09)60117-3

Dicks, L. M. T., Dreyer, L., Smith, C. and van Staden, A. D. 2018. A review: the fate of bacteriocins in the human gastro-intestinal tract: do they cross the gut-blood barrier? *Front Microbiol* 9, 2297. https://doi.org/10.3389/fmicb.2018.02297

Dowley, A., Sweeney, T., Conway, E., Maher, S., Rajauria, G., Yadav, S., Wilson, J., Gabrielli, W. and O'Doherty, J. V. 2022. The effects of dietary supplementation with mushroom or vitamin D enriched mushroom powders on finisher pig performance and meat quality. *Anim Feed Sci Tech* 288. https://doi.org/10.1016/j.anifeedsci.2022.115313

Drochner, W., Kerler, A. and Zacharias, B. 2004. Pectin in pig nutrition, a comparative review. *J Anim Physiol Anim Nutr (Berl)* 88(11–12), 367–380. https://doi.org/10.1111/j.1439-0396.2004.00490.x

Falck, P., Aronsson, A., Grey, C., Stalbrand, H., Nordberg Karlsson, E. and Adlercreutz, P. 2014. Production of arabinoxylan-oligosaccharide mixtures of varying composition from rye bran by a combination of process conditions and type of xylanase. *Bioresour Technol* 174, 118–125. https://doi.org/10.1016/j.biortech.2014.09.139

Fiesel, A., Gessner, D. K., Most, E. and Eder, K. 2014. Effects of dietary polyphenol-rich plant products from grape or hop on pro-inflammatory gene expression in the intestine, nutrient digestibility and faecal microbiota of weaned pigs. *Bmc Vet Res* 10. https://doi.org/10.1186/s12917-014-0196-5

Flint, H. J. 2004. Polysaccharide breakdown by anaerobic microorganisms inhabiting the mammalian gut. *Adv Appl Microbiol* 56, 89. https://doi.org/10.1016/S0065-2164(04)56003-3

Flint, H. J., Bayer, E. A., Rincon, M. T., Lamed, R. and White, B. A. 2008. Polysaccharide utilization by gut bacteria: potential for new insights from genomic analysis. *Nat Rev Microbiol* 6(2), 121–131. https://doi.org/10.1038/nrmicro1817

Flint, H. J., Scott, K. P., Duncan, S. H., Louis, P. and Forano, E. 2012a. Microbial degradation of complex carbohydrates in the gut. *Gut Microbes* 3(4), 289–306. https://doi.org/10.4161/gmic.19897

Flint, H. J., Scott, K. P., Louis, P. and Duncan, S. H. 2012b. The role of the gut microbiota in nutrition and health. *Nat Rev Gastroenterol Hepatol* 9(10), 577.

FAO & WHO. 2006. *Probiotics in Food: Health and Nutrition Properties and Guidelines for Evaluation*. FAO. http://www.fao.org/3/a-a0512e.pdf

Fouhse, J. M., Ganzle, M. G., Regmi, P. R., van Kempen, T. A. and Zijlstra, R. T. 2015. High amylose starch with low in vitro digestibility stimulates hindgut fermentation and has a bifidogenic effect in weaned pigs. *J Nut* 145(11), 2464–2470. https://doi.org/10.3945/jn.115.214353

Fransen, C. T., Van Laere, K. M., van Wijk, A. A., Brull, L. P., Dignum, M., Thomas-Oates, J. E., Haverkamp, J., Schols, H. A., Voragen, A. G., Kamerling, J. P. and Vliegenthart, J. F. 1998. alpha-D-Glcp-(1<-->1)-beta-D-Galp-containing oligosaccharides, novel products from lactose by the action of beta-galactosidase. *Carbohydr Res* 314(1–2), 101–114. https://doi.org/10.1016/s0008-6215(98)00285-7

Giavasis, I. 2014. Bioactive fungal polysaccharides as potential functional ingredients in food and nutraceuticals. *Curr Opin Biotech* 26, 162–173. https://doi.org/10.1016/j.copbio.2014.01.010

Gibson, G. R., Hutkins, R., Sanders, M. E., Prescott, S. L., Reimer, R. A., Salminen, S. J., Scott, K., Stanton, C., Swanson, K. S., Cani, P. D., Verbeke, K. and Reid, G. 2017. The International Scientific Association for Probiotics and Prebiotics (ISAPP) consensus statement on the definition and scope of prebiotics. *Nat Rev Gastro Hepat* 14(8), 491–502. https://doi.org/10.1038/nrgastro.2017.75

Gibson, G. R. and Roberfroid, M. B. 1995. Dietary modulation of the human colonic microbiota: introducing the concept of prebiotics. *J Nut* 125(6), 1401–1412.

Goh, Y. J. and Klaenhammer, T. R. 2015. Genetic mechanisms of prebiotic oligosaccharide metabolism in probiotic microbes. *Annu Rev Food Sci Technol* 6(6), 137–156. https://doi.org/10.1146/annurev-food-022814-015706

Gomez, B., Gullon, B., Remoroza, C., Schols, H. A., Parajo, J. C. and Alonso, J. L. 2014. Purification, characterization, and prebiotic properties of pectic oligosaccharides from orange peel wastes. *J Agric Food Chem* 62(40), 9769–9782. https://doi.org/10.1021/jf503475b

Gresse, R., Chaucheyras-Durand, F., Fleury, M. A., Van de Wiele, T., Forano, E. and Blanquet-Diot, S. 2017. Gut microbiota dysbiosis in postweaning piglets: understanding the keys to health. *Trends Microbiol* 25(10), 851–873. https://doi.org/10.1016/j.tim.2017.05.004

Hansen, C. F., Phillips, N. D., La, T., Hernandez, A., Mansfield, J., Kim, J. C., Mullan, B. P., Hampson, D. J. and Pluske, J. R. 2010. Diets containing inulin but not lupins help to prevent swine dysentery in experimentally challenged pigs. *J Anim Sci* 88(10), 3327–3336. https://doi.org/10.2527/jas.2009-2719

Hansson, G. C. 2012. Role of mucus layers in gut infection and inflammation. *Curr Opin Microbiol* 15(1), 57–62. https://doi.org/10.1016/j.mib.2011.11.002

Hedemann, M. S. and Bach Knudsen, K. E. 2010. A diet containing dried chicory root does not protect against post-weaning diarrhoea in an E. coli challenge model using piglets weaned at 7 weeks of age. *Livest Sci* 133, 232–235.

Hill, C., Guarner, F., Reid, G., Gibson, G. R., Merenstein, D. J., Pot, B., Morelli, L., Canani, R. B., Flint, H. J., Salminen, S., Calder, P. C. and Sanders, M. E. 2014. The international scientific association for probiotics and prebiotics consensus statement on the scope and appropriate use of the term probiotic. *Nat Rev Gastro Hepat* 11(8), 506-514. https://doi.org/10.1038/nrgastro.2014.66

Hlalukana, N., Magengelele, M., Malgas, S. and Pletschke, B. I. 2021. Enzymatic conversion of mannan-rich plant waste biomass into prebiotic mannooligosaccharides. *Foods* 10(9). https://doi.org/10.3390/foods10092010

Houdijk, J. G. M., Bosch, M. W., Tamminga, S., Verstegen, M. W. A., Berenpas, E. B. and Knoop, H. 1999. Apparent ileal and total-tract nutrient digestion by pigs as affected by dietary nondigestible oligosaccharides. *J Anim Sci* 77, 148-158.

Houdijk, J. G. M., Hartemink, R., Verstegen, M. W. A. and Bosch, M. W. 2002. Effects of dietary non-digestible oligosaccharides on microbial characteristics of ileal chyme and faeces in weaner pigs. *Arch Anim Nutr* 56(4), 297-307. https://doi.org/10.1080/00039420214346

Hu, Y., Ketabi, A., Buchko, A. and Ganzle, M. G. 2013. Metabolism of isomalto-oligosaccharides by Lactobacillus reuteri and bifidobacteria. *Lett Appl Microbiol* 57(2), 108-114. https://doi.org/10.1111/lam.12076

Isaacson, R. and Kim, H. B. 2012. The intestinal microbiome of the pig. *Anim Health Res Rev* 13, 100-109.

Izydorczyk, M. S. and Dexter, J. E. 2008. Barley β-glucans and arabinoxylans: molecular structure, physicochemical properties, and uses in food products-a review. *Int Food Res* 41(9), 850-868. https://doi.org/10.1016/j.foodres.2008.04.001

Janczyk, P., Pieper, R., Smidt, H. and Souffrant, W. B. 2007. Changes in the diversity of pig ileal lactobacilli around weaning determined by means of 16S rRNA gene amplification and denaturing gradient gel electrophoresis. *FEMS Microbiol Ecol* 61(1), 132-140.

Janeway, C. A. and Medzhitov, R. 2002. Innate immune recognition. *Annu Rev Immunol* 20, 197-216. https://doi.org/10.1146/annurev.immunol.20.083001.084359

Jang, S., Sun, J. H., Chen, P., Lakshman, S., Molokin, A., Harnly, J. M., Vinyard, B. T., Urban, J., Davis, C. D. and Solano-Aguilar, G. 2016. Flavanol-enriched cocoa powder alters the intestinal microbiota, tissue and fluid metabolite profiles, and intestinal gene expression in pigs. *J Nut* 146(4), 673-680. https://doi.org/10.3945/jn.115.222968

Jiao, L. F., Song, Z. H., Ke, Y. L., Xiao, K., Hu, C. H. and Shi, B. 2014. Cello-oligosaccharide influences intestinal microflora, mucosal architecture and nutrient transport in weaned pigs. *Anim Feed Sci Tech* 195, 85-91. https://doi.org/10.1016/j.anifeedsci.2014.05.014

Kamada, N., Seo, S. U., Chen, G. Y. and Núñez, G. 2013. Role of the gut microbiota in immunity and inflammatory disease. *Nat Rev Immunol* 13(5), 321-335. https://doi.org/10.1038/nri3430

Kelly, G. 2009. Inulin-type prebiotics: a review (part 2). *Altern Med Rev* 14(1), 36-55.

Kiernan, D. P., O'Doherty, J. V. and Sweeney, T. 2023. The effect of prebiotic supplements on the gastrointestinal microbiota and associated health parameters in pigs. *Animals (Basel)* 13(19). https://doi.org/10.3390/ani13193012

Kim, H. B., Borewicz, K., White, B. A., Singer, R. S., Sreevatsan, S., Tu, Z. J. and Isaacson, R. E. 2011. Longitudinal investigation of the age-related bacterial diversity in the feces of commercial pigs. *Vet Microbiol* 153(1-2), 124-133. https://doi.org/10.1016/j.vetmic.2011.05.021

Kim, H. B. and Isaacson, R. E. 2015. The pig gut microbial diversity: understanding the pig gut microbial ecology through the next generation high throughput sequencing. *Vet Microbiol* 177(3-4), 242-251. https://doi.org/10.1016/j.vetmic.2015.03.014

Lam, K.-L. and Chi-Keung Cheung, P. 2013. Non-digestible long chain beta-glucans as novel prebiotics. *Bioact Carbohydr Diet Fibre* 2, 45–64.

Leemhuis, H., Dobruchowska, J. M., Ebbelaar, M., Faber, F., Buwalda, P. L., van der Maarel, M. J., Kamerling, J. P. and Dijkhuizen, L. 2014. Isomalto/malto-polysaccharide, a novel soluble dietary fiber made via enzymatic conversion of starch. *J Agric Food Chem* 62(49), 12034–12044. https://doi.org/10.1021/jf503970a

Luo, J. Q., Chen, D. W., Mao, X. B., He, J., Yu, B., Cheng, L. and Zeng, D. F. 2019. Purified β-glucans of different molecular weights enhance growth performance of LPS-challenged piglets via improved gut barrier function and microbiota. *Animals-Basel* 9(9). https://doi.org/10.3390/ani9090602

Manners, D. J., Masson, A. J., Patterson, J. C., Bjorndal, H. and Lindberg, B. 1973. The structure of a beta-(1--6)-D-glucan from yeast cell walls. *Biochem J* 135(1), 31–36. https://doi.org/10.1042/bj1350031

Martinez, F. A. C., Balciunas, E. M., Converti, A., Cotter, P. D. and Oliveira, R. P. D. 2013a. Bacteriocin production by Bifidobacterium spp: a review. *Biotechnol Adv* 31(4), 482–488. https://doi.org/10.1016/j.biotechadv.2013.01.010

Martinez, R. C. R., Wachsman, M., Torres, N. I., LeBlanc, J. G., Todorov, S. D. and Franco, G. D. 2013b. Biochemical, antimicrobial and molecular characterization of a noncytotoxic bacteriocin produced by ST71KS. *Food Microbiol* 34(2), 376–381. https://doi.org/10.1016/j.fm.2013.01.011

Metzler-Zebeli, B. U., Canibe, N., Montagne, L., Freire, J., Bosi, P., Prates, J. A. M., Tanghe, S. and Trevisi, P. 2019. Resistant starch reduces large intestinal pH and promotes fecal lactobacilli and bifidobacteria in pigs. *Animal* 13(1), 64–73. https://doi.org/10.1017/S1751731118001003

Metzler-Zebeli, B. U., Trevisi, P., Prates, J. A. M., Tanghe, S., Bosi, P., Canibe, N., Montagne, L., Freire, J. and Zebeli, Q. 2017. Assessing the effect of dietary inulin supplementation on gastrointestinal fermentation, digestibility and growth in pigs: A meta-analysis. *Anim Feed Sci Tech* 233, 120–132. https://doi.org/10.1016/j.anifeedsci.2017.05.010

Mikkelsen, L. L., Jakobsen, M. and Jensen, B. B. 2003. Effects of dietary oligosaccharides on microbial diversity and fructo-oligosaccharide degrading bacteria in faeces of piglets post-weaning. *Anim Feed Sci Tech* 109(1–4), 133–150. https://doi.org/10.1016/S0377-8401(03)00172-X

Mikkelsen, L. L. and Jensen, B. B. 2004. Effect of fructo-oligosaccharides and transgalacto-oligosaccharides on microbial populations and microbial activity in the gastrointestinal tract of piglets post-weaning. *Anim Feed Sci Tech* 117(1–2), 107–119. https://doi.org/10.1016/j.anifeedsci.2004.07.015

Moniz, P., Ho, A. L., Duarte, L. C., Kolida, S., Rastall, R. A., Pereira, H. and Carvalheiro, F. 2016. Assessment of the bifidogenic effect of substituted xylo-oligosaccharides obtained from corn straw. *Carbohydr Polym* 136, 466–473. https://doi.org/10.1016/j.carbpol.2015.09.046

Mølbak, L., Thomsen, L. E., Jensen, T. K., Bach Knudsen, K. E. and Boye, M. 2007. Increased amount of bifidobacterium thermacidophilum and megasphaera elsdenii in the colonic microbiota of pigs fed a swine dysentery preventive diet containing chicory roots and sweet lupine. *J Appl Microbiol* 103(5), 1853–1867. https://doi.org/10.1111/j.1365-2672.2007.03430.x

Oerlemans, M. M. P., Akkerman, R., Ferrari, M., Walvoort, M. T. C. and de Vos, P. 2021. Benefits of bacteria-derived exopolysaccharides on gastrointestinal microbiota, immunity and health. *J Funct Foods* 76. https://doi.org/10.1016/j.jff.2020.104289

Onumpai, C., Kolida, S., Bonnin, E. and Rastall, R. A. 2011. Microbial utilization and selectivity of pectin fractions with various structures. *Appl Environ Microbiol* 77(16), 5747–5754. https://doi.org/10.1128/AEM.00179-11

Pandey, S., Kim, E. S., Cho, J. H., Song, M., Doo, H., Kim, S., Keum, G. B., Kwak, J., Ryu, S., Choi, Y., Kang, J., Lee, J. J. and Kim, H. B. 2023. Swine gut microbiome associated with non-digestible carbohydrate utilization. *Front Vet Sci* 10, 1231072. https://doi.org/10.3389/fvets.2023.1231072

Patterson, J. K., Yasuda, K., Welch, R. M., Miller, D. D. and Lei, X. G. 2010. Supplemental dietary inulin of variable chain lengths alters intestinal bacterial populations in young pigs. *J Nut* 140(12), 2158–2161. https://doi.org/10.3945/jn.110.130302

Phillips, G. O. and Cui, S. W. 2011. An introduction: evolution and finalisation of the regulatory definition of dietary fibre. *Food Hydrocoll* 25(2), 139–143.

Pieper, R., Janczyk, P., Zeyner, A., Smidt, H., Guiard, V. and Souffrant, W. B. 2008. Ecophysiology of the developing total bacterial and lactobacillus communities in the terminal small intestine of weaning piglets. *Microb Ecol*. https://doi.org/10.1007/s00248-00008-09366-y

Rajani, J., Dastar, B., Samadi, F., Torshizi, M. A. K., Abdulkhani, A. and Esfandyarpour, S. 2016. Effect of extracted galactoglucomannan oligosaccharides from pine wood on colonisation, growth performance and intestinal morphology in broiler chicks. *Brit Poultry Sci* 57(5), 682–692. https://doi.org/10.1080/00071668.2016.1200013

Rastall, R. A. 2010. Functional oligosaccharides: application and manufacture. *Annu Rev Food Sci T* 1, 305–339. https://doi.org/10.1146/annurev.food.080708.100746

Rastall, R. A. and Gibson, G. R. 2015. Recent developments in prebiotics to selectively impact beneficial microbes and promote intestinal health. *Curr Opin Biotechnol* 32, 42–46. https://doi.org/10.1016/j.copbio.2014.11.002

Regmi, P. R., Metzler-Zebeli, B. U., Ganzle, M. G., van Kempen, T. A. and Zijlstra, R. T. 2011. Starch with high amylose content and low in vitro digestibility increases intestinal nutrient flow and microbial fermentation and selectively promotes bifidobacteria in pigs. *J Nut* 141(7), 1273–1280. https://doi.org/10.3945/jn.111.140509

Rios-Rios, K. L., Dejonghe, W., Vanbroekhoven, K., Rakotoarivonina, H. and Remond, C. 2021. Enzymatic production of xylo-oligosaccharides from destarched wheat bran and the impact of their degree of polymerization and substituents on their utilization as a carbon source by probiotic bacteria. *J Agric Food Chem* 69(44), 13217–13226. https://doi.org/10.1021/acs.jafc.1c02888

Roberfroid, M., Gibson, G. R., Hoyles, L., McCartney, A. L., Rastall, R., Rowland, I., Wolvers, D., Watzl, B., Szajewska, H., Stahl, B., Guarner, F., Respondek, F., Whelan, K., Coxam, V., Davicco, M. J., Leotoing, L., Wittrant, Y., Delzenne, N. M., Cani, P. D., Neyrinck, A. M. and Meheust, A. 2010. Prebiotic effects: metabolic and health benefits. *Brit J Nutr* 104 Suppl 2, S1–63. https://doi.org/10.1017/S0007114510003363

Shim, S. B., Verstegen, M. W., Kim, I. H., Kwon, O. S. and Verdonk, J. M. 2005. Effects of feeding antibiotic-free creep feed supplemented with oligofructose, probiotics or synbiotics to suckling piglets increases the preweaning weight gain and composition of intestinal microbiota. *Arch Anim Nutr* 59(6), 419–427. https://doi.org/10.1080/17450390500353234

Singh, R. S. and Singh, R. P. 2010. Production of fructooligosaccharides from inulin by endoinulinases and their prebiotic potential. *Food Technol Biotech* 48(4), 435–450.

Spring, P., Wenk, C., Connolly, A. and Kiers, A. 2015. A review of 733 published trials on Bio-Mos®, a mannan oligosaccharide, and Actigen®, a second generation mannose rich fraction, on farm and companion animals. *J Appl Anim Nut* 3, 1–11. https://doi.org/10.1017/jan.2015.6

Sun, F. Z., Li, H. H., Sun, Z. Q., Liu, L., Zhang, X. J. and Zhao, J. B. 2023. Effect of arabinoxylan and xylo-oligosaccharide on growth performance and intestinal barrier function in weaned piglets. *Animals-Basel* 13(6). https://doi.org/10.3390/ani13060964

Tan, F. P. Y., Beltranena, E. and Zijlstra, R. T. 2021. Resistant starch: implications of dietary inclusion on gut health and growth in pigs: a review. *J Anim Sci Biotechno* 12(1). https://doi.org/10.1186/s40104-021-00644-5

Thomsen, L. E., Bach Knudsen, K. E., Jensen, T. K., Christensen, A. S., Moller, K. and Roepstorff, A. 2007. The effect of fermentable carbohydrates on experimental swine dysentery and whip worm infections in pigs. *Vet Microbiol* 119(2–4), 152–163. https://doi.org/10.1016/j.vetmic.2006.09.004

Tian, S., Wang, J., Gao, R., Zhao, F., Wang, J. and Zhu, W. 2023. Galacto-oligosaccharides alleviate LPS-induced immune imbalance in small intestine through regulating gut microbe composition and bile acid pool. *J Agric Food Chem* 71(46), 17615–17626. https://doi.org/10.1021/acs.jafc.3c00419

Tomás-Barberán, F. A., González-Sarrías, A. and García-Villalba, R. 2021. *Dietary Polyphenols: Metabolism and Health Effects*. JohnWiley & Sons, Inc.

Torres, D. P. M., Goncalves, M., Teixeira, J. A. and Rodrigues, L. R. 2010. Galacto-oligosaccharides: production, properties, applications, and significance as prebiotics. *Compr Rev Food Sci Food Saf* 9(5), 438–454. https://doi.org/10.1111/j.1541-4337.2010.00119.x

Van Loo, J., Cummings, J. H., Delzenne, N., Englyst, H., Franck, A., Hopkins, M., Kok, N., Macfarlane, G. T., Newton, D., Quigley, M., Roberfroid, M., van Vliet, T. and van den Heuvel, E. 1999. Functional food properties of non-digestible oligosaccharides: a consensus report from the ENDO project (DGXII AIRII-CT94-1095). *Brit J Nutr* 81(2), 121–132. https://doi.org/10.1017/S0007114599000252

Wellock, I. J., Fortomaris, P. D., Houdijk, J. G., Wiseman, J. and Kyriazakis, I. 2008. The consequences of non-starch polysaccharide solubility and inclusion level on the health and performance of weaned pigs challenged with enterotoxigenic Escherichia coli. *Brit J Nut* 99(3), 520–530. https://doi.org/10.1017/S0007114507819167

Wiese, M. 2019. The potential of pectin to impact pig nutrition and health: feeding the animal and its microbiome. *Fems Microbiol Lett* 366(4). https://doi.org/10.1093/femsle/fnz029

Wood, P. J., Weisz, J., Beer, M. U., Newman, C. W. and Newman, R. K. 2003. Structure of (1→3)(1→4)-β-d-Glucan in waxy and nonwaxy barley. *Cereal Chem* 80(3), 329–332. https://doi.org/10.1094/CCHEM.2003.80.3.329

Wu, W. D., Zhang, L., Xia, B., Tang, S. L., Xie, J. J. and Zhang, H. F. 2020. Modulation of pectin on mucosal innate immune function in pigs mediated by gut microbiota. *Microorganisms* 8(4). https://doi.org/10.3390/microorganisms8040535

Xing, Y. Y., Li, K. N., Xu, Y. Q., Wu, Y. Z., Shi, L. L., Guo, S. W., Yan, S. M., Jin, X. and Shi, B. L. 2020. Effects of galacto-oligosaccharide on growth performance, feacal microbiota, immune response and antioxidant capability in weaned piglets. *J Appl Anim Res* 48(1), 63–69. https://doi.org/10.1080/09712119.2020.1732394

Xu, R., Feng, N., Li, Q., Wang, H., Su, Y. and Zhu, W. 2023. Short-term supplementation of pectin alters substrate dynamics and modulates microbial carbohydrate metabolism in the gut of a pig model. *J Agric Food Chem* 71(27), 10470–10482. https://doi.org/10.1021/acs.jafc.3c02562

Xu, X., Hua, H. W., Wang, L. M., He, P. W., Zhang, L., Qin, Q., Yu, C., Wang, X. Y., Zhang, G. L. and Liu, Y. L. 2020. Holly polyphenols alleviate intestinal inflammation and alter microbiota composition in lipopolysaccharide-challenged pigs. *Brit J Nutr* 123(8), 881–891. https://doi.org/10.1017/S0007114520000082

Yan, B., Huang, C., Lai, C., Ling, Z. and Yong, Q. 2022. Production of prebiotic xylooligosaccharides from industrial-derived xylan residue by organic acid treatment. *Carbohydr Polym* 292, 119641. https://doi.org/10.1016/j.carbpol.2022.119641

Yasuda, K., Dawson, H. D., Wasmuth, E. V., Roneker, C. A., Chen, C., Urban, J. F., Welch, R. M., Miller, D. D. and Lei, X. G. 2009. Supplemental dietary inulin influences expression of iron and inflammation related genes in young pigs. *J Nut* 139(11), 2018–2023. https://doi.org/10.3945/jn.109.110528

Zdunek, A., Pieczywek, P. M. and Cybulska, J. 2021. The primary, secondary, and structures of higher levels of pectin polysaccharides. *Compr Rev Food Sci F* 20(1), 1101–1117. https://doi.org/10.1111/1541-4337.12689

Zhao, P. Y., Jung, J. H. and Kim, I. H. 2012. Effect of mannan oligosaccharides and fructan on growth performance, nutrient digestibility, blood profile, and diarrhea score in weanling pigs. *J Anim Sci* 90(3), 833–839. https://doi.org/10.2527/jas.2011-3921

Zhao, W., Wang, Y., Liu, S., Huang, J., Zhai, Z., He, C., Ding, J., Wang, J., Wang, H., Fan, W., Zhao, J. and Meng, H. 2015. The dynamic distribution of porcine microbiota across different ages and gastrointestinal tract segments. *PLoS One* 10(2), e0117441. https://doi.org/10.1371/journal.pone.0117441

Zhou, D. T., Mudhluli, T. E., Hall, L. J., Manasa, J. and Munyati, S. 2023. A scoping review of gut microbiome and bifidobacterium research in Zimbabwe: implications for future studies. *Pediatric Health Med Ther* 14, 483–496. https://doi.org/10.2147/PHMT.S414766

Chapter 9

Understanding and optimising the use of exogenous enzymes in pig nutrition

M. R. Bedford and G. Cordero, AB Vista, UK

1 Introduction

The response to any feed enzyme is maximised when the problem that it addresses is responsible for constraining the rate and efficiency of growth. Even with the most diligent efforts, there will always be variation in the exposure to such constraints between trials and between individuals within any trial. This explains the variation in performance responses noted when the data on a particular feed enzyme are reviewed (Torres-Pitarch et al., 2017, 2019) and is supported by the universal finding in any holo-analysis of feed enzyme datasets, viz. the effect of the enzyme is greater when the performance of the control-fed animals is poorer (Rosen, 2002, 2003, 2006).

Given this background, it is evident that when the problem the feed enzyme addresses is constantly present and reasonably significant, then most studies will return significant responses, whereas when the problem is less evident the responses will be less frequent and harder to detect. Further complicating the evaluation of the literature is the fact that in many cases, the responses achieved may well be real and commercially valuable, but the lack of precision in the experiments means the dataset is not deemed significant which undoubtedly delays commercial uptake. Early feed enzyme development (e.g.

http://dx.doi.org/10.19103/AS.2024.0140.13

β-glucanases and phytases) fortunately addressed relatively pressing problems and thus their uptake in these specific fields was rapid.

Feed enzymes (specifically β-glucanases) were first used commercially in the early to mid-1980s when they were employed to enable the use of higher levels of barley in poultry and pig rations. Prior to the development of β-glucanases, high levels of inclusion of barley, particularly newly harvested barley, were avoided in younger animal rations due to the associated poor performance and high moisture manure due to the presence of high viscosity β-glucans. Given the significant price differential between barley and competing grains during the barley harvest, it was economic to feed rations based almost entirely on barley despite the very significant costs of these initial enzyme offerings, and as a result, their uptake was rapid. The nascent industry very rapidly began to search for additional dietary 'problems' that could be solved with enzymatic solutions and subsequently generated xylanases, amylases, phytases, other non-starch polysaccharidases (NSPases) and proteases. Other minor classes of enzymes such as lipases exist but are not commonplace and thus not the focus of this chapter and have been discussed in the previous edition of this chapter (Bedford and Walk, 2017). Here, each enzyme is considered under the title of the principal condition or antinutritional factor (ANF) that they target.

2 Augmenting neonate and mature pig enzymatic capacity

The digestive capacity of the piglet at birth is limited and weaning places a further stress that temporarily but significantly challenges the ability of the piglet to efficiently digest feed (Lindemann et al., 1986; Kelly et al., 1991). Pancreatic and gastric mucosal enzyme activities fall 2 days post-weaning, indicating more is being secreted than is being synthesised. For some (e.g. chymotrypsin, lipase and carboxyl ester hydrolase), the subsequent digesta levels also fall to a minimum approximately 5 days post-weaning after which they slowly recover (Hedemann and Jensen, 2004). For others (e.g. pepsin, trypsin and amylase), digesta levels do not fall, indicating that although secretion exceeded synthesis, sufficient quantities were being released from the pancreatic reserves to maintain digesta levels (Hedemann and Jensen, 2004).

Regardless, it is important to realise that the replacement of a milk-based diet with solid feed places far greater demands on the digestive system and hence simply maintaining the levels of digestive enzymes post-weaning may not be enough to provide the digestive capacity required. Indeed, this capacity is likely exceeded, especially as the transition encountered at weaning is so abrupt and adaptation clearly takes some time. As a result, the weaned piglet grows more slowly than the sow-reared piglet for the first week post-weaning despite the fact that the pancreas as a proportion of body weight increases more rapidly in the former, presumably in response to the perceived lack of endogenous enzyme output. This, therefore, suggests there is an opportunity to improve piglet performance through the exogenous addition of those endogenous enzymes that are not produced in sufficient quantities to optimise digestion. Such enzymes would include proteases, amylase and lipase as the major activities to consider.

2.1 Protease

As noted in the previous edition of this book, proteases were first shown to be effective in weaning piglets 70 years ago but subsequent work has been equivocal, with protease use often benefiting digestibility of amino acids but in some cases depressing intake and thus performance (Bedford and Walk, 2017).

More recent work has shown that protease addition to markedly deficient weaner diets (10% reduction in crude protein (CP) and amino acid concentration compared with the positive control) resulted in significant improvements in performance, nutrient digestibility and intestinal structure (villus height to crypt depth ratio) to the point that performance was restored to that of the positive, high protein control (Park et al., 2020). The conclusion from this work would be that in correctly formulated diets (i.e. protein reduced), there is potential for the value of an exogenous protease to be exploited. Further work by the same group demonstrated that the same protease reduced the inflammatory responses apparent at weaning (Lee et al., 2020) and reduced white blood cell counts (although such effects were found by Kim et al. (2021) using a different protease) by means of an as yet unknown mechanism. The performance and inflammatory responses are most probably linked, suggesting that responses to such products could be greater in stressed rather than unstressed individuals, but care is needed in extrapolating from the results obtained from one protease to another. However, the premise that proteases are only effective in low-protein diets was recently challenged by the work of Kim et al. (2021) who noted that the highest protease level employed in their study improved weight gain and gain-to-feed (G:F) ratio in high protein diets fed to piglets up to 4 weeks after weaning. Interestingly, no further responses to the protease were noted up to 16 weeks of age, but between 16 and 20 weeks of age, a large and significant increase in weight gain (25%) and G:F (14%) were noted, but only at the higher dose (300 ppm vs 150 ppm) employed (Kim et al., 2021). In an unrelated study, again using a different protease, beneficial effects on the average daily gain of weaning piglets were only noted in the presence but not the absence of a xylanase (Duarte et al., 2019). The equivocal nature of these data likely is a result of the use of different proteases under different circumstances.

The current commercial proteases are quite distinct from each other and more so from the endogenous enzymes produced by the piglet and a better understanding of how these enzymes work in concert is required to predict outcomes more effectively. From the responses noted earlier, it is important that proteases are tested in diets which are deficient in the amino acids that the protease can influence if responses are to be seen (Bedford and Walk, 2017).

The age of the animal, the presence of other exogenous feed enzymes, dietary nutrient content, and the identity and dose of the enzyme employed seem to be critical parameters in determining the response to this enzyme class.

2.2 Amylase

Whether amylase per se is the limiting factor for starch digestion in the weaning pig is not known, but in 23kg pigs fed barley, maize or high-amylose maize, starch dissolution by amylase did not seem to be the rate-limiting step in the reduction of starch to maltose and glucose (Martens et al., 2019). In the first half of the small intestine, it seems there is a significant accumulation of maltotriose, maltose and glucose (Martens et al., 2019). This suggests that the rate-limiting step in young piglets in assimilating glucose from starch is more to do with glucose and maltose absorption than starch digestion per se. However, a recent review suggested that exogenous amylase may be of greater benefit in older animals due to an age-related reduction in the output of endogenous enzymes when expressed per kg of body weight (Cowieson et al., 2019). Whilst an interesting proposition, the digestive capacity of the intestine relates to the amount of enzyme per g of digesta, not the weight of the animal. As a result, such a hypothesis is open to challenge. Indeed, although evidence is scant in older pigs, a relatively large study where the maize in the control diet was replaced by a GMO maize expressing amylase suggested that even very high levels of amylase were of little benefit in terms of performance or carcass composition (Ochonski et al., 2021).

Nevertheless, supplementing weaner diets with amylase-rich enzymes has been shown to increase growth rates and improve feed conversion ratio (FCR) in young piglets (Fernandez, 2009; Dzierva et al., 2023) and further work suggests both amylases and lipases may aid piglet performance during the weaning period (Portillo et al., 2021). However, the products used in all studies probably also contained significant amounts of fibre-degrading enzyme activity (a point raised in a review of amylases (Cowieson et al., 2019)) or the fibre-degrading enzymes were not characterised at all, which challenges the allocation of the response solely to the amylase or lipase, respectively. The responses noted are interesting to the industry, however, and do need further scrutiny. It was recently noted that *in vitro* starch digestion underestimated *in vivo* small intestinal digestion and that this was due to starch hydrolysis in the stomach being overlooked (Martens et al., 2019). Thus, the focus on augmenting small intestinal amylase in weaning piglets may require some redirection if gastric rather than small intestinal dissolution is the bottleneck.

2.3 Absorptive capacity rather than enzyme output

With regard to the constraints on ingredient digestion and subsequent nutrient retrieval, the absorptive capacity of the intestine has to be considered alongside enzyme activity to fully understand where the limitations exist. As noted earlier, starch digestion and subsequent glucose uptake may be more constrained by uptake capacity than enzyme output (Martens et al., 2019). Furthermore, one notable difference between weaned and sow-fed piglets of the same age is the marked reduction in mucosal protein content and villus height coupled with an increase in crypt depth and brush border glucoamylase along the entire length of the small intestine of the former. This further suggests that absorptive capacity is indeed compromised in weaning piglets and possibly to a greater extent than digestive capacity (Kelly et al., 1991). If this is the case, then the addition of exogenous enzymes will likely not aid nutrient extraction rate to the extent expected.

2.4 Market considerations

Although the use of enzymes during the weaning period can be beneficial, the short period during which the deficit in endogenous enzymes allows such a supplementation to improve performance limits the size of this market.

3 Phytic acid

3.1 Hydrolysis of phytic acid and lower esters

Phytic acid (inositol hexa-kis phosphate, InsP6) is present in most plant-based ingredients and is a significant ANF due to the negative charges carried by the 6 phosphate groups arranged around the inositol ring. Figure 1 highlights the content and variation in the content of phytic acid phosphorus (which is 28% of total InsP6 content) in a range of ingredients, highlighting that the choice of ingredients in a diet can have a significant impact on final dietary phytic acid concentration.

In the absence of added exogenous phytase, the extent of InsP6 hydrolysis in pigs at the terminal ileum is quite limited (Schlemmer et al., 2001). The small intestinal mucosa does house phytases and some dietary ingredients do bring exogenous phytase activity with them (e.g. wheat and its bran fractions) but neither of these sources provides a consistent benefit due to the conditions of the intestinal tract with regard to the former (substrate solubility, calcium (Ca) levels) and heat and pepsin stability for the latter respectively. At the faecal

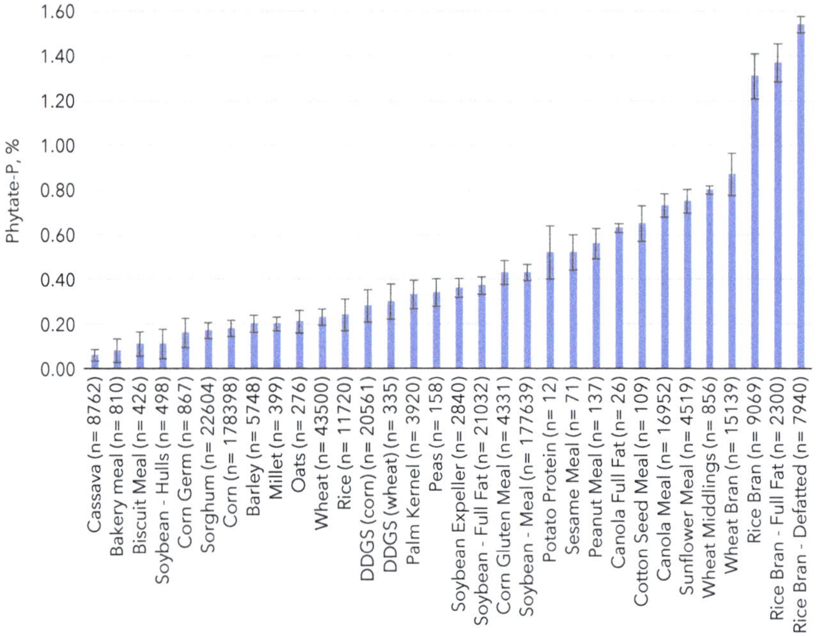

Figure 1 Phytate phosphorus content (± SD) of a range of ingredients (*n* = number of samples analysed) commonly used in pig rations Source: AB Vista (2023).

level, phytate and its lower esters are almost absent (Rosenfelder-Kuon et al., 2020) which suggests that large intestinal bacteria are very effective at InsP6 hydrolysis, but the host seems to gain no advantage from the inositol and phosphate released (Rosenfelder-Kuon et al., 2020). When phytase is added, there is a dose-dependent reduction in ileal InsP6 and InsP5 concentrations but an accumulation of lower esters (InsP4 and InsP3) which becomes so onerous, it seems to limit phosphorous (P) digestibility to a maximum of approximately 65% P in pigs (Rosenfelder-Kuon et al., 2019). The inability of exogenous phytases to address these lower esters before the digesta reaches the terminal ileum was suggested as the reason why this asymptote in P digestibility exists. In this meta-analysis (Rosenfelder-Kuon et al., 2019), however, the maximum dose of phytase employed was 2500 Phytase Units (FTU)/kg, which is at the upper extreme of doses employed commercially.

Recent work where InsP6 hydrolysis has been followed using even higher doses of phytase (8000 FTU/kg feed) has shown that the extent of InsP6 and lower ester hydrolysis can be further increased and with it, presumably, P digestibility by considering log increments rather than linear increments in phytase dosage (Laird et al., 2016). Indeed, in a recent study, a significant increment in P digestibility from 69% to 75% was noted by increasing the phytase dose from 2000 to 4000 FTU/kg (Espinosa et al., 2021). Regardless, in comparison to poultry, the ability of a given dose of phytase to hydrolyse InsP6 in pigs seems compromised suggesting that conditions in the gastrointestinal tract (GIT) are not as conducive to phytase activity. This may relate to the lack of a crop and less than ideal gastric conditions for phytase activity. Access to the substrate may also play a role as much of the phytate in small grain cereals is located in the thick cell-walled aleurone layer and this may be less accessible to pigs due to less effective grinding by teeth compared with that of the gizzard in birds. Regardless, it remains to be seen whether much higher doses of phytase can improve P digestibility above the ~65% ceiling proposed by Rosenfelder-Kuon et al. (2019).

3.2 Anti-nutritive effects of phytic acid and lower esters

Some understanding of InsP6 solubility and its anti-nutritional effects is needed to place the use of exogenous phytases in context. The extent of the charge carried by IP6 depends upon pH (Evans et al., 1982), the higher the pH the greater the charge. Even under the acidic conditions of the stomach, it carries sufficient charge to bind positively charged minerals, proteins and any other such nutrients, thus sterically hindering their digestion and absorption (Angel et al., 2002). As digesta enters the small intestine, the pH increases as does the InsP6 charge (and that of lower esters, InsP5, 4, 3 and 2) and thus so does its ability to bind positively charged ions and precipitate with its ligand (Schlemmer et al., 2001). Once the InsP(n) ester and nutrient precipitate, it is effectively unavailable for digestion/absorption and further InsP(n) hydrolysis. Phytases can, however, reduce and ultimately eliminate this ANF effect by sequential removal of phosphates prior to entry into the small intestine, and with each phosphate removal, the charge of the remaining IP ester is reduced and with it its ANF potential (Persson et al., 1998).

Despite the presumed inert (from an ANF viewpoint) nature of InsP1, it is soluble and charged and as a result, is not absorbed in the small intestine (SI) until it is hydrolysed to inositol and phosphate by intestinal phosphatases. The inositol can then be absorbed by small intestinal where it has a value in its own right as a nutrient. The quantitative reduction of InsP6 to InsP1 in the stomach is therefore highly desirable if all ANF effects are to be removed prior to entry into the small intestine, but as discussed next, such a goal is not achieved at the levels of phytase currently employed commercially.

3.3 Optimising phytate hydrolysis

In all dose-response work to date where intestinal InsP ester and inositol contents are determined, it is clear that a plateau in response is not achieved. For example, plasma and intestinal concentrations of inositol continue to increase with increased dosage of phytase (in one study, this was the case up to 16 000

FTU/kg), indicating that phytate and lower ester hydrolysis is not complete even at such extreme levels of inclusion (Lagos et al., 2021). Improvements in the rate and in particular the efficiency of gain in the first 14 days post-weaning accompanied this rise in plasma inositol (Lagos et al., 2021), suggesting that such high doses continue to bring benefit (although they were not commercially viable) and that the early weaned piglet is particularly susceptible to relatively low concentrations of InsP6 and lower esters, or responsive to the inositol released. As a result, it is unclear what the target maximum gastric InsP6–InsP1 concentration should be to ensure optimum health and performance of the pig and whether this changes with age. Indeed, the same is true for inositol, what is the target amount required, where should it be generated in the GIT, and does this change with age?

In many respects, the current market is confused as it considers the FTU added as an index of how much phytate hydrolysis takes place but this is misconceived. Two studies may have used identical FTU but the outcome, in terms of how much P is released, how much InsP6-1 remains and how much inositol is generated can be radically different. Indeed, the review by Rosenfelder-Kuon et al. (2019) highlights the enormous variation in P digestibility (approx. 15–85%) between studies when the diets contained 500 FTU of phytase. In addition to the animal (age, sex, breed, physiological status), ingredient identity and sample (endogenous phytase, phytate solubility/accessibility) and microbiome influence on the extent of phytate hydrolysis (Rosenfelder-Kuon et al., 2019), intrinsic differences exist between phytases which makes the FTU/kg declaration of little value in predicting response. Phytases differ markedly in their ability to function at gastric pH (whereas the FTU is declared at pH 5.5), survive pepsin hydrolysis, process lower esters (Menezes-Blackburn et al., 2015), resist the inhibitory effects of excess Ca and P (Sommerfeld et al., 2018; Lei et al., 1994; Adeola et al., 2006; Letourneau-Montminy et al., 2010) and thus deliver the degree of phytate hydrolysis and inositol generation required. The negative effects of these detracting factors on InsP6 hydrolysis through to InsP1 become less evident as the dose of phytase increases (Lagos et al., 2021) and the consistency of response therefore improves.

The dose required to deliver such consistency and effects is not economically viable at present. However, as new and better phytases enter the market, the viability of such high doses may be realised resulting in more efficient use of resources and less pollution. The benefits noted here go well beyond the P and Ca releasing ability of the enzyme and encompass what has often been described as the extra-phosphoric activities of phytase. Some of these benefits are described here.

3.4 Benefits of extra-phosphoric activities

3.4.1 Reduction of endogenous losses

It has long been suggested that phytic acid interferes with gastric digestion and in response the animal increases both pepsin and acid production in an attempt to restore proteolysis. This increases the hostility of the gastric environment which results in greater demands for mucin production for preservation of the gastric mucosa (Selle and Ravindran, 2008; Humer et al., 2015). In addition, compromised small intestinal digestion necessitates increased pancreatic, biliary and small intestinal secretions along with increased enterocyte replication and turnover and small intestinal mucin production in an attempt to increase the digestive and absorptive capacity of the small intestine. Unfortunately, of all the secretions, mucin proteins by their very design are most resistant to proteolysis and a large proportion thereby are excreted and comprise an increasing proportion of endogenous amino acid losses as the challenge augments. The application of phytases in both pigs and chickens has been shown to markedly reduce mucin losses (Cowieson and Bedford, 2009; Humer et al., 2015; Selle and Ravindran, 2008) although the data in pigs are more equivocal (Mesina et al., 2019) than in chickens (Cowieson et al., 2004; Cowieson et al., 2006).

InsP6 is implicated as the main culprit with regard to stimulating mucin production but work in poultry has suggested that much lower esters may be just as important. In broilers, ileal mucin production was reduced by approximately 20% when 500 or 7500 FTU/kg of phytase was employed compared with the negative control, but increasing the dose further to 15 000 FTU/kg resulted in much greater reductions, almost 50% (Persia, 2010), suggesting that the lower esters may be just as damaging if not more so than InsP6. If phytase is relatively less effective in pigs than poultry in reducing InsP6 to InsP1, possibly due to significant phytate hydrolysis in the crop (Novotny et al., 2023), and very high levels of phytase are required to significantly reduce endogenous losses, then this may explain why the endogenous loss reductions recorded in pigs to date are not so convincing.

Regardless, if a reduction in endogenous losses was responsible for the bulk of the amino acids spared, then the use of a phytase should favour the recovery of cysteine, proline, serine, threonine, aspartate, glycine and glutamate, and this seems to be the case (Cowieson and Bedford, 2009). In the future, a much greater recovery of these amino acids may be realised once higher doses of phytase become economically viable.

3.4.2 Micromineral absorption

Incremental doses of phytase initially reduce InsP6 concentrations and increase InsP3 and 4 concentrations, then go on to finally reduce the concentrations of the lower esters. This exchange of high concentrations of InsP6 for high concentrations of InsP3 and InsP4 at intermediate phytase doses could have implications for the availability of several trace element minerals. Calcium is preferentially bound to InsP6 not because it has a greater affinity than other ions, in fact, it has quite poor affinity (Luttrell, 1993), but because it is present in very high concentrations. As InsP6 concentrations decline and the lower esters increase in concentration, the Ca is less able to occupy the binding sites despite its concentration advantage (Luttrell, 1993) and other minerals may take its place. Given the assumed molar ratios of calcium to zinc (Ca:Zn) and InsP(x):Zn would be in the range of 100–200:1 and 1–8:1 respectively, then according to the work of Xu et al. (1992), even InsP3 could precipitate 75% of the Zn in the small intestine and higher esters even more whilst at the same time leaving most of the Ca in solution. Similar results are expected for iron (Fe) and magnesium (Mg). Thus, partial hydrolysis of InsP6 to lower esters may present the trace elements with as much of a challenge to remain soluble as no hydrolysis, and it may be that processing of the lower esters quantitatively to InsP1 is required to restore their full solubility. This may explain why the digestibility of some minerals responds to moderate phytase doses whereas others do not (Arredondo et al., 2019), the latter presumably being precipitated by the lower InsP3 and even InsP2 esters, and why massive doses of phytase are required to improve the availability of others.

3.4.3 Post-weaning diarrhoea – activation of alkaline phosphatase

The exact cause of post-weaning diarrhoea (PWD) is not known but it usually, but not always, involves enterotoxigenic *E. coli* strains which take advantage of stresses encountered by the piglet at weaning due to diet, housing conditions, genetics, passive immunity and other factors (Rhouma et al., 2017). The subsequent inflammatory response against the lipopolysaccharide endotoxin is thought to contribute to diarrhoea as it increases intestinal permeability and reduces the ability of the intestine to absorb minerals which then osmotically draw water back into the digesta (Campbell et al., 2013). A reduction in mucosal alkaline phosphatase activity is also noted at weaning (Lackeyram et al., 2010) which is significant since this enzyme is involved in the dephosphorylation of the lipopolysaccharide, rendering it inert, and as a result this reduction of capability exposes the piglet to greater risk of inflammatory damage. Dietary phytase application has been shown in broilers to increase ileal alkaline phosphatase activity in a dose-dependent manner (Walk et al., 2018) possibly as a result of the generation of InsP1 which is a substrate for this enzyme and/or the fact that very high doses of phytase results in greater levels of free Zn and Mg as noted earlier, the former being crucial for alkaline phosphatase activity (Davies and

Flett, 1978). Use of phytase at doses well above that used commercially to date may therefore convey some benefit in weaning piglets in dampening down the inflammatory response that accompanies the weaning process and thereby reduce the incidence and severity of PWD.

3.4.4 Parturition effects

Farrowing and lactation periods require the expenditure of a great deal of metabolic energy, and this has increased over time as modern sow genotypes experience extended farrowing durations due to increased litter size (Tokach et al., 2019). Increased litter size has also been associated with an increase in the duration of gestation by an average of 1 day over the last 20 years, from 114.6 to 115.6 days. The increase in prolificity is likely the cause of the increase in the duration of gestation since this is needed to ensure optimum maturity of piglets at the point of farrowing and thus enhance the likelihood of their survival. Poor maturation as a result of greater litter size and insufficient gestation time can result in an increased stillborn rate (van Dijk et al., 2005).

With phytate having the ability to bind cations (e.g. Ca2+, Zn2+, Fe2+ and Cu2+), the effect of superdosing phytase on mineral utilisation is of interest in relation to mineral excretion and environmental concerns as well as animal performance. Ca demands are understandably greater in modern sow lines with larger litter sizes and greater milk production capabilities. Ca also plays an important role in the skeletal development of piglets during late pregnancy. Finally, calcium is important for muscle contraction and superdoses of phytase may improve calcium uptake through near-complete phytate degradation and thus provide sufficient calcium for optimal muscle activity.

Data regarding phytase dose-response effects on lactating sow performance are scarce (Jongbloed et al., 2004). Wealleans et al. (2015) observed decreased body weight (BW) loss for sows fed a low-P diet when it was supplemented with 2000 FTU/kg of phytase. Manu et al. (2018) observed that feeding 2500 FTU/kg of phytase in lactation diets from day 109 to weaning in adequate P diets had no effect on sow performance; however, a reduction in farrowing duration compared to the control (521.5 ± 45.24 min vs 710.4 ± 83.63 min, $P < 0.046$) tended to reduce the number of stillborn piglets (1.26 ± 0.18 vs 1.69 ± 0.23, $P = 0.08$) and numerically reduced piglet pre-weaning mortality by 1.7% units relative to the control sows. These effects are likely a result of improved absorption of a number of nutrients but clearly energy is of importance if a reduction in farrowing time is a result of improved energy reserves. Phytase superdosing can also be used in peripartum diets to improve piglet weight gain by an average of 490 g/weaned piglet ($P < 0.01$), which again is linked with greater release of nutrients from InsP6 and all lower esters and thus improved utilisation (Cordero et al., 2022).

4 Fibre

4.1 Identity and biological relevance

Fibre is a term that is used generically to describe the bulk of the indigestible plant material in the diet. As such it is an unhelpful term because it can, depending on its identity, be beneficial, inert or detrimental towards the digestive process and as such a far better sub-classification is needed. For a recent detailed description of fibre structure, composition, analysis and functionality, the reader is referred to the cited reviews (Rybicka and Medel, 2021; Bach Knudsen, 2014). In order to understand how fibre-degrading enzymes function in the pig, it is first necessary to understand how the pig's small and large intestines function and how this environment has changed since fibre-degrading enzymes were first introduced more than 30 years ago.

4.1.1 Small intestinal function

Soluble viscous fibres such as the high-molecular weight arabinoxylans and β-glucans can create problems in pigs as they effectively slow down diffusion and hence the rate of enzymatic deconstruction of particles and macro-molecules into absorbable nutrients (Bedford and Schulze, 1998). However, the quantity of such non-starch polysaccharides (NSP) needed to markedly compromise digestion seems to be significantly higher than that which would incapacitate a broiler chicken due to significantly greater dilution of ingesta in the pig's intestinal tract (Bedford and Schulze, 1998). Nevertheless, it is accepted that in the recently weaned piglet, there is a need to limit soluble and particularly viscous fibre and indeed maintain a minimum amount of insoluble fibre so as to optimise intestinal transit (Rybicka and Medel, 2021). Piglets are susceptible to viscous NSPs since they lack a functioning microbiota and they are on a digestive knife-edge due to the stress of weaning as noted earlier. During this phase, viscosity-reducing enzymes have proven effective in enhancing piglet performance and nutrient digestibility (Molist et al., 2014; Lærke et al., 2015; Chen et al., 2020). A few weeks post-weaning, pigs develop a microbiota in the small intestine capable of depolymerising and thus removing the viscous nature of these soluble fibres much earlier than poultry (de Vries et al., 2016). One added advantage of adding a fibre-degrading enzyme is that it not only degrades an anti-nutrient but in doing so produces prebiotic and stimbiotic smaller polysaccharides and oligosaccharides which stimulate the development of beneficial, fibre-degrading bacteria (Gomes et al., 2019; Cho et al., 2020; Petry et al., 2021b, c).

Insoluble fibre sources such as dried distillers grain with solubles (DDGS) are often considered as a problem when fed at substantial levels not only because they dilute other nutrients but also because they speed up passage rates of digesta through the small intestine. However, in ileal and caecal cannulated pigs inclusion of 30% DDGS in corn-soy diets fed to growing pigs had no effect on the first appearance of a marker at the terminal ileum, cecum or in the faeces (Urriola and Stein, 2010). The possible reason cited for this anomaly is that the DDGS diet contained more fat and the control, invoking a more robust ileal brake mechanism which countered the expected acceleration of passage from the fibre per se. As a result, the effects of fibre need to be taken in the context of what else changes in the diet with the addition of such ingredients. A further consideration is that the dissolution and depolymerisation of such fibre types may benefit the pig but the quantity of insoluble fibre dissolved very much depends upon the susceptibility of the fibre to enzymatic attack (Vangsøe et al., 2021), which ultimately means responses may be variable depending upon the substrate employed.

The capacity of the small intestine to ferment fibre can be great and seems to increase with the maturation of the pig and its intestinal microbiota (Hoogeveen et al., 2021). A recent observation made in several papers has suggested that this increment can be so substantial that the bulk of the rapidly fermentable fibre sources such as fructans, oligosaccharides and even soluble β-glucans are almost quantitatively removed from the digesta before it exit into the caecum and consequently the quantity of fibre digested in the colon actually falls with age (Urriola and Stein, 2010; Petry et al., 2021a). This creates pressure on the large intestinal microbiota to adapt to a seemingly never-ending reduction in 'nutrient' (i.e. fermentable fibre) supply as the small intestine effectively ferments and removes the easiest substrates. If this small intestinal transition is too rapid then there could be a consequential loss in the ability of the large intestinal bacteria to evolve rapidly enough to deal with the restricted supply of usable substrates, and protein putrefaction may ensue.

Careful management of the capacities of the small and large intestinal microbiota is therefore needed if problems are to be avoided. This raises a particular point of interest, in that the provision of fibre-degrading enzymes which simply dissolve insoluble fibre that ordinarily would have been attacked and fermented by bacteria living in the distal colon may accelerate this 'substrate starvation over time' effect noted earlier, and indeed in one experiment, this appears to have been the case when a xylanase was applied to a diet rich in corn bran (Petry et al., 2021a).

4.1.2 Large intestinal function

As indigesta exit the small intestine and enter the caecum, the vast majority of readily digestible nutrients have been removed leaving a composite of fermentable (resistant starch, fibre and protein) and inert (ash and intransigent fibres which are in large part cellulose and lignin) materials, the latter ultimately making up the non-microbial components of the faeces. The pig effectively separates the digesta in the caecum into the soluble and small particulate components, which enter the haustrae, and the larger particulates and insoluble materials into the core of the large intestine (Moran and Bedford, 2022). The haustrae are balloon-like structures which are formed as a result of the arrangements of the circular muscles but in particular the longitudinal muscle clusters which effectively create narrow entrances into the chamber of these structures. The narrow entrances restrict entry to liquids and small particulate matter which is then exposed to a specialised, mucosal resident microbiota which is adapted to ferment the material presented (Moran and Bedford, 2022). As the bacteria metabolise and replicate, the volatile fatty acids (VFA) and other compounds they produce are absorbed and the volume reduction of the haustrae is then dependent upon the relative rate of repletion of liquid and solutes from the core. Through muscular contractions, the haustrae move distally and subsequently expose new mucosal populations to the partially fermented material, such that by the time the haustrae reach the terminal colon, the material remaining has been largely fermented and there is little value left in their contents. Consequently, the microbiota must adapt to increasingly difficult to ferment materials (Bach Knudsen, 2016) as the haustrae move caudally and also rely more and more on the 'refill' of the haustra from the much less fermentable core materials that can enter.

The distal large intestinal microbiota should be considered as being poorly nourished compared with their more caudal counterparts (Bach Knudsen, 2016), and thus dietary adjustments that provide the distal hindgut with fermentable materials might be considered beneficial, especially when considering the fact that if fermentable carbohydrate in the form of fibre is not available, then the microbiota will use undigested protein if it is available, producing undesirable end products such as indoles, phenols, skatoles, amines and ammonia (Apajalahti and Rinttila, 2019). Fermented protein end products can result in significant damage to the enterocytes, a reduction in intestinal integrity which leads to leakage and translocation of intestinal bacteria to the blood and subsequently the liver and peripheral organs (Apajalahti and Rinttila, 2019). In the past, when antibiotics were routinely added to pig diets, these events were minimised or even eliminated through direct control of the protein-putrefactive bacterial species. In today's environment, however, the opportunity for such events to occur has increased manyfold. Thus, controlled, and consistent delivery of a more desirable fermentative fibre throughout the length of the large intestine should be the goal if intestinal health is to be maintained throughout its entire length in the absence of prophylactic antibiotics.

4.2 Role of fibre-degrading enzymes

Fibre-degrading enzymes clearly have a role to play here but their role may well be more significant in some sections of the intestine than others. Their predominant effects are to alter the presentation of the fibre in the finished feed which in general terms can be described as a reduction in particle size, dissolution of some insoluble material into the soluble fibre, and progressive reduction in molecular size of the soluble fibre over time.

Reduction in particle size, if it takes place in the small intestine, may benefit the digestion of nutrients as a whole as this effectively increases the surface area exposed to attack by digestive enzymes. It also enables greater inputs into the haustrae further down the intestinal tract where the most rapid fermentation takes place.

Reduction in molecular weight of the large soluble material may well benefit the digestive processes as this would likely be associated with a reduction in digesta viscosity which is sometimes (Bartelt et al., 2002; Chen et al., 2020) but not always (Duarte et al., 2019) associated with an improvement in nutrient digestibility (Jensen et al., 1998). It has been argued that in the pig, and perhaps even in the piglet, viscosity is far less of a problem than in the chicken because of the lower dry matter and hence lower concentration of viscous polymers (Bedford and Schulze, 1998) of porcine intestinal contents. Nevertheless, when digestion is compromised, as it often is at weaning, even marginal viscosity reductions may translate into performance gains.

Reduction of high-molecular weight soluble polysaccharides to lower molecular weight, and ultimately oligosaccharides effectively increases the fermentability of these fibre fractions such that they become more and more attractive to fibre-degrading species (Rybicka and Medel, 2021). In this case, the use of a xylanase can therefore result in increased energy extraction from the diet simply by increasing the proportion of fibre that is fermented (Petry et al., 2019). However, the benefits extend beyond simply increasing energy extraction. In presenting more fermentable carbohydrates, the activity and identity of the resident microbiota may change to the benefit of the health of the host as less protein will be putrefied to damaging amines and breakdown products (Apajalahti and Rinttila, 2019). This results in less resources wasted in the repair of damaged epithelial tissue and mounting an inflammatory response.

If the fermentation is augmented substantially then enterohormone signalling may well encourage the delayed release of digesta from the stomach and as a result protein digestion benefits (Ajuwon, 2019). In the younger pig, the provision of more oligosaccharides such as xylo-oligosaccharides (XOSs) by enzymatic depolymerisation of arabinoxylan substrates also seems to select for a more rapid establishment of a fibre fermenting microbiota in both the ileum and large intestine in what is termed a stimbiotic effect (Petry et al., 2021b,c), which effectively accelerates the establishment of a stable microbiota. The rationale for such a response has recently been discussed and is based on the idea that individual fibre-degrading and fibre-utilising bacteria tend to absorb specific oligosaccharides and metabolise these internally rather than degrade them to monosaccharides before absorption (Leth et al., 2023). It is the presence of these targeted oligosaccharides (the identity of which varies between species and defines their niche) which stimulate the augmentation of transcription of a number of enzymes and transporters which favour the growth of the species concerned (Leth et al., 2023). An added xylanase provided it can survive the intestinal environment, will continue to attack the insoluble and soluble larger molecular weight fibre as it transits through the small and subsequently the large intestine. In doing so it will continue to provide substrates and stimbiotic oligosaccharides for the resident microbiota as long as a suitable substrate remains.

One consideration is that fibre-degrading enzymes may in fact increase fermentation in the small intestine and caecum but then run out of substrate to the detriment of the distal colon (Chassé et al., 2022; Petry et al., 2021a). In such cases, the pig may simply respond by reducing the size of the distal colon as it is no longer providing for the host. Alternatively, it may result in substrates other than fibre being considered by the distal microbiota which could bring challenges as noted earlier. A possible solution to this problem if indeed it is real is to formulate increasingly difficult to ferment substrates into the diet as the pig ages, and ensure that enzymes are applied which can dissolve these substrates throughout the length of the colon. Insoluble fibre from soy hulls seems more intransigent than that from DDGS which in turn was slower to ferment than that from wheat middlings (Jaworski and Stein, 2017). This suggests that there may be an advantage in slowly adding soy hulls and a suitable enzyme at the expense of wheat middlings as the pig ages to maintain a substrate flow to the most distal microbiota.

4.3 Benefits noted from xylanase supplementation

Given the application of a xylanase can yield profound effects not only on the composition but also the activity of both small and large intestinal microbiota, it is perhaps not surprising that health benefits have been reported. Indeed, a recent review detailed effects such as increased intestinal integrity, reduced inflammatory responses and improved oxidative status (Petry and Patience, 2020) which all stem from intestinal effects but ultimately influence systemic health as well. This likely explains the improvements noted in livability when pigs are fed a xylanase, this being one effect of several that have been recently reported and are discussed in the following sections.

4.3.1 Livability

As noted earlier, xylanases and their arabinoxylan oligosaccharide (aka AXOS) products can accelerate the development of a fibre-degrading microbiota in the lower intestinal tract leading to better intestinal health which may have implications for systemic health. For example, Petry et al. (2020) demonstrated that the addition of a xylanase to a high-fibre corn-soy diet (high fibre as a result of inclusion of 30% corn bran) modified the microbiota in the colon, increasing butyrate producers (*Bifidobacterium* sp., *Faecalibacterium* sp.), lactate producers (*Lactobacillus* sp., Lachnospiraceae species), and fibre fermenters (e.g. *Ruminococcus* sp., *Prevotella* sp.) with a resultant increase in SCFA production (108.9 vs 82.8 mMol; $P < 0.05$; Petry et al., 2021). Butyric acid (+17%), acetate (+40%) and propionate (+17%; Ajuwon et al., 2018) increased significantly and all (Zhang et al., 2019), especially butyrate (Lin et al., 2020), are involved in improving intestinal structure and health.

One of the derived benefits observed in relation to gut health and functionality is the reduction in mortality rates in pigs. A number of studies (Fig. 2) have been conducted looking at the mortality rates in finisher pigs with/without a commercial xylanase, with each trial showing a numerical reduction in mortality when fed with the enzyme. This results in an overall significant improvement in livability (defined as 100 – mortality%, 95.2 vs 92.9; $P < 0.08$) when all results are combined and analysed in a meta-analysis. On average, xylanase reduces finisher mortality by 1/3 compared to pigs not fed xylanase.

Another more recent finding is that the incidence of ear necrosis in piglets (PEN) is reduced when a xylanase combined with a XOS was fed from weaning (Merriman et al., 2023). How this and improved longevity is brought about is

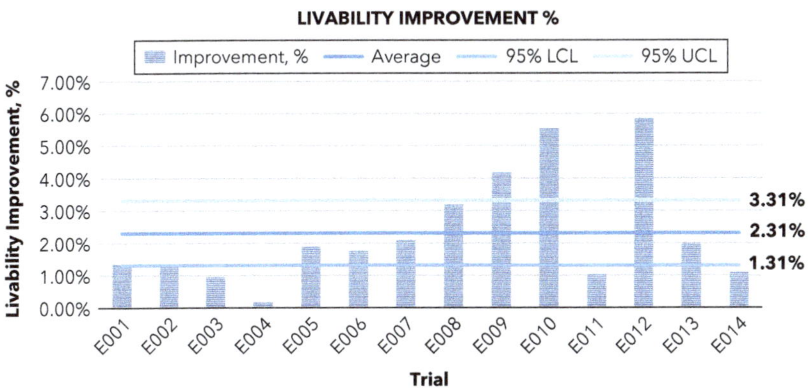

Figure 2 Summary of improved livability response with xylanase application across several trials conducted by AB Vista. Data presented are the livability improvement for each trial, the average livability improvement across all trials, and the 95% upper (UCL) and 95% lower (LCL) confidence level.

unclear but it could be that the voided faecal microbiota stimulated through use of the enzyme either presented less of a topical infection risk in the pen or the intestines were more robust and less subject to bacterial penetration and translocation to susceptible sites. Regardless, this is another piece of data that points to the link between good intestinal health and good health overall.

The main conclusion from the latest findings in use of fibre-degrading enzymes in pigs is that many of the benefits do not directly relate to improvements in rate or efficiency of gain but rather because of a generalised improvement in intestinal health, which improved the overall health of the pig. The above-mentioned metrics (liveability and PEN) were surprising but valuable for the industry, and perhaps if the net is cast wider, we may find other non-performance related benefits that accrue from long-term use of NSP-degrading enzymes in pigs. An important point to note here is that both of these benefits were recognised by the industry rather than by direct research. This is a consequence of the scale of a research trial compared with that of commercial operations. A reduction of 1% in mortality would be hard to detect in a trial where each treatment comprised 200 pigs (which would be a moderately large trial) but quite easy in a commercial operation where 50000 pigs might be processed per week. Commercial operations with good data collection and recovery for analysis are the most likely source of new discoveries in this field.

5 Constraints on enzyme functionality

In order for feed enzymes to be of value in the animal they must be able reach the parts of the GIT where they are needed which means they have to survive several stresses including the following, discussed below.

5.1 Feed manufacture

Much of the feed produced for pigs is pelleted, and specifically for this form of feed any enzyme that is added to the diet needs to be able to survive the thermal challenge of the pelleting process. This can be achieved either through discovery or design of an intrinsically thermostable molecule, through a protective coating or by applying the enzyme as a liquid spray onto the pellets after the process has completed. Issues exist with each which must be overcome in order to guarantee consistency of response. For example, selection of a naturally thermostable molecule may result in an enzyme which is not particularly good at working at body temperature. The use of coating technology is interesting but it may be susceptible to damage in storage, mixing with minerals and in can delay the release of an enzyme which could influence functionality. This is especially the true in the case of phytase as it must function as soon as possible if maximum InsP6 → InsP1 conversion is to be achieved as discussed earlier. With liquid application, the challenge is always to get uniform distribution and to maintain such delivery over time in a dusty environment. Fortunately, the more modern applicators seem to overcome many of the problems encountered with initial models.

5.1.1 Digestion stability

All enzymes are proteins and are subjected to the proteolytic activities ongoing in the stomach and small intestine. Some enzymes are more resilient than others which used to be cited as a major reason for differences in efficacy noted (Igbasan et al., 2002; Menezes-Blackburn et al., 2015), but more modern versions have been selected or evolved to be more stable and thus effective under the conditions of the intestinal tract (Menezes-Blackburn et al., 2015). As a result, proteolytic stability is less of an issue now than it was several years ago

5.1.2 Dietary ingredients/nutrients

Some dietary ingredients influence the manner in which an enzyme can function. Phytases, for example, are less effective in diets containing high levels of zinc for example (Walk et al., 2013, 2015) which means care should be taken when applying matrices to this enzyme in diets with pharmacological levels of Zn present

6 Conclusion

The use of feed enzymes may be at a turning point where the end user starts to consider the ancillary benefits of use of such products more so than they consider the potential to save costs by applying nutrient matrices. In the case of the two major enzyme classes used in pig diets, xylanase and phytase, it is becoming apparent that the benefits observed in terms of increased liveability, reduced disease incidence and improved sow productivity may be as valuable if not more so than the value that feed enzymes deliver in reducing feed cost. In some cases, this may involve using these products at inclusion rates which may not be justifiable on the basis of nutrient recovery alone. As the pig industry faces increasing challenges with regard to feed costs, welfare and health issues, the industry may start to view feed enzymes in a much broader sense than it currently does.

7 Where to look for further information

- Bedford, M. R., Partridge, G. G., Hruby, M. and Walk, C. L. eds., 2022. *Enzymes in farm animal nutrition*. Cabi.

- Kiarie E., Romero L. F. and Nyachoti, C. M. 2013. The role of added feed enzymes in promoting gut health in swine and poultry. *Nutrition Research Reviews*. 26(1):71-88. doi:10.1017/S0954422413000048.
- Moran, E.T. and Bedford, M. R. 2022. Large intestinal dynamics differ between fowl and swine: Anatomical modifications, microbial collaboration, and digestive advantages from fibrolytic enzymes. *Animal Nutrition* 11;160-170. ISSN 2405-6545, https://doi.org/10.1016/j.aninu.2022.07.004.
- Knudsen, K. E. B., Lærke, H. N. and Jørgensen, H. 2022. Carbohydrates and Carbohydrate Utilization in Swine.In Sustainable Swine Nutrition pp.151-187).

8 References

Adeola, O., Olukosi, O. A., Jendza, J. A., Dilger, R. N. and Bedford, M. R. 2006. Response of growing pigs to Peniophora lycii- and Eschericia coli-derived phytases or varying ratios of calcium to total phosphorus. *Animal Science* 82(5), 637–644.

Ajuwon, K. M. 2019. The role of fiber in the regulation of brain function: implication for welfare and appetite regulation in the pig. *MSNC* 19-29.

Angel, R., Tamin, N. M., Applegate, T. J., Dhandu, A. S. and Ellestad, L. E. 2002. Phytic acid chemistry: influence on Phytin phosphorus availability and phytase efficacy. *Journal of Applied Poultry Research* 11(4), 471–480.

Apajalahti, J. and Rinttila, T. 2019. Assessing the complex ecology of intestinal microbiome. In: Gonzalez-Ortiz, G. (Ed.). *The Value of Fibre -Engaging the Second Brain for Animal Nutrietion*. Wageningen, The Netherlands: Wageningen Academic Publishers.

Arredondo, M. A., Casas, G. A. and Stein, H. H. 2019. Increasing levels of microbial phytase increases the digestibility of energy and minerals in diets fed to pigs. *Animal Feed Science and Technology* 248, 27–36.

Bach Knudsen, K. E. 2014. Fiber and nonstarch polysaccharide content and variation in common crops used in broiler diets. *Poultry Science* 93(9), 2380–2393.

Bach Knudsen, K. E. 2016. Carbohydrates in pig nutrition: recent advances. *Journal of Animal Science* 94, 1–12.

Bartelt, J., Jadamus, A., Wiese, F., Swiech, E., Buraczewska, L. and Simon, O. 2002. Apparent precaecal digestibility of nutrients and level of endogenous nitrogen in digesta of the small intestine of growing pigs as affected by various digesta viscosities. *Archiv Fur Tierernahrung* 56(2), 93–107.

Bedford, M. R. and Schulze, H. 1998. Exogenous enzymes for pigs and poultry [Review]. *Nutrition Research Reviews* 11(1), 91–114.

Bedford, M. R. and Walk, C. L. 2017. The use of exogenous enzymes to improve feed efficiency in pigs. In: Wiseman, J. (Ed.). *Achieving Sustainable Production of Pig Meat*. Nottingham: Burleigh Dodds Science Publishing.

Campbell, J. M., Crenshaw, J. D. and Polo, J. 2013. The biological stress of early weaned piglets. *Journal of Animal Science and Biotechnology* 4(1), 19.

Chassé, É., Toft Vangsøe, C. T., Bach Knudsen, K. E., Guay, F. and Létourneau-Montminy, M.-P. 2022. Exogenous enzymes, meal size, and meal frequency: effect on ileal and total tract digestibility of carbohydrates, and energy and fibre degradation in growing pigs fed a wheat-barley grain-based high-fibre diet. *Journal of Animal Science* 100.

Chen, H., Zhang, S. and Kim, S. W. 2020. Effects of supplemental xylanase on health of the small intestine in nursery pigs fed diets with corn distillers' dried grains with solubles. *Journal of Animal Science* 98(6), skaa185.

Cho, H. M., González-Ortiz, G., Melo-Durán, D., Heo, J. M., Cordero, G., Bedford, M. R. and Kim, J. C. 2020. Stimbiotic supplementation improved performance and reduced inflammatory response via stimulating fiber fermenting microbiome in weaner pigs housed in a poor sanitary environment and fed an antibiotic-free low zinc oxide diet. *PLOS ONE* 15(11), e0240264.

Cowieson, A. J., Acamovic, T. and Bedford, M. R. 2004. The effects of phytase and phytic acid on the loss of endogenous amino acids and minerals from broiler chickens. *British Poultry Science* 45, 101–108.

Cowieson, A. J., Acamovic, T. and Bedford, M. R. 2006. Phytic acid and Phytase: implications for Protein Utilization by Poultry. *Poultry Science* 85(5), 878–885.

Cowieson, A. J. and Bedford, M. R. 2009. The effect of phytase and carbohydrase on ileal amino acid digestibility in monogastric diets: complimentary mode of action? *World's Poultry Science Journal* 65(4), 609–624.

Cowieson, A. J., Vieira, S. L. and Stefanello, C. 2019. Exogenous microbial amylase in the diets of poultry: what do we know? *Journal of Applied Poultry Research* 28(3), 556–565.

Davies, N. T. and Flett, A. A. 1978. The similarity between alkaline phosphatase (EC 3.1.3.1) and phytase (EC 3.1.3.8) activities in rat intestine and their importance in phytate-induced zinc deficiency. *British Journal of Nutrition* 39(2), 307–316.

De Vries, S., Gerrits, W. J. J., Kabel, M. A., Vasanthan, T. and Zijlstra, R. T. 2016. B-glucans and resistant starch alter the fermentation of recalcitrant fibers in growing pigs. *PLOS ONE* 11(12), e0167624.

Duarte, M. E., Zhou, F. X., Dutra, W. M. and Kim, S. W. 2019. Dietary supplementation of xylanase and protease on growth performance, digesta viscosity, nutrient digestibility, immune and oxidative stress status, and gut health of newly weaned pigs. *Animal Nutrition* 5(4), 351–358.

Dzierva, L., Conceição Dos Santos, M., De Almeida, L. M., Da Rocha, C., Bassi, L. S., Muramatsu, K. and Maiorka, A. 2023. Feed particle size and amylase supplementation in diets for nursery piglets: digestibility and growth performance. *Livestock Science*, 105211.

Espinosa, C. D., Oliveira, M. S. F., Velayudhan, D. E., Dersjant-Li, Y. and Stein, H. H. 2021. Influence of a novel consensus bacterial 6-phytase variant on mineral digestibility and bone ash in young growing pigs fed diets with different concentrations of phytate-bound phosphorus. *Journal of Animal Science* 99(8), 1-2.

Evans, W. J., Mccourtney, E. J. and Shrager, R. I. 1982. Titration studies of phytic acid. *Journal of the American Oil Chemists' Society* 59(4), 189–191.

Fernandez, J. I. 2009. Boosting growth in young piglets with amylase based enzymes. *International Pig Topics* 24(7), 11-13.

Gomes, G. A., Dos Santos, T. T., Cordero, G. and Rousseau, X. 2019. "Stimbiotics" may be a new approach to beneficial fibre fermentation. *Feedstuffs* 91, 2-3.

Hedemann, M. S. and Jensen, B. B. 2004. Variations in enzyme activity in stomach and pancreatic tissue and digesta in piglets around weaning. *Archives of Animal Nutrition* 58(1), 47–59.

Hoogeveen, A. M. E., Moughan, P. J., Henare, S. J., Schulze, P., Mcnabb, W. C. and Montoya, C. A. 2021. Type of dietary fiber is associated with changes in ileal and hindgut microbial communities in growing pigs and influences in vitro ileal and hindgut fermentation. *The Journal of Nutrition* 151(10), 2976–2985.

Humer, E., Schwarz, C. and Schedle, K. 2015. Phytate in pig and poultry nutrition. *Journal of Animal Physiology and Animal Nutrition* 99(4), 605–625.

Igbasan, F. A., Simon, O. and Manner, K. 2002. A comparative study of the stability of microbial phytases and phytate phosphorus hydrolysis in the gastrointestinal tract of chickens. *Journal of Sustainable Agriculture and the Environment* 4, 8–16.

Jaworski, N. W. and Stein, H. H. 2017. Disappearance of nutrients and energy in the stomach and small intestine, cecum, and colon of pigs fed corn-soybean meal diets containing distillers dried grains with solubles, wheat middlings, or soybean hulls. *Journal of Animal Science* 95(2), 727–739.

Jensen, M. S., Knudsen, K. E., Inborr, J. and Jakobsen, K. 1998. Effect of β glucanase supplementation on pancreatic enzyme activity and nutrient digestibility in piglets fed diets based on hulled and hulless barley varieties. *Animal Feed Science and Technology* 72(3–4), 329–345.

4Jongbloed, A. W., van Diepen, J. T. M., Kemme, P. A. and Broz, J. (2004). Efficacy of microbial phytase on mineral digestibility in diets for gestating and lactating sows. *Livestock Production Science* 91(1–2), 143–155. https://doi.org/10.1016/j.livprodsci.2004.07.017.

Kelly, D., Smyth, J. A. and Mccracken, K. J. 1991. Digestive development of the early-weaned pig: 1. Effect of continuous nutrient supply on the development of the digestive tract and on changes in digestive enzyme activity during the first week post-weaning. *British Journal of Nutrition* 65(2), 169–180.

Kim, Y. J., Lee, J. H., Kim, T. H., Song, M. H., Yun, W., Oh, H. J., Lee, J. S., Kim, H. B. and Cho, J. H. 2021. Effect of low protein diets added with protease on growth performance, nutrient digestibility of weaned piglets and growing-finishing pigs. *Journal of Animal Science and Technology* 63(3), 491–500.

Lackeyram, D., Yang, C., Archbold, T., Swanson, K. C. and Fan, M. Z. 2010. Early weaning reduces small intestinal alkaline phosphatase expression in pigs. *Journal of Nutrition* 140(3), 461–468.

Lærke, H. N., Arent, S., Dalsgaard, S. and Bach Knudsen, K. E. 2015. Effect of xylanases on ileal viscosity, intestinal fiber modification, and apparent ileal fiber and nutrient digestibility of rye and wheat in growing pigs1,2. *Journal of Animal Science* 93(9), 4323–4335.

Lagos, L. V., Bedford, M. R. and Stein, H. H. 2021. Increased microbial phytase increased phytate destruction, plasma inositol, and feed efficiency of weanling pigs, but reduced dietary calcium and phosphorus did not affect gastric pH or fecal score and reduced growth performance and bone ash. *Journal of Animal Science* 99(12), 1–13.

Laird, S., Brearley, C., Kuhn, I., Miller, H. and Miller, H. M. 2016. The effect of high phytase doses on phytate degradation and myo-inositol release in the grower pig. *BSAS* 2016, 105-105.

Lee, J. J., Kang, J., Park, S., Cho, J. H., Oh, S., Park, D. J., Perez-Maldonado, R., Cho, J. Y., Park, I. H., Kim, H. B. and Song, M. 2020. Effects of dietary protease on immune responses of weaned pigs. *Journal of Animal Science and Technology* 62(2), 174–179.

Lei, X. G., Ku, P. K., Miller, E. R., Yokoyama, M. T. and Ullrey, D. E. 1994. Calcium level affects the efficacy of supplemental microbial phytase in corn soybean meal diets of weanling pigs. *Journal of Animal Science* 72(1), 139–143.

Leth, M. L., Pichler, M. J. and Abou Hachem, M. 2023. Butyrate-producing colonic clostridia: picky glycan utilization specialists. *Essays in Biochemistry* 67(3), 415–428.

Letourneau-Montminy, M. P., Narcy, A., Magnin, M., Sauvant, D., Bernier, J. F., Pomar, C. and Jondreville, C. 2010. Effect of reduced dietary calcium concentration and phytase supplementation on calcium and phosphorus utilization in weanling pigs with modified mineral status. *Journal of Animal Science* 88(5), 1706–1717.

Lin, F., Li, X., Wen, J., Wang, C., Peng, Y., Feng, J. and Hu, C. 2020. Effects of coated sodium butyrate on performance, diarrhea, intestinal microflora and barrier function of pigs during the first 2-week post-weaning. *Animal Feed Science and Technology* 263, 114464.

Lindemann, M. D., Cornelius, S. G., El Kandelgy, S. M., Moser, R. L. and Pettigrew, J. E. 1986. Effect of age, weaning and diet on digestive enzyme levels in the piglet. *Journal of Animal Science* 62(5), 1298–1307.

Luttrell, B. M. 1993. The biological relevance of the binding of calcium ions by inositol phosphates. *Journal of Biological Chemistry* 268(3), 1521–1524.

Martens, B. M. J., Flécher, T., De Vries, S., Schols, H. A., Bruininx, E. M. A. M. and Gerrits, W. J. J. 2019. Starch digestion kinetics and mechanisms of hydrolysing enzymes in growing pigs fed processed and native cereal-based diets. *British Journal of Nutrition* 121(10), 1124–1136.

Menezes-Blackburn, D., Gabler, S. and Greiner, R. 2015. Performance of seven commercial phytases in an in vitro simulation of poultry digestive tract. *Journal of Agricultural and Food Chemistry* 63(27), 6142–6149.

Merriman, L. A., Becker, S. and Hasan, S. 2023. Use of a stimbiotic to ameliorate ear necrosis in nursery pigs. 2023 APSA Conference, Nov13–16, 2023. Brisbane: Australasian Pig Science Association, 1.

Mesina, V. G. R., Lagos, L. V., Sulabo, R. C., Walk, C. L. and Stein, H. H. 2019. Effects of microbial phytase on mucin synthesis, gastric protein hydrolysis, and degradation of phytate along the gastrointestinal tract of growing pigs. *Journal of Animal Science* 97(2), 756–767.

Molist, F., Van Oostrum, M., Perez, J. F., Mateos, G. G., Nyachoti, C. M. and Van Der Aar, P. J. 2014. Relevance of functional properties of dietary fibre in diets for weanling pigs. *Animal Feed Science and Technology* 189, 1–10.

Moran, E. T. and Bedford, M. R. 2022. Large intestinal dynamics differ between fowl and swine: anatomical modifications, microbial collaboration, and digestive advantages from fibrolytic enzymes. *Animal Nutrition* 11, 160–170.

Novotny, M., Sommerfeld, V., Krieg, J., Kühn, I., Huber, K. and Rodehutscord, M. 2023. Comparison of mucosal phosphatase activity, phytate degradation, and nutrient digestibility in 3-week-old turkeys and broilers at different dietary levels of phosphorus and phytase. *Poultry Science* 102, 102457.

Ochonski, P., Wu, F., Tokach, M. D., Derouchey, J. M., Dritz, S. S., Goodband, R. D., Woodworth, J. C. and Lattimer, J. M. 2021. Evaluation of Enogen Feed Corn on growth performance and carcass characteristics of finishing pigs. *Translational Animal Science* 5(2), yxab052.

Park, S., Lee, J. J., Yang, B. M., Cho, J. H., Kim, S., Kang, J., Oh, S., Park, D. J., Perez-Maldonado, R., Cho, J. Y., Park, I. H., Kim, H. B. and Song, M. 2020. Dietary protease improves growth performance, nutrient digestibility, and intestinal morphology of weaned pigs. *Journal of Animal Science and Technology* 62(1), 21–30.

Persia, M. E. 2010. An enzymatic approach to dealing with the negative consequences of phytate. US Multistate Poultry Meeting 2010, 1–9.

Persson, H., Turk, M., Nyman, M. and Sandberg, A.-S. 1998. Binding of Cu^{2+}, Zn^{2+} and Cd^{2+} to inositol tri-, tetra-, penta- and hexa-phosphates. *Journal of Agricultural and Food Chemistry* 46(8), 3194–3200.

Petry, A. L., Masey O'neill, H. V. and Patience, J. F. 2019. Xylanase, and the role of digestibility and hindgut fermentation in pigs on energetic differences among high and low energy corn samples1. *Journal of Animal Science* 97(10), 4293–4297.

Petry, A. L. and Patience, J. F. 2020. Xylanase supplementation in corn-based swine diets: a review with emphasis on potential mechanisms of action. *Journal of Animal Science* 98(11), 1-12; skaa318.

Petry, A. L., Huntley, N. F., Bedford, M. R. and Patience, J. F. 2021a. The influence of xylanase on the fermentability, digestibility, and physicochemical properties of insoluble corn-based fiber along the gastrointestinal tract of growing pigs. *Journal of Animal Science* 99(7), skab159.

Petry, A. L., Patience, J. F., Huntley, N. F., Koester, L. R., Bedford, M. R. and Schmitz-Esser, S. 2021b. Xylanase supplementation modulates the microbiota of the large intestine of pigs fed corn-based fiber by means of a stimbiotic mechanism of action. *Frontiers in Microbiology* 12, 619970.

Petry, A. L., Patience, J. F., Koester, L. R., Huntley, N. F., Bedford, M. R. and Schmitz-Esser, S. 2021c. Xylanase modulates the microbiota of ileal mucosa and digesta of pigs fed corn-based arabinoxylans likely through both a stimbiotic and prebiotic mechanism. *PLOS ONE* 16(1), e0246144.

Portillo, G. A., Renaut Aquino, J. E. and Silva Mazacotte, E. E. 2021. Efectos de la utilizaciÃ³n de enzimas ±-amilasa y lipasa sobre el desempeÃ±o productivo de lechones reciÃ©n destetados. *InvestigaciÃ³N Agraria* 23, 111–116.

Rhouma, M., Fairbrother, J. M., Beaudry, F. and Letellier, A. 2017. Post weaning diarrhea in pigs: risk factors and non-colistin-based control strategies. *Acta Veterinaria Scandinavica* 59(1), 31.

Rosen, G. D. 2002. Exogenous enzymes as pro-nutrients in broiler diets. In: Garnsworthy, P. C. and Wiseman, J. (Eds). *Recent Advances in Animal Nutrition 2002* (1st edn.). Nottingham: Nottingham University Press.

Rosen, G. D. 2003. Setting and meeting standards for the efficient replacement of pronutrient antibiotics in broiler, turley and pig nutrition. In: Kreuzer, M., Wenk, C. and Lanzini, T. (Eds). *Gesunde Nutztiere. Heutiger Stellewert der Futterzusatzstoffe in der Tierernaehrung*. Tagungsbericht: Band 24, Schriftenreihe aus den Institut fuer Nutztierwissenschaften, Ernaehrung-Produkte-Umwelt, ETH-Zentrum, CH-8092 Zuerich.

Rosen, G. D. 2006. Holo-analysis. *Poultry Science* 85(6), 957–959.

5Rosenfelder-Kuon, P., Siegert, W. and Rodehutscord, M. (2019). Effect of microbial phytase supplementation on P digestibility in pigs: a meta-analysis. *Archives of Animal Nutrition* 74(1), 1–18. https://doi.org/10.1080/1745039X.2019.1687249.

Rosenfelder-Kuon, P., Klein, N., Zegowitz, B., Schollenberger, M., Kühn, I., Thuringer, L., Seifert, J. and Rodehutscord, M. 2020a. Phytate degradation cascade in pigs as affected by phytase supplementation and rapeseed cake inclusion in corn–soybean meal-based diets. *Journal of Animal Science* 98(3).

Rybicka, A. and Medel, P. 2021. Fibre in swine nutrition. In: FEDNA (Ed.). *Curso de Especializacion Fedna*. Madrid: Fedna.

Schlemmer, U., Jany, K. D., Berk, A., Schulz, E. and Rechkemmer, G. 2001. Degradation of phytate in the gut of pigs: pathway of gastro-intestinal inositol phosphate hydrolysis and enzymes involved. *Archiv Fur Tierernahrung* 55(4), 255–280.

Selle, P. H. and Ravindran, V. 2008. Phytate degrading enzymes in pig nutrition. *Livestock Science* 113(2–3), 99–122.

Sommerfeld, V., Schollenberger, M., Kuhn, I. and Rodehutscord, M. 2018. Interactive effects of phosphorus, calcium and phytase supplements on products of phytate degradation in the digestive tract of broiler chickens. *Poultry Science*. https://doi.org/10.3382/ps/pex404.

8Tokach, M. D., Menegat, M. B., Gourley, K. M. and Goodband, R. D. (2019). Review: nutrient requirements of the modern high-producing lactating sow, with an emphasis on amino acid requirements. *Animal* 13(12), 2967–2977. https://doi.org/10.1017/S1751731119001253.

Torres-Pitarch, A., Hermans, D., Manzanilla, E. G., Bindelle, J., Everaert, N., Beckers, Y., Torrallardona, D., Bruggeman, G., Gardniner, G. E. and Lawlor, P. G. 2017. Effect of feed enzymes on digestibility and growth in weaned pigs; A systematic review and meta-analysis. *Animal Feed Science and Technology*, in PRESS, 233, 145-159.

Torres-Pitarch, A., Manzanilla, E. G., Gardiner, G. E., O'doherty, J. V. and Lawlor, P. G. 2019. Systematic review and meta-analysis of the effect of feed enzymes on growth and nutrient digestibility in grow-finisher pigs: effect of enzyme type and cereal source. *Animal Feed Science and Technology* 251, 153–165.

Urriola, P. E. and Stein, H. H. 2010. Effects of distillers dried grains with solubles on amino acid, energy, and fiber digestibility and on hindgut fermentation of dietary fiber in a corn-soybean meal diet fed to growing pigs. *Journal of Animal Science* 88(4), 1454–1462.

6van Dijk, A. J., van Rens, B. T. T. M., van der Lende, T. and Taverne, M. A. M. (2005). Factors affecting duration of the expulsive stage of parturition and piglet birth intervals in sows with uncomplicated, spontaneous farrowings. *Theriogenology* 64(7), 1573–1590. https://doi.org/10.1016/j.theriogenology.2005.03.017.

Vangsøe, C. T., Bonnin, E., Joseph-Aime, M., Saulnier, L., Neugnot-Roux, V. and Knudsen, K. E. B. 2021. Improving the digestibility of cereal fractions of wheat, maize, rice by a carbohydrase complex rich in xylanases and arabinofuranosidases: an in vitro digestion study. *Journal of the Science of Food and Agriculture* 101(5), 1910–1919.

Walk, C. L., Bedford, M. R. and Olukosi, O. A. 2018. Effect of phytase on growth performance, phytate degradation and gene expression of myo- inositol transporters in the small intestine, liver and kidney of 21 day old broilers. *Poultry Science* 97(4), 1155–1162.

Walk, C. L., Srinongkote, S. and Wilcock, P. 2013. Influence of a microbial phytase and zinc oxide on young pig growth performance and serum minerals. *Journal of Animal Science* 91(1), 286-291.

Walk, C. L., Wilcock, P. and Magowan, E. 2015. Evaluation of pharmacological zinc oxide, phytase and phosphorus on weaned piglet growth performance, plasma minerals and faecal mineral digestibility and excretion. *Animal*. 5

7Wealleans, A. L., Bold, R. M., Dersjant-Li, Y. and Awati, A. (2015). The addition of a Buttiauxella sp. phytase to lactating sow diets deficient in phosphorus and calcium reduces weight loss and improves nutrient digestibility. *Journal of Animal Science* 93(11), 5283–5290. https://doi.org/10.2527/jas.2015-9317.

Xu, P., Price, J., Wise, A. and Aggett, P. J. 1992. Interaction of inositol phosphates with calcium, zinc and histidine. *Journal of Inorganic Biochemistry* 47(2), 119–130.

Zhang, Y., Chen, H., Zhu, W. and Yu, K. 2019. Cecal infusion of sodium propionate promotes intestinal development and jejunal barrier function in growing pigs. *Animals* 9(6), 284.

Part 4

Alternative feed sources

Chapter 10

Co-products and other alternative feedstuffs for sustainable pig production: an overview

Ruurd T. Zijlstra[1], *Li Fang Wang and Eduardo Beltranena, University of Alberta, Canada*

1 Introduction

Achieving long-term sustainability in swine production requires the consideration of issues such as the economics, societal acceptance and environmental footprint of the sector. In most of North America, there is a strong tradition of feeding pigs diets containing high levels of feed grains. However, the growing global human population, coupled with increased crop production for biofuel due to government incentives (Martin and Lazarevic, 2015) and increasing fluctuations in weather patterns have all affected crop cultivation. This raises questions about how reliable, abundant and cheap feed grain supply for livestock will be in the future. Moreover, questions have been raised about the impact of human consumption of animal-based diets based on grain-fed diets on both human health and the environment (Willett et al., 2019). Nonetheless, a minimum contribution of animal-derived food is

1 Corresponding author: ruurd.zijlstra@ualberta.ca.

http://dx.doi.org/10.19103/AS.2024.0140.16

important for human nutrition and health, especially during periods of high nutritional demands such as growth, gestation and lactation (FAO, 2023).

A more sustainable food system involves using grains and oilseeds mainly for direct human consumption (the food industry) and then feeding more co-products to pigs and other livestock to produce animal protein for human nutrition within a circular food system (Van Zanten et al., 2019). A large array of alternative feedstuffs exists (Schedle et al., 2013; Florou-Paneri et al., 2014; Woyengo et al., 2014) including pulse grains (Wang et al., 2022). However, the present chapter is focused on the co-products of cereal grains and oilseeds. Co-products can originate from a wide range of sources including, for example, rice bran from rice (Casas et al., 2015), wheat millrun from wheat (Garcia et al., 2015) or corn distillers dried grains with solubles (DDGS) from corn (Stein and Shurson, 2009).

A major advantage of using co-products in feeding growing-finishing pigs, which consume around 85% of the feed in most swine operations, is in mitigating feed cost (Woyengo et al., 2014). However, a shift towards feeding diets that contain more co-products and less feed grains also poses a range of challenges. Plant-based co-products generally contain more fibre than the parent cereal or oilseed (NRC, 2012). As starch is removed from cereals, or oil is removed from oilseeds (representing the most valuable components for human food or renewable energy sources), the remaining components, including fibre, are left. Consequently, co-products have lower energy and protein digestibility than the parent crop (Sauvant et al., 2004; NRC, 2012). Co-products also tend to be more variable in nutritional value, in part because of processing methods and conditions, than the parent crop (NRC, 2012). This makes it more challenging to achieve constant energy values and digestible nutrient content in achieving consistent animal growth through diet. These issues and challenges are discussed in detail in the following sections.

2 Feed evaluation, animal growth and carcass characteristics

Diets containing co-products generally contain more fibre than diets without co-products. As a result, more undigested residue remains within the lumen of the gastrointestinal tract when fed to pigs, thus altering digestive physiology (Zijlstra et al., 2012). Feeding diets containing high-fibre co-products is associated with increased fermentation of the undigested residue within the gastrointestinal tract and with changes in the rate of digesta flow, depending on the type and functionality of the fibre (Zijlstra et al., 2012). The extent of excretion of the undigested co-product residue depends on cell wall components and physico-chemical properties of the fibre such as viscosity and fermentability (Dikeman and Fahey, 2006; Williams et al., 2019). Challenges to the animal from a greater fermentable fibre content include, for example, greater heat increment (difference between metabolizable energy and net energy value), lower feed conversion, increased enterocyte proliferation resulting in greater mass of the gastrointestinal tract and increased endogenous amino acid losses (Libao-Mercado et al., 2007).

Increased dietary inclusion of co-products has historically raised concerns. For decades, fibre was regarded as a negative dietary factor causing the dilution of dietary energy and nutrients, and reducing energy and nutrient digestibility (Fairbairn et al., 1999). In addition, excessive dietary fibre (with a high water-holding capacity) was associated with reduced voluntary feed intake in pigs (Kyriazakis and Emmans, 1995). High dietary fibre has been seen as a concern in maintaining the high growth performance of nursery pigs due to an immature gut (Wang et al., 2016) compared with growing-finishing pigs or sows. However, these challenges can now be managed nutritionally by formulating diets based on the net energy system. This involves formulating diets based on digestible amino acids and protein with additional threonine to optimise protein synthesis (Zhu et al., 2005; NRC, 2012).

Variation in nutritional value among batches of the same co-products can be managed using existing technologies. For example, large numbers of samples of the same ingredient category can be characterized using *in vitro* digestibility technologies. The resulting digestibility coefficients and data from laboratory analyses can then serve to develop near-infrared reflectance spectroscopy calibrations to predict the nutritional value of individual ingredient samples (Wang et al., 2021).

Alternative feedstuffs may also contain an array of anti-nutritional factors (ANF) (Woyengo et al., 2017; Wang et al., 2022). However, the ANF phytate is partially degraded during the production of the co-product corn DDGS, resulting in greater phosphorus digestibility in corn DDGS than in the parent grain (Widyaratne and Zijlstra, 2007). Mycotoxins in particular may concentrate in co-products during the production process. For example, deoxynivalenol (vomitoxin) survives fermentation and drying and becomes concentrated in corn DDGS because of the loss of starch that is converted into ethanol and CO_2 (Khatibi et al., 2014). ANF and mycotoxins content may place limits on the inclusion of alternative feedstuffs in swine diets (Frobose et al., 2015).

Using modern feed evaluation systems, such as the net energy system and standardized ileal digestible amino acids for feed formulation, the growth performance of growing-finishing pigs from a body mass perspective can likely be maintained with a relatively high dietary inclusion of co-products (Jha et al., 2013; Woyengo et al., 2014). This is on the assumptions that dietary energy values are allowed to vary and that the cost of maintaining these values is not too high. In the absence of disease or heat challenge, feeding reduced energy diets means that feed cost per unit of weight gain is lower with increased feed intake (Beltranena, 2022). The reduced feed cost may outweigh the increased cost of extending days to reach slaughter weight due to slower nursery growth plus the consequential reduction in housing utilization. Feeding diets containing fibrous co-products increases visceral mass and thus reduces carcass weight, requiring careful management of dressing percentage (Agyekum et al., 2012; Jha et al., 2013; Beltranena, 2022).

3 Meat quality and animal health

In addition to growth performance and carcass characteristics, feeding co-products can help address key aspects of pig production such as: nutrient management (including greenhouse gas emissions),animal health and welfare, as well as pork quality.

In the case of feeding pigs co-products with high remaining oil content, such as original, high-fat corn DDGS (Stein and Shurson, 2009), dietary unsaturated lipid content reduces carcass fat firmness (Jha et al., 2013). This produces loins with less firmness as well as processed pork products with a higher unsaturated fatty acid content, poorer sensory quality and higher risk of rancidity. This requires anticipating and managing potential reductions in pork quality. For example, carcass fatty composition can be managed by withdrawing dietary corn DDGS inclusions within weeks prior to slaughter (Hilbrands et al., 2013; Beltranena 2022).

For sows, feeding dietary co-products may provide additional benefits. Isocaloric diets containing wheat bran, sugar beet pulp and oat grain increased diet crude fibre content from 38 to 70 g/kg, which provided enteral stimulation and drastically reduced constipation from 5 days before to 5 days after farrowing (Oliviero et al., 2009; Peltoniemi et al., 2016; Lu et al., 2022). Similarly, feeding sows during late gestation a diet containing sugar beet pulp, soybean hulls, wheat bran and sunflower meal increased diet crude fibre content from 28 to 110 g/kg, which increased subsequent voluntary feed intake of a common lactation diet during the first 3 weeks of lactation and may thereby prevent postpartum dysgalactia syndrome (Quesnel et al., 2009).

Due to the increased feed intake and thus nutrient intake of lactating sows, suckling piglets increase their growth during lactation due to increased sow milk production resulting in heavier weaning weight. For sows, fibre source and quantity of fibre included in the diet are important. Gestating sows housed in stalls and fed diets with increased fibre showed a reduced incidence of behavioural stereotypies (Bergeron et al., 2000).

Feeding an adequate mass of fibre-rich co-products to nursery pigs directly after weaning may provide gut health and developmental benefits resulting in reduced incidence of post-weaning diarrhoea and increased survival (Pluske et al., 2018). In an array of studies challenging weaned pigs with *E. coli*, the dietary inclusion of 40 g/kg of coarsely ground wheat bran improved faeces consistency, reduced proliferation of *E. coli* in the large intestine and increased concentration of short-chain fatty acids in the colon (Molist et al., 2009, 2014). Resistant starch may also provide benefits to gut health in pigs after weaning (Tan et al., 2021). For example, resistant starch may function as a fermentable fibre resulting in increased butyrate concentration in the large intestine (Fouhse et al., 2017). Increased carbohydrate fermentation is associated with increased *Bifidobacteria*, i.e. bacteria that are positively associated with gut health (Fouhse et al., 2017).

However, if there is too much dietary soluble, fermentable fibre, for example 60 g sugar beet pulp plus 20 g/kg soybean hulls, diarrhoea incidence in nursery pigs housed under poor sanitary conditions will increase (Montagne et al., 2012; Molist et al., 2014). A dietary inclusion of up to 240 g sugar beet pulp/kg was found to not cause diarrhoea but instead cause loose faeces in the first week and marginally reduced growth in weaned pigs (Wang et al., 2016).

Finally, changes in fibre characteristics may alter the physical characteristics of digesta, especially the water-retention capacity of colonic digesta. For example, 80 g/kg dietary wheat bran and 60 g/kg sugar beet pulp increased the water-retention capacity of colonic digesta that may partly explain the loosening of faeces (Molist et al., 2010). Consequently, an optimum but not a maximum amount of dietary fibre with specific properties is likely required to enhance gut health.

4 Sustainability: ammonia and greenhouse gas emissions

In growing-finishing pigs, the dietary inclusion of co-products can be used to alter nitrogen and carbon excretion patterns (Jarret et al., 2011). For example, the dietary inclusion of soy hulls or sugar beet pulp increases dietary fermentable fibre content. Fermentable fibre serves as an energy source for microbes in the large intestine (Zervas and Zijlstra, 2002). Consequently, these microbes use excess nitrogen in the gut lumen to synthesize microbial protein that will be in turn be excreted as faecal nitrogen thereby reducing urinary nitrogen excretion.

The reduction in urinary nitrogen excretion due to feeding fermentable fibre reduces ammonia emissions from pigs (Canh et al., 1998). Reduction of ammonia emissions from livestock production system is presently an important political issue, for example in The Netherlands (Ge et al., 2023). Conversely, increased intestinal microbial fermentation due to increased dietary fermentable fibre increases methane emissions. For example, increasing dietary sugar beet pulp from 0 to 15 g/kg doubled methane emissions in pigs (Schrama et al., 1998). Methane emissions are an important component of greenhouse gas emissions from livestock production systems.

During the current century, emissions of greenhouse gases (CO_2, CH_4 and N_2O), including those from livestock production, have attracted more attention. Based on standard diets, the production of feedstuffs used in feed contributes 60–70% of overall emissions from pig production; in contrast, manure production and storage contributes 15–20% (Kebreab et al., 2016). Particle size reduction, mixing and pelleting of feed only contribute minimally to carbon emissions associated with feed production. Thus, changing feedstuff selection and thereby changing dietary feedstuff composition provides an opportunity to alter the carbon and nitrogen emissions of feed production (Table 1).

As an example, dietary inclusion of cake from macaúba (*Acrocomia aculeata*; an oilseed) kernels or a mixture of co-products reduced the greenhouse warming potential of a diet based on corn and soybean meal (SBM) in Brazil by 10% (Ali et al., 2017). Diet ingredient composition also alters the carbon emission of manure because the undigested residue accumulates in manure and can subsequently be fermented into methane and carbon dioxide. For example, dietary inclusion of 200 g/kg of wheat bran and 100 g/kg of bakery meal instead of corn and SBM increased greenhouse warming potential by 54% (Jaworkski et al., 2022).

An important component of these life cycle analyses is the allocation of greenhouse warming potential between the main product and co-product, for example, soybean oil and SBM, respectively (Table 2). A mass basis approach allocates a disproportionate amount of greenhouse warming potential to co-products relative to their lower economic value compared with the main product (Blonk et al., 2009). The allocation of greenhouse warming potential between main and co-product should instead be based on the economic value of each product. This calculation does not excessively increase calculated carbon emissions (Blonk et al., 2009). For example, the economic value of the canola oil produced for culinary food use is far greater than the that of canola meal, even though the mass of canola meal is greater than the mass of canola oil (600 vs. 400 g/kg; Zijlstra and Beltranena, 2023). Greenhouse gas

Table 1 Effect of wheat quality category included in livestock feed on the human-edible protein index

Livestock category	Total protein	Edible protein	
		Food wheat	Feed wheat
Fattening pigs	0.43	0.87	1.36
Broilers	0.54	0.61	0.96
Layers	0.34	0.86	1.30
Dairy	0.22	1.71	1.74

Source: Adapted from: De Cuyper et al. (2022). The index is a simple equation that describes efficiency as the total or edible amount of protein produced divided by the amount of total or edible protein consumed by different livestock categories.

Table 2 Calculation of allocation of fractions of soybean oil and soybean meal based on mass and economic value

Variable	Oil	Meal	Sum
Output, kg/kg input	0.175	0.800	0.975
Mass fraction allocation	0.179	0.821	1.000
Price of (co-)products, USD/kg output	700	230	–
Revenue (price of (co-)product × output), USD/kg output	122.5	184	306.5
Economic allocation fraction	0.400	0.600	1.000

Source: Adapted from: Blonk et al. (2009).

emissions are generally estimated using models and not directly evaluated using continuous measurements. Therefore, to evaluate contributions from the inclusion of co-products in swine diets, life cycle analyses including modelling are a useful tool. However, a fundamental understanding of the inputs of the models is required and consideration of disposal of waste streams that are not fed as co-products.

5 Sustainability: human-edible protein balance and land use

From a food security perspective, a concern is the amount of human edible ingredients, such as grains, that are being consumed by pigs to produce pork (Zijlstra and Beltranena, 2013, 2022). However, only about 50% of crops harvested can be directly utilized by humans (Van Zanten et al., 2018). The human edible protein index is a simple equation that describes efficiency as the total or edible amount of protein produced divided by the amount of total or edible protein consumed by livestock, for example pigs (FAO, 2011). Therefore, a value greater than 1 indicates that more human-edible protein is produced than what is consumed by livestock, and a value less than 1 indicates that less human-edible protein is produced than what is consumed.

An important consideration described in a recent study from Belgium is the quality of grain that is included in the feed (De Cuyper et al., 2022). Dietary inclusion of low-quality grains and co-products from food and biofuel production will considerably improve the human-edible protein balance (edible protein output/input) of swine production and will reduce waste streams from the food and bio-industry.

A second social consideration is the amount of land required to produce human food protein (Van Zanten et al., 2018). The current Western diet mandates the greatest land use, because livestock-producing meat requires a large land base. In contrast, a strict vegan diet to feed the human population requires a lower land base. However, the lowest land base is required by a combination of plant and animal protein in the human diet. The underlying reason is that crop production produces food-grade protein but also human-inedible co-products such as SBM and canola meal. Livestock including pigs can upcycle these products into high-quality animal protein for human consumption (Van Zanten et al., 2019). Therefore, dietary inclusion of co-products results in the lowest footprint on arable land. Based on land use, the most sustainable human diet would be 60 g protein/day comprised of 20 g/day from animals and 40 g/day from plants (Van Zanten et al., 2018).

6 Case study: canola co-products as feedstuffs for pigs

Co-product production and utilization are important parts of the food system. For example, annually more than 40 million metric tons of corn DDGS is produced in the USA resulting from the production of ethanol from corn grain (US Grains Council, 2023). In Canada, an important case study is canola co-products: 5.2 million metric tonnes of canola meal result from the production of canola oil mostly for culinary use by crushing canola seed (Canola Council of Canada, 2023). These co-products are produced in solvent-extraction plants with human food grade oil production and safety standards. However, the co-product canola meal aren't human edible but are suitable as feedstuff for livestock, including pigs, that convert this inedible co-product into human-edible animal protein.

Crop breeding turned rapeseed that yielded historically industrial oil and low-quality co-product rapeseed meal into canola seed yielding edible (low erucic acid) canola oil and the higher-quality co-product canola meal (Bell, 1993). In the past decades, feeding young pigs diets containing canola meal caused reduced growth performance compared with pigs fed diets containing dehulled SBM as a protein source (Baidoo et al., 1986, 1987). However, in these previous experiments, diets were not formulated using modern nutritional concepts that are based on net energy and standardized ileal digestible amino acids. In addition, the initially developed canola meal contained more residual glucosinolates (>35 μmol/g) than modern canola meal produced currently (<10 μmol/g).

During the last decade, an array of studies conducted at the University of Alberta demonstrated that modern canola meal included into diets using modern feed evaluation systems resulted in equal growth performance in nursery pigs (Landero et al., 2011). A meta-analysis across multiple studies was also in agreement with our research outcomes (Hansen et al., 2020). Similarly, equal growth performance in starter pigs is possible with dietary inclusion of canola expeller (Landero et al., 2012) and canola cake (Zhou et al., 2016). Equal growth of pigs fed canola meal instead of SBM is possible even though young pigs preferred diets containing SBM (Landero et al., 2018). Similar lack of performance changes by feeding canola meal instead of SBM have been observed in growing-finishing pigs (Smit et al., 2014a,b) and gestating and lactating sows (Velayudhan and Nyachoti, 2017; Velayudhan et al., 2018), provided that the diets were balanced well for energy and amino acids.

Apart from growth performance, other factors, such as gut health, are important in pig production. Canola co-product has indeed a lower nutrient digestibility than dehulled SBM (Woyengo et al., 2016). Consequently, feeding diets containing canola meal instead of SBM result in more undigested protein and fibre within the lumen of the gastrointestinal tract. While this change may not result in changes in gut health for pigs managed under high health scenarios, increased undigested protein may increase the risk of post-weaning diarrhoea (Hong et al., 2020), mainly attributed to protein fermentation in the hindgut. Therefore, initiatives to lower the fibre content of canola meal using crop breeding will increase the protein content and energy digestibility of canola meal, with a greater amino acid digestibility and less undigested protein in digesta (Berrocoso et al., 2015).

Fractionation technologies can also be applied to commodity feedstuffs to generate novel fractions that contain not only less fibre but also more protein (Mejicanos et al., 2016; Zijlstra et al., 2023). For example, fine grinding followed by air classification resulted in separation of protein and fibre by particle size and density. Consequently, canola meal can be separated into light, small particles (393 g/L) containing less acid-detergent fibre (ADF) (138 vs. 198 g/kg) and more crude protein (CP) (410 vs. 381 g/kg), and heavy, large particles (610 g/L) containing more ADF (230 vs. 198 g/kg) than the parent canola meal (Zhou et al., 2013). The light fraction had greater digestibility of energy and greater net energy value than the parent canola meal (Zhou et al., 2015). At 200 g/kg dietary inclusion, reducing fibre and increasing protein content in the light fraction increased gain-to-feed of weaned pigs compared with feeding parent canola meal (Zhou et al., 2013). However, simple heat and pressure using extrusion or steam-explosion did not change the nutritional value of canola meal (Heyer et al., 2021; Sanchez-Zannatta et al., 2023). Combined, co-products should not be considered as static entities. The nutritional value including nutrient digestibility can be improved over time using novel plant breeding, novel feed processing such as fermentation technologies (Scholten et al., 1999) and novel feed additive technologies such as feed enzymes (Zijlstra et al., 2010).

7 Conclusion and future trends

Feeding more co-products and less grains to pigs is paramount to reach sustainable food production systems. Pigs as omnivores can readily convert human-inedible co-products from food and bio-industry production into wholesome pork for human consumption, thereby considerably improving the human-edible protein balance (edible protein output/input) of swine production. Compared with traditional diets fed to pigs based on cereal grains as an energy source and SBM as a protein source, feeding high inclusion levels of co-products poses greater risk. This risk can be managed using modern feed formulation, feed evaluation, feed enzymes and feed processing to attain predictable swine growth performance, carcass characteristics, and pork quality. Co-product quality characteristics will continue to change overtime due to crop improvement and diversification, and evolving processing methods. Thus, feed evaluation of novel and evolved co-products is paramount to support proper diet formulation resulting in predictable pig performance. Apart from risks, feeding co-products also presents opportunities for animal health, nutrient management, carcass traits and pork quality that must be managed carefully. Feed enzymes have substantially increased P availability and the soluble fibre digestibility of co-products with further gains to be realized using new feed enzyme and processing technologies. Reducing excess supply of amino acids and protein fermentation in the gut, and mitigating N excretion related to feeding co-products is a challenge ahead. Dietary inclusion of co-products may reduce feed cost per unit of pork produced. Fast adoption of feeding co-products is now driven by unprecedented high prices of corn and SBM during the last couple of years.

The knowledge of the nutritional value of co-products continues to expand. As nutritional and anti-nutritional properties of crops change over time due to crop breeding, changing agronomic conditions and industrial processes that improve over time, the nutritional value of co-products requires regular updating (Sauvant et al., 2004; NRC, 2012), especially when major changes do occur. An excellent example is the crude fat content of corn DDGS that was commonly above 100 g/kg with the initial expansion of the bioethanol industry during 2000–2010 in the USA (Stein and Shurson, 2009). However, after the high value of corn oil for the generation of biodiesel was realized, most ethanol plants introduced partial separation of corn oil resulting in corn DDGS with a fat content closer to 50 g/kg and thus a lower energy value and concurrent greater protein content (Cristobal et al., 2020). Due to the lower digestibility of plant-based co-products, their dietary inclusion generally increases excretion of organic matter, P and N by pigs. Advancement in technologies such as tail-end processing, feed enzymes and feed fermentation are thus required to increase nutrient digestibility (Jha et al., 2015; Xu et al., 2019; Zijlstra et al., 2023), especially in cases that the limitation in digestibility or palatability also cause limits of dietary inclusion. Simple tail-end sieving or air classification of canola meal can generate low-fibre and high-fibre fractions (Zhou et al., 2015; Mejicanos et al., 2017). With the evolution of phytase enzyme technology, tremendous strides have been made to increase P digestibility and reduce

P excretion. Similar efforts are now required to reduce N excretion. These technologies can be combined to create increasingly advanced, highly digestible plant-based protein sources (Ton Nu et al., 2020).

To reduce the environmental footprint related to feeding co-products to pigs, reducing excess dietary supply of amino acids and mitigating N excretion including mitigation of protein fermentation are challenges ahead. Excretion of N and especially N volatilization as NH_3 is an issue of major concern in areas with high livestock density, such as the Netherlands. The production of NH_3 and H_2S from manure stored in enclosed barns that house pigs in northern latitudes is a worker health and safety issue. However, technologies can be enhanced to reduce greenhouse gas emissions and better capture the value that pig manure provides. For example, bacterial denitrification of manure can reduce emissions of N_2O, which has a 10 times greater global warming potential than methane (Wang et al., 2013; Luo et al., 2020). Manure slurry with high organic matter content can be bio-digested so that produced methane can be captured and used as an energy source instead of being emitted into the atmosphere (Lymperatou et al., 2021).

Finally, a whole system approach is required so that major aspects of sustainability are considered. For example, human food waste must be considered as a source of nutrients for pigs if food safety can be carefully managed (Shurson, 2020; Urriola, 2022). Pork production should be profitable as a part of sustainable food production. Produced pork needs to be nutritious, tasty and safe, as well as contribute to food security, especially for people with high nutrition demands (FAO, 2023) in a diet properly balanced for other concerns including human health (Willett et al., 2019). Pork needs to be produced using the smallest environmental footprint possible, taking into consideration the use of land, water and feed grains, excretion of nutrients, and emissions of greenhouse gases and ammonia. For each, feeding co-products to pigs is an important piece of the puzzle to ensure that the net impact of pork production on society is positive.

8 Where to look for further information

A plethora of information exists about feeding co-products to pigs. Some examples of information on the topic are listed in the references. By far, the strongest information set is the technical information about nutritional value of co-products and their effect on an array of biological measurements in the pig including growth and carcass composition. Information about key components of sustainability is more difficult to find. Economic analyses are generally excluded from scientific journal articles containing technical information. Economic information is included in some review articles (e.g. Woyengo et al., 2014), proceedings of industry conferences (e.g. Beltranena,

2022) or in articles published in magazines for pork producers such as the Canadian Hog Journal (e.g. Wang et al., 2017). Using the inverse of nutrient digestibility, information about the environmental footprint can be modelled. Recently, information about social aspect of feeding co-products started to be published, partly due to increasing interest in reaching circular food systems (Van Zanten et al., 2019).

9 References

Agyekum, A. K., Slominski, B. A. and Nyachoti, C. M. (2012). Organ weight, intestinal morphology, and fasting whole-body oxygen consumption in growing pigs fed diets containing distillers dried grains with solubles alone or in combination with a multienzyme supplement. *J. Anim. Sci.* 90(9): 3032–3040. doi: 10.2527/jas.2011-4380.

Ali, B. M., van Zanten, H. H. E., Berentsen, P., Bastiaansen, J. W. M., Bikker, P. and Oude Lansink, A. O. (2017). Environmental and economic impacts of using co-products in the diets of finishing pigs in Brazil. *J. Clean. Prod.* 162: 247–259. doi: 10.1016/j.jclepro.2017.06.041.

Baidoo, S. K., McIntosh, M. K. and Aherne, F. X. (1986). Selection preference of starter pigs fed canola meal and soybean meal supplemented diets. *Can. J. Anim. Sci.* 66(4): 1039–1049. doi: 10.4141/cjas86-114.

Baidoo, S. K., Mitaru, B. N., Aherne, F. X. and Blair, R. (1987). The nutritional value of canola meal for early-weaned pigs. *Anim. Feed Sci. Technol.* 18(1): 45–53. doi: 10.1016/0377-8401(87)90028-9.

Bell, J. M. (1993). Factors affecting the nutritional value of canola meal: a review. *Can. J. Anim. Sci.* 73(4): 689–697. doi: 10.4141/cjas93-075.

Beltranena, E. (2022). Reducing feed cost in grow-finish: alternative feedstuffs and production strategies. *Adv. Pork Prod.* Proceedings of the Banff Pork Seminar 33: 95–105. University of Alberta, Edmonton, AB, Canada.

Bergeron, R., Bolduc, J., Ramonet, Y., Meunier-Salaün, M. C. and Robert, S. (2000). Feeding motivation and stereotypies in pregnant sows fed increasing levels of fibre and/or food. *Appl. Anim. Behav. Sci.* 70(1): 27–40. doi: 10.1016/S0168-1591(00)00142-8.

Berrocoso, J. D., Rojas, O. J., Liu, Y., Shoulders, J., Gonzalez-Vega, J. C. and Stein, H. H. (2015). Energy concentration and amino acid digestibility in high-protein canola meal, conventional canola meal, and soybean meal fed to growing pigs. *J. Anim. Sci.* 93(5): 2208–2217. doi: 10.2527/jas.2014-8528.

Blonk, H., Ponsioen, T., Henne, W., van Kernebeek, H. and Waarts, Y. (2009). *Towards a Tool for Assessing Carbon Footprints of Animal Feed*. Product Board Animal Feed, The Netherlands.

Casas, G. A., Almeida, J. A. S. and Stein, H. H. (2015). Amino acid digestibility in rice co-products fed to growing pigs. *Anim. Feed Sci. Technol.* 207: 150–158. doi: 10.1016/j.anifeedsci.2015.05.024.

Canh, T. T., Schrama, J. W., Aarnink, A. J. A., Verstegen, M. W. A., van't Klooster, C. E. and Heetkamp, M. J. W. (1998). Effect of dietary fermentable fibre from pressed sugar-beet pulp silage on ammonia emission from slurry of growing-finishing pigs. *Anim. Sci.* 67(3): 583–590. doi: 10.1017/S1357729800033026.

Canola Council of Canada (2023). Canola industry: at a glance. Available at: https://www.canolacouncil.org/markets-stats/ (accessed 2023 July 1).

Cristobal, M., Acosta, J. P., Lee, S. A. and Stein, H. H. (2020). A new source of high-protein distillers dried grains with solubles (DDGS) has greater digestibility of amino acids and energy, but less digestibility of phosphorus, than de-oiled DDGS when fed to growing pigs. *J. Anim. Sci.* 98(7): skaa200. doi: 10.1093/jas/skaa200.

De Cuyper, C., Van Mullem, J., Millet, S., Delezie, E., Vandaele, L., De Boever, J. and De Campeneere, S. (2022). *Evaluation of the Net Contribution of Belgian Livestock Production to Human Protein Nutrition (in Dutch).* Instituut voor Landbouw-, Visserij- en Voedingsonderzoek (ILVO) Eenheid Dier, Melle, Belgium.

Dikeman, C. L. and Fahey, G. C., Jr. (2006). Viscosity as related to dietary fiber: a review. *Crit. Rev. Food Sci. Nutr.* 46(8): 649–663. doi: 10.1080/10408390500511862.

Fairbairn, S. L., Patience, J. F., Classen, H. L. and Zijlstra, R. T. (1999). The energy content of barley fed to growing pigs: characterizing the nature of its variability and developing prediction equations for its estimation. *J. Anim. Sci.* 77(6): 1502–1512. doi: 10.2527/1999.7761502x.

FAO (2011). *World Livestock 2011 – Livestock in Food Security*. Food and Agriculture Organization, Rome, Italy.

FAO (2023). *Contribution of Terrestrial Animal Source Food to Healthy Diets for Improved Nutrition and Health Outcomes: An Evidence and Policy Overview on the State of Knowledge and Gaps*. Food and Agriculture Organization, Rome, Italy. doi: 10.4060/cc3912en.

Florou-Paneri, P., Christaki, E., Giannenas, I., Bonos, E., Skoufos, I., Tsinas, A., Tzora, A. and Peng, J. (2014). Alternative protein sources to soybean meal in pig diets. *J. Food Agric. Environ.* 12: 655–660. doi: 10.1234/4.2014.5214.

Fouhse, J. M., Gänzle, M. G., Beattie, A. D., Vasanthan, T. and Zijlstra, R. T. (2017). Whole grain composition modifies substrate availability in the hindgut for fermentation and thereby shifts fecal microbial profiles in pigs. *J. Nutr.* 147(11): 2031–2040. doi: 10.3945/jn.117.255851.

Frobose, H. L., Fruge, E. D., Tokach, M. D., Hansen, E. L., DeRouchey, J. M., Dritz, S. S., Goodband, R. D. and Nelssen, J. L. (2015). The effects of deoxynivalenol-contaminated corn dried distillers grains with solubles in nursery pig diets and potential for mitigation by commercially available feed additives. *J. Anim. Sci.* 93(3): 1074–1088. doi: 10.2527/jas.2013-6883.

Garcia, H., Wang, L. F., Landero, J. L., Beltranena, E., Cervantes, M., Morales, A. and Zijlstra, R. T. (2015). Effect of feeding wheat millrun on diet nutrient digestibility and growth performance in starter pigs. *Anim. Feed Sci. Technol.* 207: 283–288. doi: 10.1016/j.anifeedsci.2015.06.018.

Ge, X., Schaap, M. and de Vries, W. (2023). Improving spatial and temporal variation of ammonia emissions for the Netherlands using livestock housing information and a Sentinel-2-derived crop map. *Atmos. Environ. X* 17: 100207. doi: 10.1016/j.aeaoa.2023.100207.

Hansen, J. Ø., Øverland, M., Skrede, A., Anderson, D. M. and Collins, S. A. (2020). A meta-analysis of the effects of dietary canola / double low rapeseed meal on growth performance of weanling and growing-finishing pigs. *Anim. Feed Sci. Technol.* 259: 114302. doi: 10.1016/j.anifeedsci.2019.114302.

Heyer, C. M. E., Wang, L. F., Beltranena, E. and Zijlstra, R. T. (2021). Nutrient digestibility of extruded canola meal in ileal-cannulated growing pigs, growth performance and diet nutrient digestibility in weaned pigs fed extruded canola meal. *J. Anim. Sci.* 99(5): skab135. doi: 10.1093/jas/skab135.

Hilbrands, A. M., Johnston, L. J., McClelland, K. M., Cox, R. B., Baidoo, S. K., Souza, L. W. O. and Shurson, G. C. (2013). Effects of abrupt introduction and removal of high and low digestibility corn distillers dried grains with solubles from the diet on growth performance and carcass characteristics of growing-finishing pigs. *J. Anim. Sci.* 91(1): 248–258. doi: 10.2527/jas.2012-5162.

Hong, J., Ndou, S. P., Adams, S., Scaria, J. and Woyengo, T. A. (2020). Canola meal in nursery pig diets: growth performance and gut health. *J. Anim. Sci.* 98(11): 1–18. doi: 10.1093/jas/skaa338.

Jarret, G., Martinez, J. and Dourmad, J. Y. (2011). Effect of biofuel co-products in pig diets on the excretory patterns of N and C and on the subsequent ammonia and methane emissions from pig effluent. *Animal* 5(4): 622–631. doi: 10.1017/S1751731110002041.

Jaworkski, N. W., Ferguson, N. S., Gaasbeek, A., Lizaso, J. and Smits, C. H. M. (2022). Use of feed technologies to reduce environmental impact at the feed mill and on swine farms. In: *Book of Abstracts*. 2nd International Feed Technology Congress, Wageningen University Research, Wageningen, The Netherlands, p. 26. (Abstr.).

Jha, R., Htoo, J. K., Young, M. G., Beltranena, E. and Zijlstra, R. T. (2013). Effects of increasing co-product inclusion and reducing dietary protein on growth performance, carcass characteristics, and jowl fatty acid profile of growing–finishing pigs. *J. Anim. Sci.* 91(5): 2178–2191. doi: 10.2527/jas.2011-5065.

Jha, R., Woyengo, T. A., Li, J., Bedford, M. R., Vasanthan, T. and Zijlstra, R. T. (2015). Enzymes enhance degradation of the fiber-starch-protein matrix of distillers dried grains with solubles as revealed by a porcine in vitro fermentation model and microscopy. *J. Anim. Sci.* 93(3): 1039–1051. doi: 10.2527/jas.2014-7910.

Kebreab, E., Liedke, A., Caro, D., Deimling, S., Binder, M. and Finkbeiner, M. (2016). Environmental impact of using specialty feed ingredients in swine and poultry production: a life cycle assessment. *J. Anim. Sci.* 94(6): 2664–2681. doi: 10.2527/jas.2015-9036.

Khatibi, P. A., McMaster, N. J., Musser, R. and Schmale, D. G., III (2014). Survey of mycotoxins in corn distillers' dried grains with solubles from seventy-eight ethanol plants in twelve states in the U.S. in 2011. *Toxins* 6(4): 1155–1168. doi: 10.3390/toxins6041155.

Kyriazakis, I. and Emmans, G. C. (1995). The voluntary feed intake of pigs given feeds based on wheat bran, dried citrus pulp and grass meal, in relation to measurements of feed bulk. *Br. J. Nutr.* 73(2): 191–207. doi: 10.1079/bjn19950023.

Landero, J. L., Beltranena, E., Cervantes, M., Morales, A. and Zijlstra, R. T. (2011). The effect of feeding solvent-extracted canola meal on growth performance and diet nutrient digestibility in weaned pigs. *Anim. Feed Sci. Technol.* 170(1–2): 136–140. doi: 10.1016/j.anifeedsci.2011.08.003.

Landero, J. L., Beltranena, E., Cervantes, M., Araiza, A. B. and Zijlstra, R. T. (2012). The effect of feeding expeller-pressed canola meal on growth performance and diet nutrient digestibility in weaned pigs. *Anim. Feed Sci. Technol.* 171(2–4): 240–245. doi: 10.1016/j.anifeedsci.2011.11.004.

Landero, J. L., Wang, L. F., Beltranena, E., Bench, C. J. and Zijlstra, R. T. (2018). Feed preference of weaned pigs fed diets containing soybean meal, *Brassica napus* canola meal, or *Brassica juncea* canola meal. *J. Anim. Sci.* 96(2): 600–611. doi: 10.1093/jas/skx052.

Libao-Mercado, A. J., Zhu, C. L., Fuller, M. F., Rademacher, M., Seve, B. and de Lange, C. F. M. (2007). Effect of feeding fermentable fiber on synthesis of total and mucosal protein in the intestine of the growing pig. *Livest. Sci.* 109(1-3): 125-128. doi: 10.1016/j.livsci.2007.01.116.

Lu, D., Pi, Y., Ye, H., Wu, Y., Bai, Y., Lian, S., Han, D., Ni, D., Zou, X., Zhao, J., Zhang, S., Kemp, B., Soede, N. and Wang, J. (2022). Consumption of dietary fiber with different physicochemical properties during late pregnancy alters the gut microbiota and relieves constipation in sow model. *Nutrients* 14(12): 2511. doi: 10.3390/nu14122511.

Luo, Z., Wang, D., Yang, J. and Zeng, W. (2020). The effect of using pig manure as an internal carbon source in a traditional piggery wastewater treatment system for biological denitrification. *Ecol. Eng.* 143: 105638. doi: 10.1016/j.ecoleng.2019.105638.

Lymperatou, A., Rasmussen, N. B., Gavala, H. N. and Skiadas, I. V. (2021). Improving the anaerobic digestion of swine manure through an optimized ammonia treatment: process performance, digestate and techno-economic aspects. *Energies* 14(3): 787. doi: 10.3390/en14030787.

Martin, M. and Lazarevic, D. (2015). A review of North American biofuel production, policies and research. Report No 2015:06, f3. The Swedish Knowledge Centre for Renewable Transportation Fuels, Sweden. doi: 10.13140/RG.2.1.1946.3446.

Mejicanos, G., Sanjayan, N., Kim, I. H. and Nyachoti, C. M. (2016). Recent advances in canola meal utilization in swine nutrition. *J. Anim. Sci. Technol.* 58: 7. doi: 10.1186/s40781-016-0085-5.

Mejicanos, G. A., Rogiewicz, A., Nyachoti, C. M. and Slominski, B. A. (2017). Fractionation of canola meal using sieving technology. *Can. J. Anim. Sci.* 97: 613-621. doi: 10.1139/cjas-2016-0229.

Molist, F., Gómez de Segura, A. G., Gasa, J., Hermes, R. G., Manzanilla, E. G., Anguita, M. and Pérez, J. F. (2009). Effects of the insoluble and soluble dietary fibre on the physicochemical properties of digesta and the microbial activity in early weaned piglets. *Anim. Feed Sci. Technol.* 149(3-4): 346-353. doi: 10.1016/j.anifeedsci.2008.06.015.

Molist, F., Gómez de Segura, A., Pérez, J. F., Bhandari, S. K., Krause, D. O. and Nyachoti, C. M. (2010). Effect of wheat bran on the health and performance of weaned pigs challenged with *Escherichia coli* K88+. *Livest. Sci.* 133(1-3): 214-217. doi: 10.1016/j.livsci.2010.06.067.

Molist, F., van Oostrum, M., Pérez, J. F., Mateos, G. G., Nyachoti, C. M. and van der Aar, P. J. (2014). Relevance of functional properties of dietary fibre in diets for weanling pigs. *Anim. Feed Sci. Technol.* 189: 1-10. doi: 10.1016/j.anifeedsci.2013.12.013.

Montagne, L., Le Floc'h, N., Arturo-Schaan, M., Foret, R., Urdaci, M. C. and Le Gall, M. (2012). Comparative effects of level of dietary fiber and sanitary conditions on the growth and health of weanling pigs. *J. Anim. Sci.* 90: 2556-2569. doi: 10.2527/jas2011-4160.

NRC (2012). *Nutrient Requirements of Swine*, 11th rev. edn. National Academies Press, Washington, DC, USA.

Oliviero, C., Kokkonen, T., Heinonen, M., Sankari, S. and Peltoniemi, O. (2009). Feeding sows with high fibre diet around farrowing and early lactation: impact on intestinal activity, energy balance related parameters and litter performance. *Res. Vet. Sci.* 86(2): 314-319. doi: 10.1016/j.rvsc.2008.07.007.

Peltoniemi, O. A. T., Björkman, S. and Oliviero, C. (2016). Parturition effects on reproductive health in the gilt and sow. *Reprod. Domest. Anim.* 51 (Suppl. 2): 36–47. doi: 10.1111/rda.12798.

Pluske, J. R., Turpin, D. L. and Kim, J. C. (2018). Gastrointestinal tract (gut) health in the young pig. *Anim. Nutr.* 4(2): 187–196. doi: 10.1016/j.aninu.2017.12.004.

Quesnel, H., Meunier-Salaün, M. C., Hamard, A., Guillemet, R., Etienne, M., Farmer, C., Dourmad, J. Y. and Père, M. C. (2009). Dietary fiber for pregnant sows: influence on sow physiology and performance during lactation. *J. Anim. Sci.* 87(2): 532–543. doi: 10.2527/jas.2008-1231.

Sanchez-Zannatta, J., Le Thanh, B. V., Wang, L. F., Beltranena, E., Newkirk, R. W. and Zijlstra, R. T. (2023). Ileal nutrient and energy digestibility of steam-exploded canola meal in cannulated grower pigs and total tract nutrient digestibility and growth performance of diets containing steam-exploded canola meal in weaned pigs. *Anim. Feed Sci. Technol.* 295(Jan): 115518. doi: 10.1016/j.anifeedsci.2022.115518.

Sauvant, D., Perez, J. M. and Tran, G. (2004). Tables *of Composition and Nutritional Value of Feed Materials: Pigs, Poultry, Cattle, Sheep, Goats, Rabbits, Horses and Fish*. Wageningen Academic Publishers, The Netherlands and INRA Editions, Wageningen/Versailles, France.

Schedle, K., Humer, E. and Mair, C. (2013). Alternative protein sources to soybean meal: potentials and limitations in diets for growing-finishing pigs: a review. *Tieraerztl. Umsch.* 68: 336–342.

Scholten, R. H. J., van der Peet-Schwering, C. M. C., Verstegen, M. W. A., den Hartog, L. A., Schrama, J. W. and Vesseur, P. C. (1999). Fermented co-products and fermented compound diets for pigs: a review. *Anim. Feed Sci. Technol.* 82(1–2): 1–19. doi: 10.1016/S0377-8401(99)00096-6.

Schrama, J. W., Bosch, M. W., Verstegen, M. W. A., Vorselaars, A. H. P. M., Haaksma, J. and Heetkamp, M. J. W. (1998). The energetic value of nonstarch polysaccharides in relation to physical activity in group-housed, growing pigs. *J. Anim. Sci.* 76(12): 3016–3023. doi: 10.2527/1998.76123016x.

Shurson, G. C. (2020). "What a waste" – can we improve sustainability of food animal production systems by recycling food waste streams into animal feed in an era of health, climate, and economic crises? *Sustainability* 12(17): 7071. doi: 10.3390/su12177071.

Smit, M. N., Seneviratne, R. W., Young, M. G., Lanz, G., Zijlstra, R. T. and Beltranena, E. (2014a). Feeding increasing inclusions of canola meal with distillers dried grains and solubles to growing-finishing barrows and gilts. *Anim. Feed Sci. Technol.* 189: 107–116. doi: 10.1016/j.anifeedsci.2013.12.012.

Smit, M. N., Seneviratne, R. W., Young, M. G., Lanz, G., Zijlstra, R. T. and Beltranena, E. (2014b). Feeding *Brassica juncea* or *Brassica napus* canola meal at increasing dietary inclusions to growing-finishing gilts and barrows. *Anim. Feed Sci. Technol.* 198: 176–185. doi: 10.1016/j.anifeedsci.2014.09.010.

Stein, H. H. and Shurson, G. C. (2009). Board-invited review: the use and application of distillers dried grains with solubles in swine diets. *J. Anim. Sci.* 87(4): 1292–1303. doi: 10.2527/jas.2008-1290.

Tan, F. P. Y., Beltranena, E. and Zijlstra, R. T. (2021). Resistant starch: implications of dietary inclusion on gut health and growth in pigs: a review. *J. Anim. Sci. Biotechnol.* 12(1): 124. doi: 10.1186/s40104-021-00644-5.

Ton Nu, M. A., Lupatsch, I., Zannatta, J. S., Schulze, H. and Zijlstra, R. T. (2020). Thermo-mechanical and enzyme-facilitated processing of soybean meal enhanced in vitro kinetics of protein digestion and protein and amino acid digestibility in weaned pigs. *J. Anim. Sci.* 98(8): skaa224. doi: 10.1093/jas/skaa224.

Urriola, P. E. (2022). Optimizing resource utilization for future feeds: evaluation of agricultural resources for optimal upcycling in pigs. *Animal - Science Proceedings* 13: 123–126.

US Grains Council (2023). DDGS production and exports. Available at: https://grains.org/buying-selling/ddgs/ (accessed 2023 June 29).

Van Zanten, H. H. E., Herrero, M., Van Hal, O., Röös, E., Muller, A., Garnett, T., Gerber, P. J., Schader, C. and De Boer, I. J. M. (2018). Defining a land boundary for sustainable livestock consumption. *Glob. Change Biol.* 24(9): 4185–4194. doi: 10.1111/gcb.14321.

Van Zanten, H. H. E., Van Ittersum, M. K. and De Boer, I. J. M. (2019). The role of farm animals in a circular food system. *Glob. Feed Sec.* 21: 18–22. doi: 10.1016/j.gfs.2019.06.003.

Velayudhan, D. E. and Nyachoti, C. M. (2017). Effect of increasing dietary canola meal inclusion on lactation performance, milk composition, and nutrient digestibility of lactating sows. *J. Anim. Sci.* 95(7): 3129–3135. doi: 10.2527/jas.2016.1191.

Velayudhan, D. E., Hossain, M. M., Regassa, A. and Nyachoti, C. M. (2018). Effect of canola meal inclusion as a major protein source in gestation and lactation sow diets with or without enzymes on reproductive performance, milk composition, fecal bacterial profile and nutrient digestibility. *Anim. Feed Sci. Technol.* 241: 141–150. doi: 10.1016/j.anifeedsci.2018.05.001.

Wang, C., Lu, H., Dong, D., Deng, H., Strong, P. J., Wang, H. and Wu, W. (2013). Insight into the effects of biochar on manure composting: evidence supporting the relationship between N_2O emission and denitrifying community. *Environ. Sci. Technol.* 47(13): 7341–7349. doi: 10.1021/es305293h.

Wang, L. F., Beltranena, E. and Zijlstra, R. T. (2016). Diet nutrient digestibility and growth performance of weaned pigs fed sugar beet pulp. *Anim. Feed Sci. Technol.* 211: 145–152. doi: 10.1016/j.anifeedsci.2015.11.005.

Wang, L. F., Beltranena, E. and Zijlstra, R. T. (2017). Feed nutrient digestibility and growth performance of weaned pigs fed canola meal varying in nutritive quality. *Can. Hog J.* 38(5): 46–48.

Wang, L. F. and Zijlstra, R. T. (2018). Prediction of bioavailable nutrients and energy. In: Moughan, P. J. and Hendriks, W. H. (Eds) *Feed Evaluation Science*. Wageningen Academic Publishers, Wageningen, The Netherlands, pp. 337–386.

Wang, L. F., Beltranena, E. and Zijlstra, R. T. (2022). Chapter 13. Pulse grains and their coproducts in swine diets. In: Chiba, L. (Ed.) *Sustainable Swine Nutrition* (2nd edn.), John Wiley & Sons, Hoboken, NJ, USA, pp. 343–374.

Widyaratne, G. P. and Zijlstra, R. T. (2007). Nutritional value of wheat and corn distiller's dried grain with solubles: digestibility and digestible contents of energy, amino acids and phosphorus, nutrient excretion and growth performance of grower-finisher pigs. *Can. J. Anim. Sci.* 87(1): 103–114. doi: 10.4141/A05-070.

Willett, W., Rockström, J., Loken, B., Springmann, M., Lang, T., Vermeulen, S., Garnett, T., Tilman, D., DeClerck, F., Wood, A., Jonell, M., Clark, M., Gordon, L. J., Fanzo, J., Hawkes, C., Zurayk, R., Rivera, J. A., De Vries, W., Sibanda, L. M., Afshin, A., Chaudhary, A., Herrero, M., Agustina, R., Branca, F., Lartey, A., Fan, S., Crona, B., Fox, E., Bignet, V., Troell, M., Lindahl, T., Singh, S., Cornell, S. E., Srinath Reddy, K., Narain, S., Nishtar, S. and Murray, C. J. L. (2019). Food in the Anthropocene: the EAT-Lancet Commission on healthy diets from sustainable food systems. *Lancet* 393(10170): 447–492. doi: 10.1016/S0140-6736(18)31788-4.

Williams, B. A., Mikkelsen, D., Flanagan, B. M. and Gidley, M. J. (2019). "Dietary fibre": moving beyond the "soluble/insoluble" classification for monogastric nutrition, with an emphasis on humans and pigs. *J. Anim. Sci. Biotechnol.* 10: 45. doi: 10.1186/s40104-019-0350-9.

Woyengo, T. A., Beltranena, E. and Zijlstra, R. T. (2014). Controlling feed cost by including alternative ingredients into pig diets: a review. *J. Anim. Sci.* 92(4): 1293–1305. doi: 10.2527/jas.2013-7169.

Woyengo, T. A., Jha, R., Beltranena, E. and Zijlstra, R. T. (2016). *In vitro* digestion and fermentation characteristics of canola co-products simulate their digestion in the pig intestine. *Animal* 10(6): 911–918. doi: 10.1017/S1751731115002566.

Woyengo, T. A., Beltranena, E. and Zijlstra, R. T. (2017). Effect of anti-nutritional factors of oilseed co-products on feed intake of pigs and poultry. *Anim. Feed Sci. Technol.* 233: 76–86. doi: 10.1016/j.anifeedsci.2016.05.006.

Xu, B., Zhu, L., Fu, J., Li, Z., Wang, Y. and Jin, M. (2019). Overall assessment of fermented feed for pigs: a series of meta-analyses. *J. Anim. Sci.* 97(12): 4810–4821. doi: 10.1093/jas/skz350.

Zervas, S. and Zijlstra, R. T. (2002). Effects of dietary protein and fermentable fiber on nitrogen excretion patterns and plasma urea in grower pigs. *J. Anim. Sci.* 80(12): 3247–3256. doi: 10.2527/2002.80123247x.

Zhou, X., Oryschak, M. A., Zijlstra, R. T. and Beltranena, E. (2013). Effects of feeding high- and low-fibre fractions of air-classified, solvent-extracted canola meal on diet nutrient digestibility and growth performance of weaned pigs. *Anim. Feed Sci. Technol.* 179(1–4): 112–120. doi: 10.1016/j.anifeedsci.2012.12.002.

Zhou, X., Zijlstra, R. T. and Beltranena, E. (2015). Nutrient digestibility of solvent-extracted *Brassica napus* and *Brassica juncea* canola meals and their air-classified fractions fed to ileal-cannulated grower pigs. *J. Anim. Sci.* 93(1): 217–228. doi: 10.2527/jas.2014-7451.

Zhou, X., Beltranena, E. and Zijlstra, R. T. (2016). Effects of feeding canola press-cake on diet nutrient digestibility and growth performance of weaned pigs. *Anim. Feed Sci. Technol.* 211: 208–215. doi: 10.1016/j.anifeedsci.2015.12.001.

Zhu, C. L., Rademacher, M. and de Lange, C. F. M. (2005). Increasing dietary pectin level reduces utilization of digestible threonine intake, but not lysine intake, for body protein deposition in growing pigs. *J. Anim. Sci.* 83(5): 1044–1053. doi: 10.2527/2005.8351044x.

Zijlstra, R. T., Owusu-Asiedu, A. and Simmins, P. H. (2010). Future of NSP-degrading enzymes to improve nutrient utilization of co-products and gut health in pigs. *Livest. Sci.* 134(1–3): 255–257. doi: 10.1016/j.livsci.2010.07.017.

Zijlstra, R. T., Jha, R., Woodward, A. D., Fouhse, J. and van Kempen, T. A. T. G. (2012). Starch and fiber properties affect their kinetics of digestion and thereby digestive physiology in pigs. *J. Anim. Sci.* 90 (Suppl. 4): 49–58. doi: 10.2527/jas.53718.

Zijlstra, R. T. and Beltranena, E. (2013). Swine convert co-products from food and biofuel industries into animal protein for food. *Anim. Front.* 3(2): 48–53. doi: 10.2527/af.2013-0014.

Zijlstra, R. T. and Beltranena, E. (2022). Feeding co-products to pigs to reduce feed cost and reach sustainable food production. *Anim. Front.* 12(6): 18–22. doi: 10.1093/af/vfac067.

Zijlstra, R. T. and Beltranena, E. (2023). Ingredient fractionation for monogastric animal nutrition: the worth of sum of parts versus the whole. *Anim. Feed Sci. Technol.* 304: 115737. doi: 10.1016/j.anifeedsci.2023.115737.

Chapter 11

Developing alternative sources of protein in pig nutrition: insects

Kristy DiGiacomo, The University of Melbourne, Australia

1 Introduction

There is an increased global interest in the production of insects for animal feed and human food. This interest is driven by a desire to reduce environmental footprints and increase sustainability, reduce reliance on traditional proteins (such as soybeans) and imported feed ingredients, reduce feed costs, utilize waste, and create alternative revenue streams (or a combination of each of these). Global pork production is an optimized, efficient and integrated system that utilizes breeding, nutrition, housing, and management to optimize production. An increase in the production of pork and pork products is occurring globally as human populations increase and diets shift to include larger portions of meat, particularly in developing countries. The predicted global pig meat production for 2030 is approx. 45 000 tonnes, which is a sustained increase from the approx. 40 000 tonnes produced in 2010 (OECD and Food

http://dx.doi.org/10.19103/AS.2024.0140.17

Agriculture Organization of the United Nations, 2022). Given that poultry meat production is predicted to increase by approximately 14 000 tonnes in the same period, there is increasing competition for animal feed ingredients (OECD and Food Agriculture Organization of the United Nations, 2022). The recent global coronavirus disease 2019 (COVID-19) pandemic highlighted that pork production systems are vulnerable to external factors, particularly global supply chains, further increasing the need to diversify diet ingredients.

Providing protein in the production of animal feed is a major contributor to the environmental impact of pork production systems, although pig production industries do commonly utilize waste co- and by-products (e.g. tallows and oils or oilseed mills) as feed sources in diets. Some production systems are also designed to facilitate direct feeding of waste products from human food production systems, such as waste confectionary and dairy products. Thus, the pig production industry is well placed structurally and likely open to utilizing alternate feed protein sources such as insects. Improving sustainability in pig production systems requires a reduced footprint (e.g. by using more sustainable sources of feed, requiring less use of resources such as land or water) or a reduced input (feed choices and animal genetic selection pressures are based on using local feeds and selecting robust animals that can perform efficiently in a changing climate) or both. Insects can contribute to sustainability goals by both making less from more and by potentially improving animal growth efficiency and health. This chapter will review some of the current knowledge on the use of insects in pig diets, highlighting the current research trends and knowledge gaps. Where relevant, this chapter will highlight reviews and directions for further reading but will not cover the engineering or processing elements of insect production.

2 Why insects

Insects are rapidly growing, cold-blooded, require minimal space, and contribute to the natural diet of many species including pigs and poultry. Insects can bioconvert many organic (and potentially inorganic) substrates including vegetables and fruits, grains, manure, and animal remains into a product high in protein and lipids. While the feed conversion ratio of production animals ranges from 5.0 for pork (i.e. 1 kg meat gain requires 5 kg feed input) to 10.0 for beef, these ratios are approaching 2.0 for insects (van Huis, 2013). Fly larvae like black soldier flies (BSFL) and other insects can consume organic waste, sequester carbon, and provide a nutrient-rich biomass that can be consumed by animals or humans. One study found that bioconversion of waste released 70% less carbon into the atmosphere than composting (Perednia, 2017). Compared to commonly fed protein sources such as soybean meal, insects require considerably less land (and potentially water and energy) use. To produce a similar quantity of pork or chicken protein, enterprises would require 2–3.5 ha of land compared to only 1 ha of land to produce equivalent insect (mealworm) protein (van Huis et al., 2013).

Insect species being investigated for production as human feed ingredients include crickets, mealworms, and locusts while the animal feed sector is more focused on the production of BSFL, mealworms, silkworms, common houseflies, and crickets (van Huis, 2016). Of these, BSFLs appear to be the most suited for larger scale production and unlike other species, they are not disease vectors as they do not lay their eggs on decaying organic materials and the adults do not eat. The feed provided during rearing changes the chemistry of the resultant insects, which means the rearing substrate composition is imperative and can also be manipulated to somewhat specify the appropriate composition of the insects. For example, the fat content of BSFL was greatest in those reared on dairy products (DiGiacomo et al., 2019) followed by fruits and then vegetables (Jucker et al., 2017). While being a valuable source of high-quality protein and overall have an excellent balance of amino acids (Veldkamp and Vernooij, 2021), insects also provide a source of lipids, vitamins, and minerals but can also possess bactericidal, antifungal, antiviral, and antioxidant compounds that may be of use as nutraceutical additives in pig diets (Sogari et al., 2023).

3 Current state of insect production

Alternative protein sources being developed as animal feed are currently focused on insects, novel plant proteins, seaweed, algae, and other single-cell sources. Of these, insects are receiving global interest. While the commercial use of insects in pig diets is relatively new, pigs will naturally consume insects, particularly when dietary protein is scarce (Kierończyk et al., 2022). While the insect farming industry is in its infancy, large-scale research and investment are underway globally to confirm the safety, reliability, sustainability, and potential contribution of insects as an animal feed source. This increased interest in insects from a commercial perspective has been supported by an increase in academic research. For example, the European Association of Animal Production (EAAP) has been hosting sessions on insects in animal feed since 2017, and in 2019, the insect study commission was established.

The use of insects in pig diets is not a new idea, with the first publications examining this topic published in the late 1970s (Newton et al., 1977). There has been a steady increase in the number of publications in the last two decades which has been linked to an increase in research funding globally, and many of these publications are in the agricultural sector. According to Baiano (2020), 637 manuscripts with 'edible insect' in the title were published from 1975 until January 2020. In recent years, numerous reviews of the use of insects in pig diets have been published. I previously reviewed the role of insects in commercial pig diets (DiGiacomo and Leury, 2019) and summarized the findings for their use in pig diets. An updated version of these data is provided in Table 1, although this is by no means a complete list. In an invited review, Hong and Kim (2022) provide a comprehensive summary and reference list of current data from feeding insects to pigs, and this data will not be repeated in depth in this chapter. Given there is a large variation in the experimental methods including the type of insect and inclusion rate fed, age of pigs, number of animals fed, and nutrient content/processing of the insects, and the definitive conclusions regarding the use of insects as pig feed are not yet possible.

While Western countries are often averse to consuming insects, countries such as Papua New Guinea, Thailand, African countries, Mexico, India, China, and many others have traditionally consumed insects in human diets (DeFoliart, 1999). It is predicted that approx. 2 billion people consume insects regularly in over 130 countries (Costa-Neto and Dunkel, 2016). Indigenous populations, such as those in Mexico and Australia, consumed insects as a valuable and highly sought out food source, often associated with celebrations and special occasions (Meyer-Rochow and Changkija, 1997; DeFoliart, 1999). A significant barrier to the uptake of insects as human feed is the aversion to eating insects by Western consumers, likely driven by the association of insects with disease (Bartrim, 2017). Conversely, there is increased current interest in encouraging the consumption of insects in Western societies. As the flavor of insects changes with each species and can also be manipulated by the growth medium and preparation method, there is increased interest in the production of high-end insects for human consumption. For animal feed producers, the lack of uptake is primarily driven by the limited volumes and high prices of presently available products although fear and lack of understanding likely also contribute. Further, most farmers' previous experience with insects will be in dealing with insects as disease vectors or pests (or at least a nuisance), which makes it more difficult to change perceptions and acceptability of insects. According to one survey, farmers perceive insects in feed as having a better nutrient profile and improved sustainability but a lower microbiological safety (Verbeke et al., 2015).

The environmental and socioeconomic benefits of the insect industry may surpass current projections based solely on costs of input vs output given the wide variety of uses for insects and insect products. Insects can be farmed as bait, for medicinal purposes and for the provision of fertilizer. These properties are particularly beneficial to the production of small-scale farms in developing countries. Even when not feasible to grow insects in stand-alone farms, if used as additional income/product streams, there can be localized economic empowerment benefits, particularly for the development of food and economic security in developing nations. For example, producing insects in orphanages in the Democratic Republic of Congo can generate significant income (approx.

Table 1 Summary of some of the published growth responses in pigs fed diets containing insects

Animals used	Insect type fed	Inclusion rate	Production response	Reference
5-week-old barrows, $n = 6$ (Latin square)	BSFL	33% replacement of soybean meal	Increased feed intake ($P < 0.05$); reduced apparent DMD ($P < 0.05$)	Newton et al. (1977)
Early weaned pigs (*n* not reported)	BSFL	0%, 50%, or 100% replacement of dried plasma	50% diet improved performance; 100% diet decreased performance	Newton et al. (2005)
Castrated male piglets ($n = 16$); barrows ($n = 16$)	BSFL	Complete replacement of soybean meal with and without AA fortification	No effect on growth or intake; BSF combined with AA fortification improved piglet protein quality ($P < 0.005$); BSF improved apparent N digestibility	Neumann et al. (2018)
Barrows ($n = 48$)	BSFL	50%, 75%, and 100% replacement of soybean meal	No effect on base meat quality measures, increased juiciness ($P < 0.05$); higher backfat polyunsaturated fatty acid (PUFA) contents ($P < 0.05$)	Altmann et al. (2019)
Weaned female pigs ($n = 48$)	BSFL	0%, 30%, and 60% replacement of soybean meal	Linear increase in ADFI; no effect on growth	Biasato et al. (2019)
Crossbred female finishing pigs ($n = 72$)	BSFL	BSFL at 0%, 4%, or 8% of the diet for 46 days	4% BSFL reduced colonic *Streptococcus* and increased *Lactobacillus* ($P < 0.05$); increased short-chain fatty acids and decreased protein fermentation ($P < 0.05$); downregulated pro-inflammatory cytokines and upregulated anti-inflammatory cytokines ($P < 0.05$). 8% BSFL increased *Clostridium* clusters ($P < 0.05$) and increased butyrate ($P < 0.05$)	Yu et al. (2019)
Crossbred weaner pigs ($n = 40$)	BSFL	BSFL as a replacement of fish meal at 0%, 25%, 50%, 75%, or 100%	No effect of diet on feed intake, weight gain, or blood parameters except for platelet counts that were reduced in pigs fed 25%, 75%, or 100% BSFL ($P = 0.042$).	Chia et al. (2019)
Weaning piglets ($n = 128$)	BSFL	Full-fat BSF at 0%, 1%, 2%, or 4% (isoenergetic and isonitrogenous) over two feeding phases (1–14 days and 15–28 days)	Linear increase in BW, ADG, organ weights (liver, pancreas, small intestine) from 0 day to 14 days with BSFL inclusion ($P < 0.05$), linear and quadratic decrease in feed:gain and CP/CF digestibility ($P < 0.05$). Changes to metabolism and immune status and intestinal morphology	Yu et al. (2020)

(*Continued*)

Table 1 (*Continued*)

Animals used	Insect type fed	Inclusion rate	Production response	Reference
Growing pigs ($n = 10$, ileal cannulated)	BSFL and HFL	BSFL or houseflies (HF) as the primary source of nitrogen (vs. nitrogen-free diet) fed for 7 days. Titanium dioxide marker	The CSID of all AA for HF and BSFL were more than 0.726 and 0.641, respectively. The values for the CSID of all AA in HF were greater ($P < 0.05$) than in BSFL. The values for the CSID of all AA except methionine and cysteine in HF are greater ($P < 0.05$) than in BSFL	Tan et al. (2020)
Crossbred (Large White × Landrace) finisher pigs ($n = 40$)	BSFL	Fish meal replacement with BSFL at 0%, 25%, 50%, 75%, or 100% for 98 days	Carcass weight of pigs fed diets with BSFL replacing FM by 50%, 75%, or 100% (w/w) was higher ($P < 0.001$) than for pigs fed control diet with 100% FM as protein source. Mean FCR was decreased with diets containing ≥50% BSFL ($P < 0.05$). CP content of pork tissues was high (65–93% on dry-matter basis) across all dietary groups	Chia et al. (2021)
Castrated male pigs (ileal cannulated; $n = 8$)	BSFL	Latin square (four diets, five periods: nine replicates). Fish meal vs. defatted BSFL meal vs. adult BSF vs. nitrogen-free diet. Chromium oxide marker	The coefficient of standardized ileal digestibility (CSID) of CP in defatted BSFL meal was lower (0.738 vs. 0.883; $P < 0.05$) than that in fish meal but was higher ($P < 0.05$) than that in adult BSF (0.561). The CSID of all indispensable AA except for methionine and phenylalanine in defatted BSFL meal was lower ($P < 0.05$) than that in fish meal. The CSID of all AA except for proline in defatted BSFL meal was higher ($P < 0.05$) than that in adult BSF	Kim et al. (2023)
In vitro (pig GI tract)	Crickets, mealworms	Plant proteins vs. insects cooked four ways: raw, autoclaved (25 min), or oven cooked (150°C for 30 min or 200°C for 10 min)	Insect IVCPD decreased with thermal treatments ($P < 0.05$); mealworms were more digestible than crickets ($P < 0.01$).	Poelaert et al. (2016)
Castrated male piglets ($n = 21$)	Crickets	Three iso-nitrogenous diets (18.4% CP) including either fish meal (control), whole cricket meal, or body cricket meal (legs removed) fed *ad libitum*	Cricket diets improved DM intake ($P < 0.001$) and digestibility ($P < 0.05$), nutrient intake ($P < 0.001$), piglets were heavier ($P < 0.001$) and FCR ($P < 0.01$) was lower and N retention is greater ($P < 0.05$).	Miech et al. (2017)

Crossbred weaning pigs (n = 100)	Crickets	Replacement of fish meal with crickets on a lysine basis (100%), replacing fish meal a kg/kg basis (100%) or total replacement of soybean and fish meal for 28 days	Cricket supplementation decreased the incidence of diarrhea ($P < 0.05$), increased bodyweight and ADG ($P < 0.05$). Crickets increased CP digestibility ($P = 0.041$) and crude fat digestibility ($P = 0.024$). Crickets increased jejunal villus height ($P = 0.009$) and serum immunity markers ($P < 0.05$)	Boontiam et al. (2022)
Weaning pigs (n = 120)	Mealworm	Mealworm at 0%, 1.5%, 3%, 4.5%, or 6% inclusion (isoenergetic and isonitrogenous) for 35 days (2 phases, 0–14 days and 14–35 days)	Linear increase in intake, bodyweight, ADG during phase 1 ($P < 0.01$). During phase 2 ADG tended to increase ($P = 0.08$), increased ADG ($P < 0.01$) and feed intake ($P < 0.05$). Feed:gain tended to increase across the whole period ($P = 0.07$). Linear increase in N retention and DMD and digestibility of CP ($P = 0.05$). Linear decrease in blood urea N ($P = 0.05$) and increase in IGF-1 ($P = 0.03$)	Jin et al. (2016)
Crossbred weaned pigs (n = 240)	Mealworm	Fish meal portion of the diet replaced at 0%, 50%, or 100% for 35 days	100% mealworm increased feed:gain during the first week ($P < 0.05$). During the second week, 50% mealworm increased ADG ($P < 0.05$), although across the whole period 50% replacement decreased ADG and final BW ($P < 0.05$)	Ao et al. (2020)
Weaning pigs (n = 180)	Mealworm	Fish meal portion of the diet replaced at 0%, 50%, or 100% for 28 days (two phases, 0–14 days and 15–28 days).	Feed to gain was increased in the 100% compared to 50% treatment in phase 1 ($P = 0.024$). The ADG was greater in 50% compared to 100% treatments ($P = 0.01$). Plasma IgG increased with mealworm diets in phase 1 ($P = 0.048$).	Ko et al. (2020)
Male 5-week-old, crossbred pigs (n = 30)	Mealworm	Mealworm at 0%, 5%, or 10% inclusion (isoenergetic and isonitrogenous) for 4 weeks	Decrease in AA digestibility ($P < 0.05$), weak impact on intermediary metabolism (using omics analysis)	Meyer et al. (2020)
Crossbred weaned pigs (n = 180)	Mealworm	Mealworm at 0%, 1%, or 2% in the diet (replacement of fish meal)	Pigs fed 1% mealworm had reduced final BW compared to control ($P < 0.05$). During days 0–7, feed:gain ratio was higher in mealworm diets ($P<0.05$). Feeding 1% mealworm decreased ($P < 0.05$) ADG during days 8–21 and 0–35. The 1% mealworm treatment had lower apparent total tract digestibility of DM and nitrogen ($P < 0.05$)	Ao and Kim (2019)

(*Continued*)

Table 1 (*Continued*)

Animals used	Insect type fed	Inclusion rate	Production response	Reference
Piglets (n = 144)	Mealworm, HFL, superworms	(Companion papers) Soybean-based diet with 5% of plasma protein powder (control), 5% mealworm powder, 5% HFL powder, or 5% superworm powder. Fed in two phases (1–28 days and 29–56 days)	ADFI was decreased by mealworm and HFL on day 7 and increased in superworm diets on days 28 and 56 ($P < 0.05$). Insect diets decreased diarrhea from days 15–28 ($P < 0.05$) and decreased plasma ammonia ($P < 0.05$) and increased the methionine AID ($P < 0.05$)	Ji et al. (2016)
	Mealworm, HFL, superworms	(As earlier, companion papers)	Plasma lysine concentrations were reduced in mealworm and superworm diets during phase 1 ($P < 0.05$). Insect powders overall altered AA transporters and improved the metabolism of AAs	Liu et al. (2020)

Please note, this list is not reflective of all current published data.
AA: amino acid; ADG: Average daily gain; ADFI: Average daily feed intake; AID: apparent ileal digestibility; BSFL: Black soldier fly larvae; BW: body weight; CF: crude fat; CP: crude protein; CSID: coefficient of standardized ileal digestibility; DM: dry matter; DMD: dry matter digestibility; FCR: feed conversion ratio; HFL: housefly larvae; IgG: immunoglobulin G; IGF-1: insulin-like growth factor 1.

$300 US/month) while also producing a sustainable and nutritious food and feed source (Franklin et al., 2018).

4 Potential insects and current production volumes

When selecting suitable insects, producers need to be sure they do not contain toxins, venoms, or unpalatable exoskeletons. Selection also needs to ensure that risks are avoided or minimized. For example, mealworms are a pest of grains and flower products, while houseflies are a pest to humans and other animals. There is a large variety of insects that can be reared for animal and human consumption. Those most investigated to date include crickets, flies (black soldier and house flies), silkworms, crickets, mealworms (giant, lesser, etc.), moths (wax moths, etc.), locusts, and grasshoppers. The selection of an appropriate insect species for individual production systems or target consumers will need to consider multiple factors including feed substrate available, environmental conditions, facilities, reproductive rate, size, behavior, disease risk, storage potential, and marketability. The lifecycle of insects will differ between species and some insects are consumed during their larval phase (i.e. flies), while others are consumed as adults (i.e. crickets). Thus, the length of time and input required to rear and farm these insects will be highly varied as it is when comparing the rearing of pigs to beef cows for example.

In 2019, Mancuso et al. (2019) reported that the European insect industry was producing low volumes of around 3000 tonnes of protein and BSFs were retailing for €2-9 per kg and mealworms retailing for €10–32 per kg. While these prices are not current, they do demonstrate the high cost of the protein driven by the low production capacity of the industry. It is predicted that the European industry will be producing 1.2 million tonnes by 2025 (IPIFF, 2019), although the COVID-19 pandemic may have stalled some growth in the sector. An increase in production volume should see a reduction in prices, but this will depend upon input costs and market pressures. Production may begin to increase more rapidly given that regulatory changes occurring in 2021 approved 'processed insect protein' for incorporation into animal diets (including pigs) in Europe. This is a positive step forward given that in the EU insects were previously prohibited from use in animal feed based on the risks (real or perceived) associated with transmissible spongiform encephalopathy (TSE).

Currently most insect production companies remain in the start-up phase and are not yet producing at a large or consistent scale. This lack of scale is driven by numerous factors including technology and engineering controls, access to capital investment, lack of customer base given the small-scale volumes presently produced, and lack of knowledge or research specific to each production system and environment. Given most insect producers were motivated to enter the industry because of their interest in contributing to sustainability goals, and while engineering and processing systems remain under development, most producers are attempting to develop environmentally sustainable processing methods utilizing existing efficient technologies such as solar power. While this review will not discuss the technical processes of rearing insects on a commercial scale, rearing processes must be developed for efficiency in output, energy/resource usage, and greenhouse gas emissions to remain sustainable. For example, as rearing temperature can impact the conversion efficiency of insects, energy inputs may be required to maintain appropriate growth climates which would increase energy resource usage (Rumpold and Schlüter, 2013). Further, given the scale of feed required for global pig diets, it is unlikely that insects would be more than just a sidestream or small diet component rather than a major constituent or complete replacement of traditional protein sources (such as soybean), but even small improvements in sustainability, efficiency, and reductions in costs could be valuable.

5 Production and health responses in pigs

Research into the feeding of insects to commercial pigs (from piglets to finishing pigs) has increased in recent years. Numerous reviews have been published summarizing such findings and thus will not be repeated here (e.g. Hong and Kim, 2022; Veldkamp and Bosch, 2015; Veldkamp et al., 2022). In a comprehensive recent review, Hong and Kim (2022) summarized the current literature examining insect digestibility and the impacts on meat quality in pork production systems, citing >20 publications from across the globe since 2020. Overall, including insects in pig diets either improved or did not alter growth rates, health, and meat quality. For example, weaning piglets had increased weight and efficiency markers when fed BSFL (Yu et al., 2020), crickets (Boontiam et al., 2022), or mealworms (Jin et al., 2016). In barrows, feeding BSFL had no negative effect on meat quality when fed as a replacement for soybean meal (Altmann et al., 2019). Recently, Ipema et al. (2021) examined the use of live BSFL as feed and enrichment for pigs and showed that the BSFL was preferred compared to corn, pellets, or raisins. Pigs not only interacted with the feed dispersal mechanisms but also consumed all the BSFL (Ipema et al., 2021), indicating that the provision of live insects can not only contribute nutrients but also enrich and the promotion of a positive welfare state.

Although the published literature is inconsistent with experimental design and diet composition, the consensus is that insect protein nutrient digestibility is comparable with traditional protein sources. It is important to note that digestibility of house crickets and mealworm larvae decreased when comparing raw to heated products (oven baked at either 150°C or 200°C or autoclaved) in pigs using an *in vitro* model (Poelaert et al., 2016), which is the opposite to most grains which increase in digestibility when cooked. Antinutritional factors, including chitin, can also decrease the digestibility of insects when fed to pigs while the addition of a protease can reduce some of these negative responses (Go et al., 2022). The exoskeletons of insects can be rich in chitin, which can therefore decrease digestibility. Given this, the analysis of insect protein concentrations using the Kjeldahal method overestimates the available protein (Jonas-Levi and Martinez, 2017). However, chitin and its derivatives can improve immunity, act as an antioxidant and antimicrobial, and can therefore improve growth performance in growing animals (Swiatkiewicz et al., 2015). Despite this potential decline in digestibility, most studies have shown either no change or an improvement in growth performance when feeding insects. However, without optimization of rearing and processing methods and inclusion rates, it is difficult to draw complete conclusions.

6 Competition for resources

A major initiative to reduce food waste will require cooperation between producers and consumers to both reduce oversupply and efficiently utilize unavoidable waste (FAO, 2011). Insects reared by and for the pig production industry can certainly contribute to this process. While most of the published literature has focused on the specifics of insect production including nutrient content, safety, and processing practice; a few have examined the economics and production volume capabilities. Substrates investigated for rearing insects include food and feed waste, manure, horticulture waste, and other organic substrates. Forest or gardening waste could also be utilized as rearing substrates if logistical difficulties could be overcome. For example, working in conjunction with local governments to redirect organic waste from landfills or even build insect processing plants within waste processing sites. Substrates could also include agricultural waste products, such as plant stalks and leaves, condemned grains (mycotoxins, etc.), manure, hulls and skins, spent bedding, etc. Many of these substrates can be in competition with other production systems including direct competition with pork production systems who already utilize waste products in diets as much as possible. However, in most cases processing waste incurs a cost (labor, transport, disposal fees, etc.). Insects like BSF consume waste rapidly and they generally consume the most digestible portion of the substrate first, meaning valuable nutrients can be left unchanged in the unconsumed waste. This leftover feed together with insect waste (excrement and exoskeletons for example) is called frass and is a nutrient-rich fertilizer product. In sub-Saharan Africa where soils are mostly nutrient deficient, including BSFL frass as a fertilizer increased the income from the farming practice (Beesigamukama et al., 2022). Considering that for every kilogram of larvae produced 10–34 kg of frass is produced (Beesigamukama et al., 2022), frass can become a valuable income stream to accompany the insects themselves.

While some feed protein sources such as meat meals and tallow already efficiently utilize waste products from the animal production industries, they require processing to be suitable for use in monogastric animal diets. Like insect processing, drying and transport processes are likely to be the greatest contributors to energy use (and hence emissions). Like most livestock industries, the pig industry has little to no control of grain feed prices, so sourcing local and sustainably produced and price-stable alternatives is critical for industry longevity and profitability. In addition, organic waste remains a sizable problem to be addressed. For example, insects could be reared on waste or contaminated animal feed and animal bedding that would otherwise be sent to end-point waste systems like ponds and pits, contributing to a no-waste system and potentially providing additional income streams for the piggery.

Pigs, poultry, companion animals, and humans are the most suited target species for utilizing insect protein. Given the current high cost of insect protein, the more likely purchasers are going to be the industries with a higher margin and thus more money to spend on raw ingredients. Thus, those with tighter margins such as poultry and pig production systems may be unlikely to afford insect protein in its current capacity and price. Currently, there is increased interest in sourcing insects for use in companion animal diets who have larger margins and a greater ability to spend on feed ingredients, given consumers are prepared to pay premium prices for pet food products. There is a global increase in the number of producers manufacturing pet feed containing insects, including Mars Petcare in the UK who are selling an insect-based cat food 'Lovebug'. Similarly, Nestlé Purina is producing insect and plant protein-based pet food in Switzerland.

Previous research has shown that insects (BSFL and mealworms, for example) can consume feed contaminated with mycotoxins without becoming contaminated themselves (Purschke et al., 2017; van der Fels-Klerx et al., 2018). While the metabolic processes the insects utilize to achieve this remain unknown, it is an exciting possibility that insects could consume contaminated feed, thereby utilizing waste feed and also providing a safe insect product that can be utilized as feed or food (Bosch et al., 2017; Gützkow et al., 2021; Scheibelberger et al., 2017). Growing BSFL did not accumulate pesticides when fed spiked feed (Cai et al., 2018b) although heavy metals (Cd, Cr, Ni, Pb, Hg, and As) are accumulated in the BSFL in significant quantities in multiple studies (Biancarosa et al., 2018; Cai et al., 2018a). In a comprehensive review, Lievens et al. (2021) summarize the current knowledge on the accumulation of toxins and heavy metals in insects, concluding that while investigations to date have shown no negative impact on insect growth when fed toxins or heavy metals, producers, and consumers need to ensure the product safety by monitoring any potential accumulation.

Recent evidence suggests that BSFL may be able to consume microplastics (from plastics commonly found in disposable bags and containers) (Cho et al., 2020) and polystyrene waste (Peng et al., 2019). Australian research showed that feeding some types of plastic did not diminish BSFL or mealworm growth and survival rates (although porous polylactic acid blocks reduced BSFL weight by ~29%) (Beale et al., 2022). This is promising as it could increase the options for feed sources, provide a useful mechanism to reduce some of the plastic waste ending up in landfill, and potentially reduce processing requirements if some plastic could be included in insect diets (reducing the need to sort or unpack waste).

7 Rearing insects on manure

Numerous insects can consume (and thrive on) manure, including BSFL, mealworms, and houseflies, as reviewed by Cammack et al. (2021). This can assist in the processing of waste that may not always be utilized (e.g. if biogas facilities are not present on farms). In addition, the residue from corn cob fermentation used to produce biogas has been effectively used in conjunction with pig manure as a substrate for rearing BSFL (Li et al., 2015). Thus, biogas residues, that are commonly used as fertilizer, can potentially be bioconverted into a more valuable product resulting in both frass fertilizer and proteins/oils from insects. Utilizing manure as a rearing substrate may be a useful approach to utilize substrates in a meaningful way that would otherwise contribute to the pollution of soils and waterways (via runoff and leaching) and the production of greenhouse gases. This process can generate useful products such as animal feed, oils for biodiesel and cosmetics, protein for use as emulsifiers and bioplastic applications, and chitin for use in biomedical applications and frass fertilizer. Previous experiments have demonstrated that BSFL can reduce the dry weight of manure by 30–50% and reduce the nitrogen content by 30–80% (Myers et al., 2008; Newton et al., 2005; Oonincx et al., 2015). Pig manure is a suitable substrate for mealworms, beetles (*Protaetia brevitarsis sulensis*), and flies (*P. tenebrifer*); each recording an approx. 90% survival rate (Choi, 2022). Up to a 100% reduction (ranging 87–100%) in volatile organic compounds have been recorded when BSFLs are reared on manure (Beskin et al., 2018). Insects such as BSFL can also reduce the pathogenic bacterial count of pig manure (Elhag et al., 2018), the *E. coli* counts in dairy manure (Liu et al., 2008), and *E. coli* and *Salmonella enterica* counts in dairy and chicken manure (Erickson et al., 2004). Processing manure with BSFL not only creates pathogen-free fertilizer products for farm use – far safer than directly adding pathogen-rich manures to soils – but can potentially also turn manure into livestock feed.

Using manure as a substrate doesn't come without risks, the greatest of which is the risk of contamination from veterinary medicines. However, Charlton et al. (2015) showed that larvae fed with manure containing veterinary medicines generally did not accumulate the substances (except for domestic flies) and most medications tested in the larvae were below the limit of detection. However, spreading contaminated manure would also be a potential source of risk to environmental contamination, particularly for residues in soil and water. In another experiment, BSFL were shown to degrade but not accumulate three pharmaceutical compounds (carbamazepine, roxithromycin, and trimethoprim) (Lalander et al., 2016), while tetracycline was degraded more efficiently by BSFL than normal composting (Cai et al., 2018b). Thus, processing manure using insects may prove to be the most environmentally friendly option. The use of manure as a rearing substrate for insects to be fed to production animals is currently prohibited in the EU and there will likely be resistance to this changing given insect's current classification as farmed animals (Boloh, 2018). Additional research in this area may assist to persuade change, particularly when evidence suggests that BSF larvae can reduce the pathogenic bacterial counts.

8 Current research

8.1 Insects as antimicrobials

Using insects as a source of antimicrobials has received increased attention as a potential natural antibiotic alternative to use as a disease prevention and growth-promoting agent. In a review, Jozefiak and Engberg (2017) concluded that antimicrobial peptides sourced from insects have excellent potential to be effectively used in animal diets. The combined impacts of increased demand, increased pressure to improve efficiency, and a desire to reduce antibiotic use and prevent further antimicrobial resistance have emphasized the urgent need to find alternative agents. Using insects as a feed and food source of microbial may be a valuable contribution, especially given there is evidence that antimicrobial peptides (AMPs) from insects do not result in the development of resistance (Jozefiak and Engberg, 2017). As reviewed by Hadj Saadoun et al. (2022), insects should be further examined as sources of antimicrobials, and it is likely that the estimated amount of AMP in insects has been underreported to date.

8.2 Energy use and life cycle analysis

As described previously, insect-rearing processes require labor and energy inputs similar to any farm enterprise. Life cycle assessments seek to quantify total emissions from inputs (such as animal feed) to processing (such as drying) and transport. van Zanten et al. (2018) concluded that feeding BSFL as a replacement for soybean meal in pig diets reduced global warming potential and land use, though life cycle assessment demonstrated no change in global warming potential. The complexities of diet formulations, particularly for pig and poultry diets that often include multiple ingredients, mean that substitution feed products with insects or alternate proteins will not always result in a net reduction in overall environmental impact. This is due to the many steps required such as land and water use, transport, processing, storage, and so on. While water use by insect rearing processes cannot yet be fully assessed given processing methods are under development, larvae and are predicted to have lower water, and certainly land use, compared to terrestrial animals and crops.

9 Future trends in research

9.1 Alternate insect species

Native species in individual countries will potentially be more efficient farmed insect targets given they have evolved to suit local climates. Future research should involve investigating a wider range of insect species targeted to specific regions. Further, harvesting pest species from agricultural crops etc. would be an ideal scenario as it would utilize the pests in a meaningful way while also improving existing production. However, the ability to capture such pests in an efficient manner could make this concept quite complicated but likely solvable with investments into research and development of harvesting technologies.

9.2 Insect welfare

The welfare of insects is often overlooked or dismissed as unimportant in this emerging industry. Yet, if insects are classified as farmed animals, then their welfare will need to be considered. In a recent review, Lambert et al. (2021) demonstrated that there is significant scientific evidence to support the notion that insects (across a range of orders) display sentience and cognition and are also likely to feel stress and pain. Thus, the development of farming systems to produce insects on a large scale must undertake thorough research to ensure their harvesting and slaughter practices are humane and ethical. To the best of my knowledge, no such research has been undertaken to date.

9.3 Impact of coronavirus disease 2019 on global production and feed sources

The COVID-19 pandemic impacted global trade, which of course impacted the pork production industries by increasing the costs of feed freight. Feed supplies were therefore more expensive and when combined with a reduction in pork meat product prices, the pork industry profit margins for pork were reduced. Given that regular animal production and processing chains were disrupted, the global market for animal feed protein ingredients was impacted. For example, as summarized by Rahimi et al. (2022), soybean meal exports from the Americas were reduced during the pandemic. Future events such as pandemics and weather events are likely to continue to disrupt the feed supply chain, further bolstering the need for sustainable and adaptable locally produced animal feed ingredients which may include insects.

9.4 Workplace labor constraints

The insect production industry is developing automated and robotic-based systems and engineering controls to reduce the need for manual labor. While such technological advancements will ultimately improve processing and accuracy, it will not eliminate the need for labor and will shift the type of jobs required to be more control and data based rather than manual labor. Therefore, training and advancement of employees in more technological based skills (such as operating robots and computer based systems) is required to ensure a sustainable and diverse workforce can be maintained for the insect production industry. Given the relative infancy of the industry, there is a timely opportunity to target and involve workers from a diversity of backgrounds (age, location, cultural background, gender, etc.), attract a wide range of skills (e.g. digital/precision agriculture), and attract staff interested in contributing towards a more sustainable agricultural industry.

10 Barriers to uptake/challenges to production

The greatest current challenge to insect production is the lack of scale. Engineering, infrastructure, and market processes need to be developed and optimized. In an opinion piece, Aarts (2019) describes the complexities of starting an insect rearing business (based on their experience with Protix) and suggests that insect producers need to consider the science, innovate appropriately, realize that insects are not the only potential source of alternate proteins, and importantly collaborative efforts need to occur in all facets of the system. A lack of collaboration between producers is currently constraining industry advancement. This is mostly due to perceived and real risks associated with non-mutual benefits and commercial competitiveness (Ponce-Reyes and Lessard, 2021). That being said, there are numerous large-scale insect production research programs bringing together academia and industry globally, such as the insect value chain in a circular bioeconomy (inVALUABLE) in Europe, involving 10 partners and worth over €3.7 million (Heckmann, 2018). Continued investment in collaborative research projects (across insect producers, states, universities, and agricultural and government industries) is imperative.

Consumer education will also be paramount if insects are going to become a more major and consistent feed (or food). In a 2018 survey, Weinrich and Busch (2021) identified that most consumers did not feel well educated about animal feed, although negative responses towards soybean and GM feeds are often expressed. However, consumers are increasingly concerned with the environmental impact of their food choices. Highlighting the environmental and sustainability benefits of insect production will likely lead to positive consumer responses if the messaging is clear and supported by evidence.

11 Conclusion

This chapter has highlighted some of the major findings and knowledge gaps in the insect production industry. Pork production industries have the opportunity to adopt the use of insects for inclusion as a feed source but also as a tool to process waste. Insects can provide valuable nutrients to pigs including fat, protein, and antimicrobials. Despite significant investment into the upscaling of insect production, the industry remains in its infancy and extensive adoption is yet to occur. Further research is required to establish and optimize insect rearing, storage, and processing practices and develop new markets; to establish insect production emissions and energy use values; to confirm insect inclusion levels and product safety; to examine the safety of insects reared on manure; and to confirm the potential bioactive and other functional properties of insects. Pig production industries are well placed to contribute to the reduction of waste and the improvement of the sustainability of animal production industries globally by embracing the use of insects as feed and as a waste management activity.

12 Where to look for further information

Numerous reviews and book chapters on the use of insects in the production of animal diets have been published in recent years. Some suggestions are provided below, although this list is by no means exhaustive.

- Veldkamp, T. and Vernooij, A. G. (2021). Use of insect products in pig diets. *Journal of Insects as Food and Feed*, 7(5), 781–793. doi:10.3920/jiff2020.0091.
- Hong, J. and Kim, Y. Y. (2022). Insect as feed ingredients for pigs. *Animal Bioscience*, 35(2), 347–355. doi:10.5713/ab.21.0475.
- van Huis, A. and Tomberlin, J. K. (Eds.). (2017). *Insects as Food and Feed. From Production to Consumption*. https://doi.org/10.3920/978-90-8686-849-0.

- Parrini, S., Aquilani, C., Pugliese, C., Bozzi, R. and Sirtori, F. (2023). Soybean Replacement by alternative protein sources in pig nutrition and its effect on meat quality. *Animals*, 13(3), 494. doi:10.3390/ani13030494.
- Lu, S., Taethaisong, N., Meethip, W., Surakhunthod, J., Sinpru, B., Sroichak, T., Paengkoum, P. (2022). Nutritional composition of black soldier fly larvae (*Hermetia illucens* L.) and its potential uses as alternative protein sources in animal diets: a review. *Insects*, 13(9). doi:10.3390/insects13090831.

13 References

Aarts, K. (2019). How to develop insect-based ingredients for feed and food? A company's perspective. *Journal of Insects as Food and Feed* 6(1), 67–68.

Altmann, B. A., Neumann, C., Rothstein, S., Liebert, F. and Mörlein, D. (2019). Do dietary soy alternatives lead to pork quality improvements or drawbacks? A look into micro-alga and insect protein in swine diets. *Meat Science* 153, 26–34.

Ao, X. and Kim, I. H. (2019). Effects of dietary dried mealworm (*Ptecticus tenebrifer*) larvae on growth performance and nutrient digestibility in weaning pigs. *Livestock Science* 230, 103815.

Ao, X., Yoo, J. S., Wu, Z. L. and Kim, I. H. (2020). Can dried mealworm (*Tenebrio molitor*) larvae replace fish meal in weaned pigs? *Livestock Science* 239, 104103.

Baiano, A. (2020). Edible insects: an overview on nutritional characteristics, safety, farming, production technologies, regulatory framework, and socio-economic and ethical implications. *Trends in Food Science and Technology* 100, 35–50.

Bartrim, J. (2017). Insect farming and consumption in Australia - opportunities and barriers. *Australian Zoologist* 39(1), 26–30.

Beale, D. J., Shah, R. M., Marcora, A., Hulthen, A., Karpe, A. V., Pham, K., Wijffels, G. and Paull, C. (2022). Is there any biological insight (or respite) for insects exposed to plastics? Measuring the impact on an insects central carbon metabolism when exposed to a plastic feed substrate. *Science of the Total Environment* 831, 154840.

Beesigamukama, D., Mochoge, B., Korir, N., Menale, K., Muriithi, B., Kidoido, M., Kirscht, H., Diiro, G., Ghemoh, C. J., Sevgan, S., Nakimbugwe, D., Musyoka, M. W., Ekesi, S. and Tanga, C. M. (2022). Economic and ecological values of frass fertiliser from black soldier fly agro-industrial waste processing. *Journal of Insects as Food and Feed* 8(3), 245–254.

Beskin, K. V., Holcomb, C. D., Cammack, J. A., Crippen, T. L., Knap, A. H., Sweet, S. T. and Tomberlin, J. K. (2018). Larval digestion of different manure types by the black soldier fly (Diptera: Stratiomyidae) impacts associated volatile emissions. *Waste Management* 74, 213–220.

Biancarosa, I., Liland, N. S., Biemans, D., Araujo, P., Bruckner, C. G., Waagbø, R., Torstensen, B. E., Lock, E. J. and Amlund, H. (2018). Uptake of heavy metals and arsenic in black soldier fly (*Hermetia illucens*) larvae grown on seaweed-enriched media. *Journal of the Science of Food and Agriculture* 98(6), 2176–2183.

Biasato, I., Renna, M., Gai, F., Dabbou, S., Meneguz, M., Perona, G., Martinez, S., Lajusticia, A. C. B., Bergagna, S., Sardi, L., Capucchio, M. T., Bressan, E., Dama, A., Schiavone, A. and Gasco, L. (2019). Partially defatted black soldier fly larva meal inclusion in piglet diets: effects on the growth performance, nutrient digestibility, blood profile, gut

morphology and histological features. *Journal of Animal Science and Biotechnology* 10, 12.
Boloh, Y. (2018). Insect proteins inch toward approval for EU animal feed: after the authorization for insect proteins in aquafeed in July 2017, Europe may accept insect proteins for poultry and pig feeds in 2019. *Feed Strategy* 69, 16.
Boontiam, W., Hong, J., Kitipongpysan, S. and Wattanachai, S. (2022). Full-fat field cricket (*Gryllus bimaculatus*) as a substitute for fish meal and soybean meal for weaning piglets: effects on growth performance, intestinal health, and redox status. *Journal of Animal Science* 100(4), skac080.
Bosch, G., Fels-Klerx, H. J. V., Rijk, T. C. and Oonincx, D. G. A. B. (2017). Aflatoxin B1 tolerance and accumulation in black soldier fly larvae (*Hermetia illucens*) and yellow mealworms (*Tenebrio molitor*). *Toxins (Basel)* 9(6), 185.
Cai, M., Hu, R., Zhang, K., Ma, S., Zheng, L., Yu, Z. and Zhang, J. (2018a). Resistance of black soldier fly (Diptera: Stratiomyidae) larvae to combined heavy metals and potential application in municipal sewage sludge treatment. *Environmental Science and Pollution Research* 25(2), 1559–1567.
Cai, M., Ma, S., Hu, R., Tomberlin, J. K., Yu, C., Huang, Y., Zhan, S., Li, W., Zheng, L., Yu, Z. and Zhang, J. (2018b). Systematic characterization and proposed pathway of tetracycline degradation in solid waste treatment by *Hermetia illucens* with intestinal microbiota. *Environmental Pollution* 242(A), 634–642.
Cammack, J. A., Miranda, C. D., Jordan, H. R. and Tomberlin, J. K. (2021). Upcycling of manure with insects: current and future prospects. *Journal of Insects as Food and Feed* 7(5), 605–619.
Charlton, A. J., Dickinson, M., Wakefield, M. E., Fitches, E., Kenis, M., Han, R., Zhu, F., Kone, N., Grant, M., Devic, E., Bruggeman, G., Prior, R. and Smith, R. (2015). Exploring the chemical safety of fly larvae as a source of protein for animal feed. *Journal of Insects as Food and Feed* 1(1), 7–16.
Chia, S. Y., Tanga, C. M., Osuga, I. M., Alaru, A. O., Mwangi, D. M., Githinji, M., Dubois, T., Ekesi, S., Van Loon, J. J. A. and Dicke, M. (2021). Black soldier fly larval meal in feed enhances growth performance, carcass yield and meat quality of finishing pigs. *Journal of Insects as Food and Feed* 7(4), 433–447.
Chia, S. Y., Tanga, C. M., Osuga, I. M., Alaru, A. O., Mwangi, D. M., Githinji, M., Subramanian, S., Fiaboe, K. K. M., Ekesi, S., Van Loon, J. J. A. and Dicke, M. (2019). Effect of dietary replacement of fishmeal by insect meal on growth performance, blood profiles and economics of growing pigs in Kenya. *Animals* 9(10), 705.
Cho, S., Kim, C.-H., Kim, M.-J. and Chung, H. (2020). Effects of microplastics and salinity on food waste processing by black soldier fly (*Hermetia illucens*) larvae. *Journal of Ecology and Environment* 44(1), 7.
Choi, I. H. (2022). Decomposition abilities and characteristics of pig manure using three insect larvae. *Entomological Research* 52(10), 439–444.
Costa-Neto, E. M. and Dunkel, F. V. (2016). Chapter 2. Insects as food: history, culture, and modern use around the world. In: Dossey, A. T., Morales-Ramos, J. A. and Rojas, M. G. (Eds). *Insects as Sustainable Food Ingredients*. Academic Press, San Diego.
Defoliart, G. R. (1999). Insects as food: why the western attitude is important. *Annual Review of Entomology* 44, 21.
Digiacomo, K., Akit, H. and Leury, B. J. (2019). Insects: a novel animal-feed protein source for the Australian market. *Animal Production Science* 59(11), 2037–2045.

Digiacomo, K. and Leury, B. J. (2019). Review: insect meal: a future source of protein feed for pigs? *Animal* 13(12), 3022–3030.

Elhag, O. A. O., Xiao, X. P., Zheng, L. Y. and Zhang, J. B. (2018). Antibacterial activity of *Hermetia illucens* against pathogen naturally present in the pig manure and its mechanism. The 2nd International Conference 'Insects to Feed the World', 15-18 May 2018, Wuhan, China. Wageningen Academic Publishers.

Erickson, M. C., Islam, M., Sheppard, C., Liao, J. and Doyle, M. P. (2004). Reduction of Escherichia coli O157:H7 and *Salmonella enterica* serovar enteritidis in chicken manure by larvae of the black soldier fly. *Journal of Food Protection* 67(4), 685-690.

FAO (2011). *Global Food Losses and Food Waste - Extent, Causes and Prevention*. FAO, Rome, Italy.

Franklin, A., Brandt, N. and Ureda, N. (2018). Edible insect farming: a strategy for providing sustainable nutrition and economic empowerment for orphanages in the DRC. The 2nd International Conference 'Insects to Feed the World' 15–18 May 2018, Wuhan, China. Wageningen Academic Publishers.

Go, Y. B., Lee, J. H., Lee, B. K., Oh, H. J., Kim, Y. J., An, J. W., Chang, S. Y., Song, D. C., Cho, H. A., Park, H. R., Chun, J. Y. and Cho, J. H. (2022). Effect of insect protein and protease on growth performance, blood profiles, fecal microflora and gas emission in growing pig. *Journal of Animal Science and Technology* 64(6), 1063–1076.

Gützkow, K. L., Ebmeyer, J., Kröncke, N., Kampschulte, N., Böhmert, L., Schöne, C., Schebb, N. H., Benning, R., Braeuning, A. and Maul, R. (2021). Metabolic fate and toxicity reduction of aflatoxin B1 after uptake by edible *Tenebrio molitor* larvae. *Food and Chemical Toxicology* 155, 112375.

Hadj Saadoun, J., Sogari, G., Bernini, V., Camorali, C., Rossi, F., Neviani, E. and Lazzi, C. (2022). A critical review of intrinsic and extrinsic antimicrobial properties of insects. *Trends in Food Science and Technology* 122, 40–48.

Heckmann, L. H. (2018). Research and development efforts on optimizing key parameters in industrial insect production. The 69th Annual Meeting of the European Federation of Animal Science, Dubrovnik, Croatia. Wageningen Academic.

Hong, J. and Kim, Y. Y. (2022). Insect as feed ingredients for pigs. *Animal Bioscience* 35(2), 347-355.

Ipema, A. F., Gerrits, W. J. J., Bokkers, E. A. M., Kemp, B. and Bolhuis, J. E. (2021). Live black soldier fly larvae (*Hermetia illucens*) provisioning is a promising environmental enrichment for pigs as indicated by feed- and enrichment-preference tests. *Applied Animal Behaviour Science* 244, 105481.

IPIFF (2019). *The European Insect Sector Today: Challenges, Opportunities and Regulatory Landscape. IPIFF Vision Paper on the Future of the Insect Sector Towards 2030*. IPIFF, Brussels.

Ji, Y. J., Liu, H. N., Kong, X. F., Blachier, F., Geng, M. M., Liu, Y. Y. and Yin, Y. L. (2016). Use of insect powder as a source of dietary protein in early-weaned piglets. *Journal of Animal Science* 94, 111–116.

Jin, X. H., Heo, P. S., Hong, J. S., Kim, N. J. and Kim, Y. Y. (2016). Supplementation of dried mealworm (*Tenebrio molitor* larva) on growth performance, nutrient digestibility and blood profiles in weaning pigs. *Asian-Australasian Journal of Animal Sciences* 29(7), 979-986.

Jonas-Levi, A. and Martinez, J.-J. I. (2017). The high level of protein content reported in insects for food and feed is overestimated. *Journal of Food Composition and Analysis* 62, 184–188.

Jozefiak, A. and Engberg, R. M. (2017). Insect proteins as a potential source of antimicrobial peptides in livestock production: a review. *Journal of Animal and Feed Sciences* 26(2), 87.

Jucker, C., Erba, D., Leonardi, M. G., Lupi, D. and Savoldelli, S. (2017). Assessment of vegetable and fruit substrates as potential rearing media for *Hermetia illucens* (Diptera: Stratiomyidae) larvae. *Environmental Entomology* 46(6), 1415–1423.

Kierończyk, B., Rawski, M., Mikołajczak, Z., Homska, N., Jankowski, J., Ognik, K., Józefiak, A., Mazurkiewicz, J. and Józefiak, D. (2022). Available for millions of years but discovered through the last decade: insects as a source of nutrients and energy in animal diets. *Animal Nutrition* 11, 60–79.

Kim, J., Ji, S. Y. and Kim, B. G. (2023). Ileal digestibility of amino acids in defatted black soldier fly larva meal and adult black soldier fly fed to pigs. *Journal of Insects as Food and Feed*, in press, 9(10), 1345–1352.

Ko, H., Kim, Y. and Kim, J. (2020). The produced mealworm meal through organic wastes as a sustainable protein source for weanling pigs. *Journal of Animal Science and Technology* 62(3), 365–373.

Lalander, C., Senecal, J., Gros Calvo, M., Ahrens, L., Josefsson, S., Wiberg, K. and Vinnerås, B. (2016). Fate of pharmaceuticals and pesticides in fly larvae composting. *Science of the Total Environment* 565, 279–286.

Lambert, H., Elwin, A. and D'cruze, N. (2021). Wouldn't hurt a fly? A review of insect cognition and sentience in relation to their use as food and feed. *Applied Animal Behaviour Science* 243, 105432.

Li, W., Li, Q., Zheng, L., Wang, Y., Zhang, J., Yu, Z. and Zhang, Y. (2015). Potential biodiesel and biogas production from corncob by anaerobic fermentation and black soldier fly. *Bioresource Technology* 194, 276–282.

Lievens, S., Poma, G., De Smet, J., Van Campenhout, L., Covaci, A. and Van Der Borght, M. (2021). Chemical safety of black soldier fly larvae (*Hermetia illucens*), knowledge gaps and recommendations for future research: a critical review. *Journal of Insects as Food and Feed* 7(4), 383–396.

Liu, H., Tan, B., Kong, X., Li, J., Li, G., He, L., Bai, M. and Yin, Y. (2020). Dietary insect powder protein sources improve protein utilization by regulation on intestinal amino acid-chemosensing system. *Animals* 10(9), 1590.

Liu, Q., Tomberlin, J. K., Brady, J. A., Sanford, M. R. and Yu, Z. (2008). Black soldier fly (Diptera: Stratiomyidae) larvae reduce *Escherichia coli* in dairy manure. *Environmental Entomology* 37(6), 1525–1530.

Mancuso, T., Pippinato, L. and Gasco, L. (2019). The European insects sector and its role in the provision of green proteins in feed supply. *Quality – Access to Success* 20, 374–381.

Meyer, S., Gessner, D. K., Braune, M. S., Friedhoff, T., Most, E., Höring, M., Liebisch, G., Zorn, H., Eder, K. and Ringseis, R. (2020). Comprehensive evaluation of the metabolic effects of insect meal from *Tenebrio molitor* L. in growing pigs by transcriptomics, metabolomics and lipidomics. *Journal of Animal Science and Biotechnology* 11, 20.

Meyer-Rochow, V. B. and Changkija, S. (1997). Uses of insects as human food in Papua New Guinea, Australia, and North-East India: cross-cultural considerations and cautious conclusions. *Ecology of Food and Nutrition* 36(2–4), 159–185.

Miech, P., Lindberg, J. E., Berggren, Å., Chhay, T. and Jansson, A. (2017). Apparent faecal digestibility and nitrogen retention in piglets fed whole and peeled Cambodian field cricket meal. *Journal of Insects as Food and Feed* 3(4), 279–288.

Myers, H. M., Tomberlin, J. K., Lambert, B. D. and Kattes, D. (2008). Development of black soldier fly (Diptera: Stratiomyidae) larvae fed dairy manure. *Environmental Entomology* 37(1), 11–15.

Neumann, C., Velten, S. and Liebert, F. (2018). N balance studies emphasize the superior protein quality of pig diets at high inclusion level of algae meal (*Spirulina platensis*) or insect meal (*Hermetia illucens*) when adequate amino acid supplementation is ensured. *Animals* 8(10), 1–14.

Newton, G. L., Booram, C. V., Barker, R. W. and Hale, O. M. (1977). Dried *Hermetia illucens* larvae meal as a supplement for swine. *Journal of Animal Science* 44(3), 395–400.

Newton, L., Sheppard, C., Watson, D. W., Burtle, G. and Dove, R. (2005). *Using the Black Soldier Fly, Hermetia illucens, as a Value-Added Tool for the Management of Swine Manure*. Animal and Poultry Waste Management Centre, North Carolina State University, Raleigh, NC.

OECD & Food Agriculture Organization of the United Nations, R (2022). *OECD-FAO Agricultural Outlook* (Edition 2022). OECD Agriculture Statistics (database), https://doi.org/10.1787/13d66b76-en (accessed on 28 January 2024).

Oonincx, D. G. A. B., Van Huis, A. and Van Loon, J. J. A. (2015). Nutrient utilisation by black soldier flies fed with chicken, pig, or cow manure. *Journal of Insects as Food and Feed* 1(2), 131–139.

Peng, B. Y., Su, Y., Chen, Z., Chen, J., Zhou, X., Benbow, M. E., Criddle, C. S., Wu, W. M. and Zhang, Y. (2019). Biodegradation of polystyrene by dark (*Tenebrio obscurus*) and yellow (*Tenebrio molitor*) mealworms (Coleoptera: Tenebrionidae). *Environmental Science and Technology* 53(9), 5256–5265.

Perednia, D. A. (2017). A comparison of the greenhouse gas production of black soldier fly larvae versus aerobic microbial decomposition of an organic feed material. *Research and Reviews: Journal of Ecology and Environmental Sciences* 5(3), 10–16.

Poelaert, C., Beckers, Y., Despret, X., Portetelle, D., Francis, F. and Bindelle, J. (2016). In vitro evaluation of fermentation characteristics of two types of insects as potential novel protein feeds for pigs. *Journal of Animal Science* 94(Suppl. 3), 198–201.

Ponce-Reyes, R. and Lessard, B. (2021). *Edible Insects - A Roadmap for the Strategic Growth of an Emerging Australian Industry*. CSIRO, Canberra.

Purschke, B., Scheibelberger, R., Axmann, S., Adler, A. and Jäger, H. (2017). Impact of substrate contamination with mycotoxins, heavy metals and pesticides on the growth performance and composition of black soldier fly larvae (*Hermetia illucens*) for use in the feed and food value chain. *Food Additives and Contaminants: Part A, Chemistry, Analysis, Control, Exposure and Risk Assessment* 34(8), 1410–1420.

Rahimi, P., Islam, M. S., Duarte, P. M., Tazerji, S. S., Sobur, M. A., El Zowalaty, M. E., Ashour, H. M. and Rahman, M. T. (2022). Impact of the COVID-19 pandemic on food production and animal health. *Trends in Food Science and Technology* 121, 105–113.

Rumpold, B. A. and Schlüter, O. K. (2013). Potential and challenges of insects as an innovative source for food and feed production. *Innovative Food Science and Emerging Technologies* 17, 1–11.

Scheibelberger, R., Axmann, S., Adler, A. and Jäger, H. (2017). Impact of substrate contamination with mycotoxins, heavy metals and pesticides on the growth performance and composition of black soldier fly larvae (*Hermetia illucens*) for use in the feed and food value chain. *Food Additives and Contaminants: Part A* 34, 1410–1420.

Sogari, G., Amato, M., Palmieri, R., Hadj Saadoun, J., Formici, G., Verneau, F. and Mancini, S. (2023). The future is crawling: evaluating the potential of insects for food and feed security. *Current Research in Food Science* 6, 100504.

Swiatkiewicz, S., Swiatkiewicz, M., Arczewska-Wlosek, A. and Jozefiak, D. (2015). Chitosan and its oligosaccharide derivatives (chito-oligosaccharides) as feed supplements in poultry and swine nutrition. *Journal of Animal Physiology and Animal Nutrition* 99(1), 1–12.

Tan, X., Yang, H. S., Wang, M., Yi, Z. F., Ji, F. J., Li, J. Z. and Yin, Y. L. (2020). Amino acid digestibility in housefly and black soldier fly prepupae by growing pigs. *Animal Feed Science and Technology* 263, 114446.

Van Der Fels-Klerx, H. J., Camenzuli, L., Belluco, S., Meijer, N. and Ricci, A. (2018). Food safety issues related to uses of insects for feeds and foods. *Comprehensive Reviews in Food Science and Food Safety* 17(5), 1172–1183.

Van Huis, A. (2013). Potential of insects as food and feed in assuring food security. *Annual Review of Entomology* 58, 563–583.

Van Huis, A. (2016). Edible insects are the future? *Proceedings of the Nutrition Society* 75(3), 294–305.

Van Huis, A., Van Itterbeeck, J., Klunder, H., Mertens, E., Halloran, A., Muir, G. and Vantomme, P. (2013). *Edible Insects: Future Prospects for Food and Feed Security*. FAO Forestry Paper 171. Food and Agriculture Organization of the United Nations, Rome.

Van Zanten, H. H. E., Bikker, P., Meerburg, B. G. and De Boer, I. J. M. (2018). Attributional versus consequential life cycle assessment and feed optimization: alternative protein sources in pig diets. *International Journal of Life Cycle Assessment* 23(1), 1–11.

Veldkamp, T. and Bosch, G. (2015). Insects: a protein-rich feed ingredient in pig and poultry diets. *Animal Frontiers* 5, 45–50.

Veldkamp, T., Dong, L., Paul, A. and Govers, C. (2022). Bioactive properties of insect products for monogastric animals – a review. *Journal of Insects as Food and Feed* 8(9), 1027–1040.

Veldkamp, T. and Vernooij, A. G. (2021). Use of insect products in pig diets. *Journal of Insects as Food and Feed* 7(5), 781–793.

Verbeke, W., Spranghers, T., De Clercq, P., De Smet, S., Sas, B. and Eeckhout, M. (2015). Insects in animal feed: acceptance and its determinants among farmers, agriculture sector stakeholders and citizens. *Animal Feed Science and Technology* 204, 72–87.

Weinrich, R. and Busch, G. (2021). Consumer knowledge about protein sources and consumers' openness to feeding micro-algae and insects to pigs and poultry. *Future Foods* 4, 100100.

Yu, M., Li, Z., Chen, W., Rong, T., Wang, G. and Ma, X. (2019). *Hermetia illucens* larvae as a potential dietary protein source altered the microbiota and modulated mucosal immune status in the colon of finishing pigs. *Journal of Animal Science and Biotechnology* 10, 50.

Yu, M., Li, Z., Chen, W., Rong, T., Wang, G., Wang, F. and Ma, X. (2020). Evaluation of full-fat *Hermetia illucens* larvae meal as a fishmeal replacement for weanling piglets: effects on the growth performance, apparent nutrient digestibility, blood parameters and gut morphology. *Animal Feed Science and Technology* 264, 114431.

Chapter 12

Green feed in organic pig farming

Anne Grete Kongsted, Marleen Elise van der Heide, Lene Stødkilde, Jakob C. Johannsen and Maria Eskildsen, Aarhus University, Denmark

1 Introduction

Organic pork production has experienced significant growth in the European Union (EU) over the past decade, with the number of organic certified live pigs in EU28 increasing by over 50% from 2015 to 2019, reaching approximately 1.5 million animals (Eurostat, 2023). Similar trends have been observed in other regions, such as the USA, where sales from organic certified pig farms have increased from 4.6 million dollars in 2011 to 7.8 million dollars in 2021 (USDA, 2012, 2022). This global market development has heightened the demand for organically produced feed sources for pigs.

In the EU, the dependence on imported feedstuffs for organic livestock feed has been a matter of concern for several years. Früh et al. (2015) calculated the actual produced organic crude protein (CP) and dry matter (DM) relative to the total demands of bovine animals, pigs, and poultry organically produced in 11 European countries, and this revealed a self-sufficiency rate of 0.56 for CP and 0.69 for DM, on average across countries. Organic diets based on regionally and organic produced feed components is a fundamental goal in organic livestock production. However, the imbalance in organic protein supply and demand has compelled the European Commission to extend the derogation related to the use of up to 5% non-organic feed (protein sources) in the nutrition of pigs to meet their protein (amino acids) requirements (EC, 2021).

http://dx.doi.org/10.19103/AS.2024.0140.19

The ban on the use of single pure amino acids in organic livestock feeding (EC) poses a significant challenge to the formulation of diets that meet the amino acid requirements of organic pigs. This ban increases the risk of an oversupply of dietary CP, which may lead to excessive nitrogen (N) excretion and the risk of N losses, especially from outdoor areas, through nitrate leaching (Eriksen et al., 2002; Manevski et al., 2019) and ammonia volatilization (Olsson et al., 2014; Wimmler et al., 2022), thereby undermining the credibility of organic pork production. Furthermore, the protein recommendations for pigs are often expressed relative to dietary energy, and the high daily feed allowances of organic pigs to meet their extra energy requirements further reinforce the risk of protein overfeeding.

In the European Union's largest organic pork-producing countries, organic pig feeding is primarily based on concentrate, which includes processed cereals and oil seed feed sources, such as soybeans. In Denmark, compound feed comprises more than 80% of organic sows' feed supply and over 95% of organic fattening pigs' feed supply on a dry matter basis (Andersen et al., 2022). As a result of lower feed efficiencies and lower arable crop yields in organic farming, the land occupation is significantly higher than in conventional pork production. According to a recent Life Cycle Assessment (LCA) of Danish standard practice, the land occupation per kg live weight pig was 8.8 m^2 in organic production compared to 4.4 m^2 in conventional indoor production (Andersen et al., 2021; Dorca-Preda et al., 2023). Besides contributing negatively to the feed-to-food competition challenge, this extensive use of concentrate consumed within a short time may be seen as contradictory to the nature of the pig, which is highly explorative and has a diverse intake of grasses, leaves, roots, root crops, fruit, etc. depending on availability (Graves, 1984; Sandom et al., 2013). Certified organic livestock farming is guided by the four principles of Health, Ecology, Fairness, and Care formulated by The International Federation of Organic Agricultural Movement (IFOAM, 2021). The Principle of Fairness states that animals should be provided with conditions and opportunities of life that accord with their physiology, natural behavior, and well-being. In general, 'naturalness' and 'freedom of choice' are key elements in organic livestock farming and must be considered when formulating feeding strategies for organic pigs (Zollitsch, 2007).

The current challenges require a re-evaluation of organic feeding practices, taking into account land use considerations. The EU regulations for organic pig farming stipulate that roughage must be included in the daily ration (EC, 2007, 2018, 2021). Roughage sources such as grass, clover, and lucerne have higher protein yields than concentrate sources like barley, rape seeds, and soybeans under Northern European conditions (Petersen, 2010; Mogensen et al., 2018). Recent Danish studies have shown that grass-clover silage and fresh grass clover (grazed) can meet a significant portion of the amino acid requirements for free-range gestating sows during winter (Eskildsen et al., 2020c) and summer (Eskildsen et al., 2020b; Johannsen et al., 2022), respectively. These findings suggest that it may be possible to increase the use of locally produced roughage in organic pig production by replacing a portion of the concentrate with roughage. However, growing pigs and lactating sows of modern genotypes have limited voluntary roughage intake (Carlson et al., 1999; Kongsted et al., 2000), emphasizing the need for alternative sustainable concentrate components. Recently, it was demonstrated that protein extracted from grass clover ('green protein') can be added to organic fattening pig diets as a substitute for conventional concentrate components without negatively impacting daily gain (Stødkilde et al., 2021a).

The following section presents findings from recent research on the energy and protein (amino acid) requirements of organic pigs. This is followed by a discourse on the possibility of enhancing the nutritional value of organic pig diets by incorporating more forage, either through direct foraging or by allocating roughage, as well as incorporating 'green protein' sources. The emphasis of this chapter is on studies conducted under the specific conditions of Denmark and Northern Europe.

2 Energy and protein (amino acid) requirements of organic pigs

Organic pigs have distinct nutritional requirements compared to conventional pigs, as noted by Close and Poornan (1993). In accordance with European regulations for organic livestock, organic pigs have access to outdoor areas and higher space allowances than conventional production, which increases their daily energy requirements for physical activity and thermoregulation (Eskildsen et al., 2020b,c). However, this does not impact their daily amino acid requirements.

2.1 Energy requirements

Organic free-range sows' energy requirements were recently investigated during winter (first parity) and summer (second parity), respectively (Eskildsen et al., 2020b,c). The sows were highly productive with 157 kg and 205 kg weaned piglets per litter after 49 days lactation in parity 1 and 2, respectively. The huge productivity potential of outdoor sows was later confirmed in a study with sows of mixed parity (second-fifth) weaning on average 257 kg piglets per litter after 49 days lactation under similar rearing conditions during winter (Johannsen et al., 2022). To investigate the sows' energy needs during pregnancy and lactation, a factorial approach was used, and energy required for locomotor activity, thermoregulation, milk production, maintenance, and maternal retention was quantified (Eskildsen et al., 2020b,c).

2.1.1 Locomotor activity

To estimate individual sows' activity level, they were equipped with a GPS tracking system during the daytime (Fig. 1).

The data from the GPS collars revealed significant disparities among the sows in terms of the distance travelled within the observation days, such as approximately 1–3 km on day 40 in lactation (Eskildsen, 2020). The mean distances travelled were lowest in early lactation when the sows moved mainly to eliminate, eat, and drink, averaging 0.6 km and 0.8 km during winter and summer, respectively (Table 1). This pattern is consistent with findings by Buckner et al. (1998) and Schild et al. (2018), who showed that free-range sows preferred to stay inside the hut with their piglets during the first week after farrowing, representing approximately 75–95% of all daytime observations. During pregnancy, the sows travelled daily between 1.7 km and 2.8 km depending on season and pregnancy stage (Table 1). In comparison, Buckner (1996) reported a range of 0.1–3.1 km daily in outdoor pregnant sows based on pedometer values.

Close and Poornan (1993) and Edwards (2003) have suggested that energy expenditures of 7 kJ/kg body weight per km walked and a net efficiency of energy utilization of 0.8 are applicable. Based on these values, a 225-kg pregnant sow walking 2.3 km daily would dissipate an additional 3.6 MJ heat per day and require an extra 4.5 MJ ME per day to compensate for her locomotor activity. It is worth noting, however, that this has a minor impact on the total amount of energy needed by outdoor sows (Eskildsen et al., 2020b).

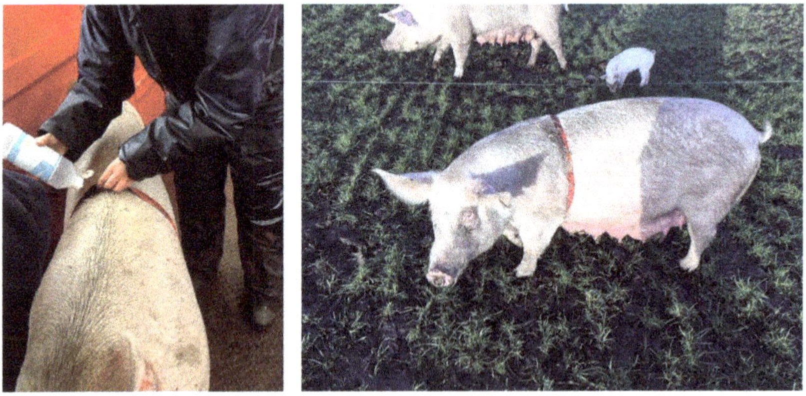

Figure 1 A GPS tracker was wrapped around the sow's belly to estimate activity level during daytime. Source: Modified after: Eskildsen (2020) and Eskildsen et al. (2020c).

Table 1 Free-range sow and litter weight, and daily distance travelled on day 60 and 100 in pregnancy and day 5, 20, and 40 in lactation

	Pregnancy		Lactation		
	D60	D100	D5	D20	D40
Winter (first parity)					
Sow weight, kg	215	242	227	206	190
Litter weight, kg			18.8	66.9	118
Milk yield, kg/day			8.0	12.5	9.3
Milk output, MJ ME/day			49.0	68.9	49.6
Daily distance travelled, km	2.78	2.46	0.64	1.54	1.68
Summer (second parity)					
Sow weight, kg	263	303	283	257	245
Litter weight, kg			27.0	78.7	144.2
Milk yield, kg/day			9.7	15.7	11.8
Milk output, MJ ME/day			51.0	76.6	55.0
Daily distance travelled, km	2.47	1.71	0.82	1.45	1.64

Source: Modified after: Eskildsen et al. 2020b,c).

Nonetheless, the activity level of sows will depend on various factors, such as feeding strategy, paddock dimensions, distribution of pasture vegetation, and other resources, as well as whether the sows are nose-ringed or not, as the ring can affect their energy-demanding rooting behavior (Eriksen et al., 2006).

2.1.2 Thermoregulation

Thermoregulation in free-range pigs is a critical aspect that requires additional daily energy demands. According to Close and Poornan (1993), the extra energy demand for thermoregulation in pigs is estimated to be 18.8 kJ ME per kg metabolic bodyweight for every degree below the lower critical temperature. In a study conducted by Verhagen et al. (1986), the lower critical temperature for sows was determined to be 18°C, and given a sow of 225 kg live weight, the daily energy spent on thermoregulation was estimated to be approximately 16 MJ ME (Eskildsen et al., 2020c). This implies that the sow would require an additional 20 MJ ME from feed, with a net efficiency of energy utilization of 0.8. During spring and summer, the energy spent on thermoregulation is generally lower, at around 5 MJ ME for a sow weighing 225 kg and an average ambient temperature of 13°C. However, as discussed in Eskildsen et al. (2020b), pregnant sows fed above their maintenance requirements generate heat due to energy retention, which likely reduces their need for additional energy for thermoregulation. Similarly, high milk production is expected to generate sufficient heat to maintain a constant body temperature in outdoor lactating sows during the spring and summer seasons.

2.1.3 Milk production

The importance of milk production in determining energy requirements during lactation cannot be overstated. However, there has been a paucity of research on organic sows' milk yield and composition. In the Danish studies on outdoor sows, milk yield was estimated based on average litter weight gain and litter size from days 2 to 40 (Table 1, Fig. 2). In peak lactation (day 20), mean daily milk yields amounted to 12.5 kg, 15.7 kg, and 16.3 kg in first, second, and mixed (second–fifth) parity sows, respectively (Eskildsen et al., 2020b,c; Johannsen et al., 2023). Consequently, milk production accounted for more than 50% of the total daily energy requirements in peak lactation. Due to an energy undersupply in early and mid-lactation, the sows in the aforementioned studies lost substantial back fat, varying from an average of 4.1 mm in mixed parity sows (Johannsen et al., 2022) to 5.8 mm in first parity sows (Eskildsen et al., 2020c) from days 5 to 40 in lactation. The under-supply was partly caused by a lower feed intake than the planned feeding curve, which underscores the need for a focus on how to improve sow appetite and optimize feed composition, especially in the sensitive period after farrowing.

Overall, the total daily energy requirements in first parity sows with access to grass-clover silage were quantified to 68 MJ ME in pregnancy and 153 MJ ME in peak lactation during temperate winter conditions (Eskildsen et al., 2020c). In

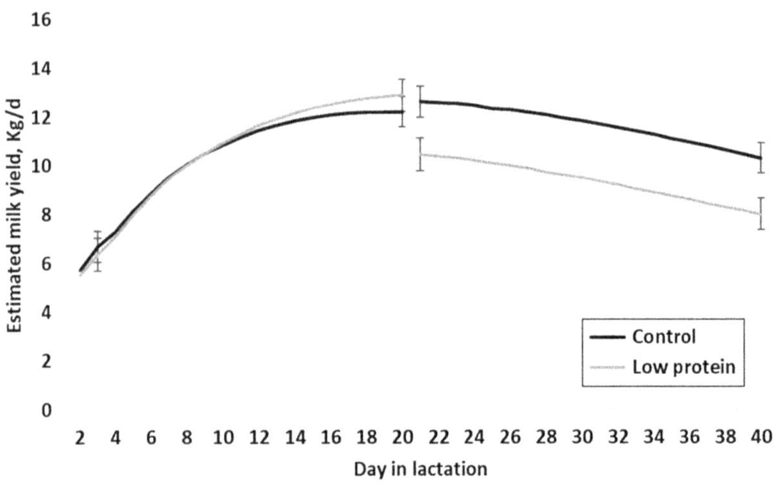

Figure 2 Estimated milk yield (day 2–40-) in second parity sows fed iso-energetic compound feed differing in proportion of protein during summer. Source: Modified after: Eskildsen et al. (2020b).

second parity sows grazing grass-clover pastures, the estimated requirements were 32 MJ ME in pregnancy and 130 MJ ME in peak lactation during summer (Eskildsen et al., 2020b).

2.2 Protein (amino acid) requirements

Amino acids are crucial for the development and maintenance of body tissues, enzyme production, and other vital functions in pigs. Therefore, like indoor pigs, organic pigs require a balanced supply of amino acids for optimal performance, with requirements depending on age, weight, and physiological stage (e.g. growth, gestation, lactation). As noted in Close and Poornan (1993) and Eskildsen (2020), the amino acid requirements of organic sows to achieve a specific output (e.g. gain or milk) are identical to those of conventional indoor sows. However, given the higher energy requirements of organic sows (2.1) and the protein contributions from pasture in the summer (3.0) and silage in the winter (4.0), a lower protein-to-energy ratio is necessary in the organic feed mixtures to avoid an oversupply of protein (Fig. 3).

Aligned with the above, recent studies have successfully demonstrated the ability to reduce the protein content in organic compound feed without compromising sow and litter productivity. Studies of free-range pregnant sows with *ad libitum* access to grass clover grazed or ensiled have shown that the protein concentration in feed mixtures can be reduced by 12-31% compared to the national recommendations for conventional indoor sows (Johannsen et al., 2022; Eskildsen et al., 2020b,c). In fact, a low-protein feeding regime (72 g vs. 88 g standardized ileal digestibale (SID) CP per kg DM) during pregnancy

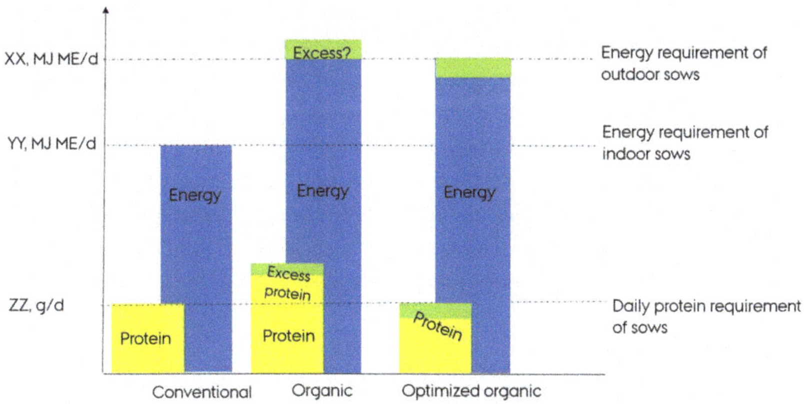

Figure 3 Protein overfeeding (middle) because the organic sows are fed the protein-to-energy ratio recommended for conventional indoor sows (left) and an optimized situation (to the right) where the protein content of the compound feed is reduced and the intakes from pasture and silage are counterbalanced. The green parts indicate contributions from grass clover Source: Modified after: Eskildsen (2020).

increased litter gain in the following lactation, with a 16% increase from 276 kg to 338 kg per litter, indicating a beneficial carry-over effect of adequate protein supply during pregnancy on lactation performance (Johannsen et al., 2022). However, the potential for reducing compound feed protein concentration is contingent upon the availability and quality of the additional protein sources, such as pasture and silage.

In lactation, the protein reduction potential is not as straightforward, particularly in high-yielding organic sows due to lower silage and pasture intakes, combined with the challenges of insufficient compound feed intakes in early and mid-lactation. In organic lactating sows housed indoors with access to outdoor runs and fed diets of 100% organic origin with varying protein content over four consecutive lactations, a 16% reduction in dietary protein content had no negative impact on sow and litter performance (Weissensteiner et al., 2018). However, in sows weaning 10 piglets or more, the low protein treatment decreased sow feed intakes and increased sow weight loss 1–2 weeks post partum. Similarly, Eskildsen et al. (2020b,c) observed that sows fed low-protein compound feed (12% less CP) had significantly greater weight loss compared to control-fed sows during lactation, and conclude that a reduction in dietary protein cannot be recommended unless a high daily compound feed intake of at least 10 kg can be achieved.

2.3 Potential

The research on organic sows highlights the importance of tailoring organic feed mixtures to specific farming conditions and management practices to meet the nutritional needs of the animals. This also applies to organic fattening pigs, for whom the intake of roughage (see sections 3.2, 4.2) contributes additional protein that must be considered in the formulation of feed mixtures during the growing-finishing stage. However, as outlined by Edwards (2022), a more precise ''precision feeding' approach requires a detailed understanding of the pigs' nutritional requirements under varying conditions, as well as the ability to handle multiple diets on a farm. Additionally, due to the significant seasonal variability in herbage quality, it is necessary to frequently analyze the available pasture and provided roughage to accurately quantify the potential contribution of nutrients from these sources, as discussed in sections 3 and 4.

3 Direct foraging

Free-range production plays a significant role in organic pig production in certain countries, such as Denmark, Sweden, and the UK, where a substantial portion of organic sows are kept in grassland-based systems (Fig. 4). These systems allow pigs to directly forage for their food. Direct foraging has the potential to improve N cycling and efficiency in organic pig production, which is

Figure 4 Pasture-based production is a central characteristics of organic pig practice in the Northern European countries (photo: K.V. Riis, Aarhus University).

a critical factor in reducing the environmental impact of this production method (Eriksen et al., 2002) and according to research by Halberg et al. (2010), organic pork production may be more environmentally competitive if direct foraging can constitute a larger proportion of the protein uptake.

3.1 Feeding value (chemical composition, digestibility)

Grass clover is the most common forage crop in intensive free-range systems in Denmark and Northern Europe. The chemical composition of grass clover may vary based on factors such as cultivar, environmental conditions, and stage of growth, as shown in Table 2. Energy levels in grass clover decrease during the growing season, with values decreasing from 12.4 MJ ME/kg ultimo April to 6.7 MJ ME/kg primo September (Eskildsen et al., 2020b). Additionally, the content of calcium, phosphorus, and most amino acids in the grass is reduced during the growing season. Although not linear, the CP content is generally higher in late stages of growth than in earlier stages, increasing from 142 g/kg DM in May-June to 194 g/kg DM in AugustSeptember (Fernández et al., 2006) and from 154 g/kg DM in August to 183-197 g/kg DM in September (Kongsted et al., 2015). As with all plant sources, the digestibility of grass clover in pigs is highly influenced by maturity stage due to differences in fiber and lignin content. In a study by Rivera Ferre et al. (2001), the apparent total tract digestibility of organic matter in ryegrass varied from 0.79 in the spring to 0.47 in the summer. In dry sows, the energy digestibility (ME/MJ) of freshly harvested grass clover was 0.68, and the apparent total tract digestibility of DM, organic matter, and N was 0.64-0.72 (Eskildsen et al., 2020b). The N intake increased linearly with increasing fresh grass-clover intake, but N deposition and N utilization were not affected by grass intake within the range investigated.

3.2 Daily intakes and impact on performance

According to reviews by Studnitz (2019) and Edwards (2003, 2022), pregnant sows' daily dry matter (DM) intake from grazed grass clover ranges from 1.3

Table 2 Chemical composition of grass clover and two silage types (see section 4) from recent Danish studies.

	Grass clover (fresh)						Grass-clover silage		Barley-pea whole-crop silage	
	April	May	June	July	August	September				
DM, g/kg	223	153	174	138	165	200	394		226	
FUsow/kg DM[1]	0.92	0.76	0.57	0.63	0.63	0.50	0.65		0.63	
ME, MJ/kg DM[2]	10.6	8.74	6.56	7.24	7.24	5.75	7.48		7.25	
CP, g/kg DM	182	236	206	194	198	190	132	(63)[3]	125	(57)[3]
Fat, g/kg DM	36	44	32	37	43	38	37		30	
Ash, g/kg DM	81	125	114	97	132	242	69		68	
Calcium, g/kg DM	7.9	12.7	8.8	7.6	6.6	4.8				
Phosphor, g/kg DM	3.4	3.5	3.4	3.7	3.7	4.1				
Amino acids, g/kg DM										
Lysine	11.0	10.3	12.0	11.0	11.9	9.6	5.13	(2.69)[3]	4.39	(2.12)[3]
Methionine	3.2	3.1	3.4	3.3	3.5	2.9	1.61		1.56	
Cystine	1.9	1.6	2.0	1.8	2.0	1.9	1.00		1.06	
Threonine	8.0	7.8	8.8	8.3	9.1	7.5	5.25		4.70	
Isoleucine	7.1	6.8	8.1	7.5	8.2	6.5	5.37		4.37	
Leucine	12.4	13.8	15.5	14.5	15.9	12.7	9.28		7.47	
Histidine	3.5	3.4	4.1	3.9	4.0	3.3	2.10		2.03	
Phenylalanine	9.0	8.9	9.9	9.6	10.3	8.5	5.90		4.52	
Tyrosine	6.6	6.6	7.4	7.4	7.1	6.2	3.15		1.32	
Valine	9.4	9.0	10.6	9.6	10.7	8.5	7.10		6.15	

Crude protein (CP) and amino acids in brackets are standardized ileal digestible (SID). Silage compositions are from Johannsen et al. (2022) and fresh grass-clover compositions from Eskildsen et al. (2020b).
[1]Danish feed units for sows (Tybirk et al., 2006).
[2]The content of metabolizable energy (ME) was calculated from the feed units (FUsow) in accordance with Theil et al. (2020).[3]The standardized ileal digestibility was calculated in accordance with Tybirk et al. (2006).

kg to 3.7 kg. Edwards (2003) further suggests that unrestricted grazing on grass clover may contribute up to 50% of the maintenance energy requirement and over 80% of the amino acid requirement for pregnant sows with a daily DM intake of 2 kg. Even a moderate mean daily intake of grass clover (428 g DM and 3.4 g/d SID lysine) during mid-gestation as observed in Eskildsen et al. (2020b) (Fig. 5) can fulfill up to 38% of the daily requirement of different digestible amino acids for pregnant sows over the growing season from April to September (Fig. 6).

Apart from N, grass clover contains phosphorus and other minerals. Approximately 3–8% of the (conventional) recommended daily intake of calcium and phosphorus of pregnant sows can be covered by grass-clover intake in summer (Eskildsen, 2020).

3.2.1 Lactating sows

Data on the intake of grazed grass clover by lactation sows are scarce but due to lactating sows' high nutrient requirements and intake limitations, the proportional nutritional contribution will be lower (Edwards, 2022). Accordingly, at peak lactation, a mean DM grass-clover intake of 574 g/d (Fig. 5) contributed up to 9 % of the estimated daily requirement of selected amino acids (Fig. 7) (Eskildsen et al., 2020a,b).

It has been observed that the intake of grazed grass clover varies widely between individuals (Sehested et al., 1999; Rivera Ferre et al., 2001; Edwards, 2003). A recent study by Eskildsen et al. (2020b) in Denmark also demonstrated individual variation in grazed grass-clover consumption among

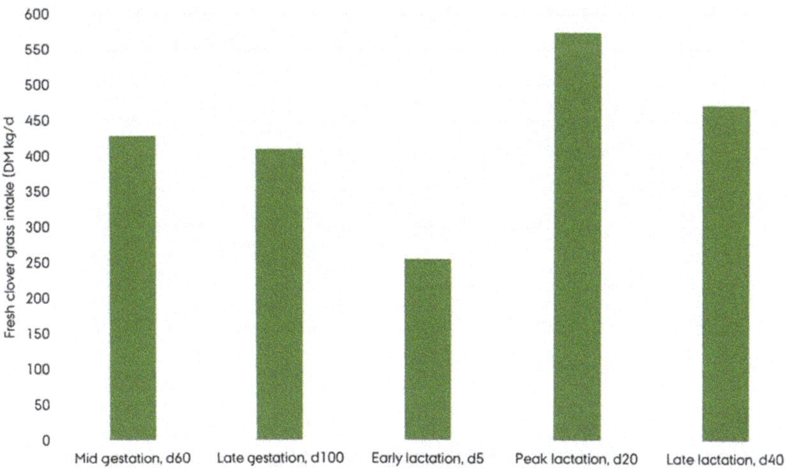

Figure 5 Average daily intake of fresh clover grass (grazed) in second parity sows during temperate summer conditions. Source: Adapted from: Eskildsen et al. (2020b).

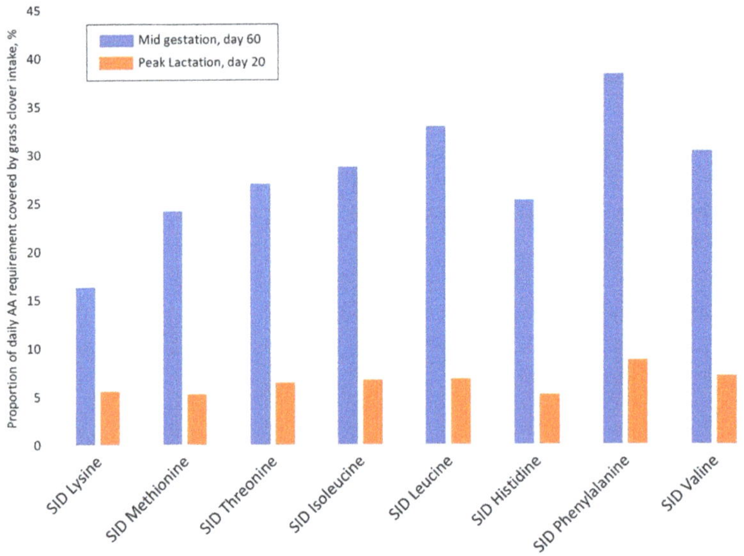

Figure 6 Percentage of SID amino acid requirement covered by the daily intake of grass clover in pregnant and lactation sows in summer (Eskildsen et al., 2020b). Requirements are based on Tybirk et al. (2019) and the digestibility of amino acids was set to 0.68% (Eskildsen et al., 2020a).

Figure 7 Growing-finishing pigs foraging fodder beets (photo: K.V. Riis, Aarhus University).

sows, particularly during peak lactation when the sows were metabolically challenged. The observed average consumption of 3.6 kg/d (as-fed) varied greatly, with 20% of the sows consuming more than 6 kg/d and some ingesting up to 11 kg/d. The reasons for this remarkable individual variation are not well understood, but it is likely influenced by several animal, dietary, environmental, and husbandry factors. However, it was found that grass-clover intake was highly correlated to live weight, with heavier sows consuming more grass clover than lighter sows, especially in late gestation and early lactation. At peak lactation, there was a positive correlation between grass-clover intake and daily distance. Additionally, sows with high grass-clover intake generally had higher compound feed intake, higher milk yield, smaller body fat pool, less back fat, higher heart rate, and a larger ash pool than sows with a low grass-clover intake (Eskildsen et al., 2020a).

3.2.2 Growing-finishing pigs

In reviewing the literature, it has been noted that while forage crops have the potential to contribute to the nutritional requirements of growing-finishing pigs, their actual intake can vary greatly depending on factors such as season, forage quality, and availability, as reported by Rivero et al. (2019) and Edwards (2022). These authors report daily intakes of grass clover or lucerne (for modern genotype pigs) ranging from 100 g of organic matter (OM) to 470 g of dry matter (DM), corresponding to up to 14% and 25% of the total energy and protein intake, respectively. Jakobsen et al. (2015a) and Kongsted and Jakobsen (2016) demonstrated that growing pigs' motivation for foraging behavior is largely influenced by feeding strategy (amount and CP content of the supplementary feed) although grazing behavior seems less affected than rooting behavior. The restriction in compound feed allowance necessary to motivate foraging behavior, and thereby forage crop intakes, has a positive impact on the lean meat percentage and compound feed efficiency but may reduce daily weight gain (Jakobsen et al., 2015a; Kongsted et al., 2015; Juul et al., 2021; Therkildsen et al., 2021). As such, the amount and nutritional composition of supplementary feed must be tailored to the nutritional value and availability of forage crops, as reported by Studnitz (2019).

3.3 Alternative forage crops and below-ground contributions

The availability of above-ground forage crops is challenged by the pigs' rooting behavior which can cause vegetation destruction (Kongsted and Jakobsen, 2015). Providing pigs with a nose-ring can help hinder this behavior, but the insertion of the ring may cause considerable pain (Horrell et al., 2001) and conflicts with organic principles of 'natural living' (Lund, 2006). As a result, alternative forage crops should be introduced, particularly during autumn and winter seasons when rooting behavior is preferred (Sandom et al., 2013). Although there has been limited focus on pigs' ability to forage below ground, Danish studies of finishing pigs (from 60 kg LW) indicate the potential for large daily intakes of root crops like Jerusalem artichoke tubers and fodder beet roots when compound feed access is restricted (Kongsted et al., 2013, 2021). In these studies, direct foraging contributed between 37% (fodder beets) and 60% (Jerusalem artichokes) of daily energy intake. While restricted access to compound feed improved forage crop intake and lean meat percentage, it reduced daily weight gain compared to *ad libitum* access (Kongsted et al., 2013). Foraging on an area with a combination of grass clover (50%) and fodder beets (50%) improved daily weight gain from 750 g to 980 g and improved the supplementary feed conversion ratio from 2.7 kg to 2.0 kg compared to foraging on grass clover alone (100%) in an autumn experiment with fattening pigs fed restrictedly (2 kg per day) (Kongsted et al., 2021). In addition to root crops, soil invertebrates, such as earthworms, can contribute to the nutritional needs of pigs in extensive systems as discussed in Jakobsen et al. (2015) and Studnitz (2019).

3.4 Potential

It is important to limit animal density in free-range systems to reduce the risk of nutrient leaching. This requires large outdoor areas (farmland) with a potential high availability of biomass to forage directly. Increasing the nutritional contribution from direct foraging will not only reduce the carbon footprint of organic pork (Jakobsen et al., 2015b) but also enhances the resilience of organic farms based on free-range pig systems (Pfeifer et al., 2022). Research has demonstrated a huge potential for meeting a significant portion of organic pigs' energy and amino acid requirements through direct foraging, but there has been limited attention on increasing the availability, palatability, and quality of forage crops. Further research is necessary to strike the right balance and optimize the financial and environmental sustainability of these systems.

4 Roughage

As discussed in Edwards (2022), the required daily access to roughage in organic production is one of the largest nutritional differences from conventional production.

4.1 Feeding value (chemical composition, digestibility)

The chemical composition of roughage is highly variable, with roughages typically containing moderate to high levels of CP (Kyntäjä et al., 2014), low energy content, and high fiber fractions, primarily consisting of cellulose, hemicellulose, and lignin. In Northern European conditions, grass-clover silage and whole crop silages are commonly utilized due to crop rotation and nutrient supply considerations. Studies conducted in Denmark have shown that the CP content in silages produced can vary between 132 g/kg DM and 169 g/kg DM for grass-clover silage and 118 g/kg DM to 125 g/kg DM for whole-crop pea barley silage (Carlson et al., 1999; Eskildsen et al., 2020c; Johannsen et al., 2022) (Table 2). Furthermore, the lysine content in these studies ranged from 3.7 g/kg DM to 4.4 g/kg DM for whole-crop silage and 5.1 to 6.3 g/kg DM for grass-clover silage.

The digestibility of roughage is influenced by its fiber composition and the presence of anti-nutritional factors and can vary not only between crops but also within crops depending on the stage of growth. Additionally, specific processing techniques such as grinding or ensiling can enhance digestibility by breaking down fibrous structures and reducing particle size. Silage digestibility for pigs has received limited attention, but Whittemore and Henderson (1977) reported moderate digestibilities of silage made from grass in the early vegetative stage (0.73 and 0.60, respectively, for protein and energy). In a Danish study of growing pigs, Carlsson et al. (1999) found high CP fecal digestibility in grass-clover (0.81) and whole-crop pea and barley (0.79) silages. Van der Peet-Scwering et al. (2010) investigated the digestibility of DM, organic matter, crude fat, NSP, and gross energy (GE) in five qualities of grass silage representing five different harvest times and corresponding to yields of 1.8, 2.2, 3.1, 3.9, and 5.0 tons of DM/hectare. The digestibility of most of the nutrients and GE increased when DM yields per ha rose from 1.8 to 2.2 tons and subsequently decreased with a further increasing DM yield per ha. Net energy (NE) per kg of DM was the highest in grass silage with a DM yield of 2.2 tons/ha (NE = 8.36 MJ) and the lowest in grass silage with a DM yield of 5.0 tons/ha (NE = 6.85 MJ).

4.2 Daily intake and impact on performance

The palatability of roughage is crucial when determining its inclusion level in pig diets. Particle size, physical form, moisture content, and processing methods can influence palatability and daily intake.

4.2.1 Pregnant sows

A recent unpublished study by Eskildsen et al. (2023) examined the intake of various types of silage among pregnant free-range sows during the winter. The results showed that the intake of clover-grass silage and barley-pea whole-crop silage was 817 g DM/d and 179 g DM/d, respectively. These intakes contributed, on average, 14.4% and 14.6% of the total intake of ME and SID lysine, respectively. Other studies have observed higher mean grass-clover silage intakes of approximately 1.5 kg DM daily during the winter season in free-range (Fernandez et al., 2006) and indoor group-housed (Bikker et al., 2014) organic sows. Additionally, a study by Van der Peet-Schwering et al. (2010) found that the NE intake from grass silage was 10.82 MJ/day in grass silage with a DM yield of 2.2 tons/ha (early spring harvest) and 7.39 MJ NE/day in grass silage with a DM yield of 5.0 tons/ha (late spring harvest). Furthermore, studies of gestating sows housed under conventional indoor conditions have shown that whole-crop barley-pea and grass silage can be consumed in large amounts without negatively impacting performance (Sørensen, 1995, 1997; Whittemore and Henderson, 1977).

4.2.2 Lactating sows

The potential to supply roughage for lactating sows is considerably lower than for gestating sows due to the large amount of compound feed required to ensure sufficient energy for milk production. Eskildsen (2023, unpublished data) found that sows fed grass-clover silage consumed an average of 99 g DM/d, while sows fed barley-pea whole-crop silage consumed an average of 161 g DM/d in lactation. Consequently, the nutritional contribution of barley-pea whole-crop silage was notably higher compared to grass-clover silage, with 1.1 % and 0.6 % for grass-clover silage and 1.6% and 0.8 % for barley-pea whole-crop silage of total ME and SID lysine intake, respectively. It appears that lactating sows' silage intake increases when access to compound feed is restricted. A reduction of 30% in compound feed allowance increased mean daily whole-crop (wheat-barley-pea) silage consumption from approximately 0.1 kg DM to 0.5 kg DM in free-range sows, with individual sow intakes of up to 1.3 kg DM per day (Kongsted et al., 2000). However, sows fed restrictedly responded with significantly lower litter gain from birth to weaning.

4.2.3 Growing-finishing pigs

The high fiber content in grass-clover silage diminishes its utility as a protein source for growing pigs of modern pig genotypes, and organic fattening pig producers often view it as a mean to enhance animal welfare rather than a significant source of energy and nutrients. Nonetheless, daily silage intakes (as-fed) ranging from 444 g (grass clover) to 656 g (whole crop pea-barley) have been observed in growing pigs (Carlsson et al., 1999), and a recent Danish study of organic finishing pigs demonstrated mean grass-clover silage intakes (as-fed) up to 0.5 kg per day (Eskildsen and Stødkilde, unpublished data). Practical challenges regarding harvesting, storage, and distribution notwithstanding, there is a growing interest among Danish farmers to allocate freshly harvested grass clover to organic growing-finishing pigs due to its higher palatability compared to grass-clover silages. This was recently demonstrated experimentally with daily fresh weight uptakes of up to 0.9 kg fresh grass clover (Fig. 8).

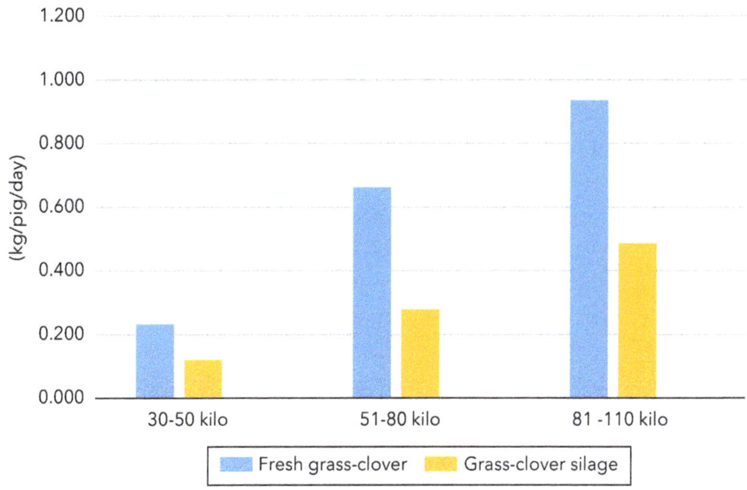

Figure 8 Daily voluntary intake of freshly harvested grass clover or grass-clover silage (from the same field) in 150 organic growing-finishing pigs from 30 kg to 110 kg. Source: Eskildsen and Stødkilde, unpublished data.

4.3 Potential

Roughage production and availability present regional variations and are contingent upon local agricultural practices and crop availability. Roughage can be sourced from crop residues, forage crops, and by-products of the food and feed industry. Achieving optimal roughage yields and availability demands efficient harvesting, storage, and processing methods, thereby ensuring a sustainable supply for pig production. The significance of roughage as a dietary component for pigs warrants careful evaluation of its feeding value, chemical composition, digestibility, palatability/intake, and impacts on performance to determine the optimal inclusion level in pig diets. Further research is needed to investigate specific roughage sources and processing techniques to maximize its potential in organic pig nutrition.

5 Green protein

Imported feed protein is a significant contributor to the negative environmental impact of pig production (Reckmann et al., 2016) and simultaneously, the dependency on imported feed compromises the adherence to organic principles. Biorefining provides a solution by utilizing local produced grass clover as a high-quality protein source for organic pigs. Utilizing grass clover as a raw material for local protein production also takes advantage of the environmental benefits associated with the cultivation of perennial forage crops, such as high carbon sequestration, low nutrient leaching, and reduced pesticide use compared to annual seed crops (Jørgensen et al., 2022). Previously, lucerne has been identified as a suitable material for biorefining (van der Heide et al., 2021), but other crops with high leaf yields, such as catch crops or paludicultures, may also be relevant (Jørgensen et al., 2021, 2022a).

The biorefinery process involves plant harvesting, transport to the biorefinery, fractionation of the plant material into a fibrous pulp and a green juice through shredding/maceration of the plant material, precipitation, and separation of the soluble proteins in the green juice into a protein paste, leaving a residual brown juice, and finally drying of the protein paste into a concentrate (green protein) (Jørgensen et al., 2022) (Figs 9 and 10). The resulting green protein, rich in soluble proteins, is suitable for use in monogastric feed.

The N in plants exists in the forms of non-protein N and protein-bound N, with varying levels of extractability depending on the method of solubilization and association with various cell components (Thers et al., 2021). Hence, not all N present in the plant will be fractionated into the green protein fraction. The recovery of N from grasses and legumes into the green protein fraction is

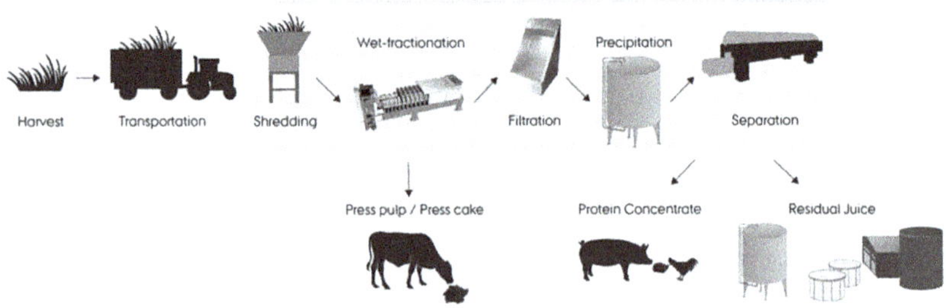

Figure 9 The green biorefinery process. Source: Adapted from: Jacobsen, (2020).

Figure 10 Fractions produced by biorefining of lucerne; lucerne plant, pulp, green juice, dried protein concentrate, and brown juice (Photo: Danish Center for Food and Agriculture, 2023).

influenced by several factors related to both plant production and the biorefinery process. These factors include cutting date (maturity), plant N content, forage species, and the choice of equipment and additives in the biorefinery process. Protein recovery in lucerne protein concentrate has been found to range from 46% to 60% of plant N (Stødkilde et al., 2020a). Additionally, Solati et al. (2018) found a higher potentially extractable true protein fraction in legume leaves compared to grass leaves. The proportion of cell-wall-bound N increases with maturity, and this N will not be recovered in the green protein fraction (Stødkilde et al., 2021b). The biorefinery process also plays a significant role in determining the degree of protein recovery. For instance, Gaffey et al. (2023) observed that the choice of equipment and additives can impact the degree of cell disruption and protein recovery. The optimization of the biorefinery process can also significantly impact the protein recovery, as evidenced by the increase in crude protein from 35% to up to 60% (Jørgensen et al., 2022b). The recovery of N will be a crucial factor in determining the economic feasibility of the production process.

5.1 Feeding value (chemical composition, digestibility)

The chemical composition of green protein is not uniform across productions. The N content in green protein is influenced by several factors, stretching from the field to the final product, as reported by Damborg et al. (2020) and Jørgensen et al. (2022b). Recently, improvements in the biorefining process have led to protein concentrates with 50–60 g CP/100 g DM (Jørgensen et al., 2022b). Along with N, the biorefined protein concentrate still contains some proportion of fiber, relative to the protein fraction. An increase in protein content is generally accompanied by a decrease in fiber content (Stødkilde et al., 2020b, 2021c). Additionally, plant lipids are prevalent in the green protein fraction and can reach concentrations of up to 14% (Jørgensen et al., 2022b). The majority of these lipids are unsaturated fatty acids, and approximately half of them is omega-3 alpha-linolenic acid (Stødkilde et al., 2021c). Unlike N content, the variation in amino acid profile among species is minimal due to the highly conserved plant proteins (Møller et al., 2021). Green protein has a high ratio of essential amino acids (EAA) to non-EAA (NEAA), with an abundance of leucine (van der Heide et al., 2021). The amino acid profile of green protein can be compared to that of soybean and other plant materials (Damborg et al., 2020). According to the protein digestibility-amino acid score (PDCAAS) using pigs as a reference, methionine and lysine are the first limiting amino acids (Stødkilde et al., 2019a).

Green protein generally has similar digestibility as other plant-based feedstuffs (van der Heide et al., 2021). Several recent studies have investigated the nutrient digestibility of green protein concentrates (Stødkilde et al., 2018,

2019b; Renaudeau et al., 2022). True fecal protein digestibility in rats fed white clover protein concentrate (451 g CP/kg DM) varied between 0.83 and 0.86 and was numerically highest in green protein from the isolated leaves compared with green protein from the whole plant (Stødkilde et al., 2018), the difference explained by a lower content of fiber-bound N from the leaves. Protein digestibility of white clover did not differ among fractions, i.e. plant, pulp, or green protein. Another study has evaluated digestibility in rats fed green protein from white clover (347 g/kg CP per DM), red clover (343 g/kg CP per DM), lucerne (388 g/kg CP per DM), and perennial ryegrass (240 g/kg CP per DM) (Stødkilde et al., 2019b). Total tract standardized digestibility of N and DM was increased in green protein compared to pulp for all plant species. Coefficients varied between 0.75 and 0.85 for the green protein with the highest digestibility observed for lucerne and the lowest for perennial ryegrass green protein. Moreover, N in lucerne green protein was more digestible than N in the whole plant.

In growing-finishing pigs, apparent ileal N digestibility was observed to be 0.81–0.86 for green protein from red clover (556 g CP/kg DM) and lucerne (538 g CP/kg DM), respectively (Renaudeau et al., 2022). This study also observed that most digestible energy was digested in the small intestine. Factors affecting green protein digestibility are fiber content and potentially saponin and preservation methods. Renaudeau et al. (2022) observed an up-concentration of saponins, which are known to depress nutrient digestibility, in the green protein fraction. Excessive heat may lead to denaturation and/or reduce protein availability due to Maillard reactions (Damborg et al., 2020).

5.2 Daily intake and impact on performance

Several *in vivo* studies with lucerne green protein (~35–60% CP per DM) have been performed earlier in growing-finishing pigs (Carr and Pearson, 1976; Hsu and Allee, 1980; Pietrzak and Grela, 2015). These studies found no adverse effects on growth rate (gain, feed intake-, or feed efficiency) of pigs when replacing soybean meal. In recent years, several studies using screw-press-derived green protein concentrate have been performed. A dose-response study using 0%, 5%, 10%, and 15% grass-clover protein concentrate (45.8 g CP/kg DM) as a substitute to primarily soybean cake was performed in organic growing-finishing pigs (Stødkilde et al., 2021a). Neither feed intake nor growth of pigs were different among diets in the weaning, grower, or finisher phase. In another study, green protein (56 g CP/100 g DM) from a grass mixture consisting of clover and lucerne was studied in grower-finisher pigs under conventional experimental conditions (Stødkilde et al., 2023). Green protein was included as part of a diet on local protein sources at a 9% inclusion rate and slightly increased feed intake as compared to soybean-based or faba bean/rapeseed cake-based diets. Daily gain was increased during the first 7 weeks for pigs fed green protein compared with soybean-based diets. Feed conversion ratio was maintained when compared with the soybean-based diets, but was slightly worse than that observed for the faba bean/rapeseed cake diet. Both studies observed no influence of feeding green protein concentrate on sensory properties, including odor, taste, or flavor, of meat from pigs. Furthermore, Stødkilde et al. (2021a) found that meat percentage increased linearly with increasing grass-clover protein concentrate inclusion. Finally, the content of n-3 fatty acids increased in adipose tissue and muscle relating to alpha-linolenic acid present in green protein (Stødkilde et al., 2021a).

5.3 Potential

Forage legumes and grasses give high yields in Northern Europe wet climates (Damborg et al., 2020). In Denmark, grass clover is a typical crop in crop rotation systems used for grass clover silage and grazing by mainly organic dairy and to a lesser extent organic pig farms (Jørgensen et al., 2021). Part of this already available biomass could be fractionated in biorefineries thus potentially increasing the value. The yield of protein in the resulting green protein will depend on its division into the different fractions during processing. The available amount of green protein depends on the amount of available, harvestable material and the capacity of processing raw material in biorefineries. Based on Danish practices and yields, Jørgensen et al. (2021) suggest that in total 40000 ha of organically managed grassland could be used for production of grass for biorefining providing in total 22 tonnes of green protein per year.

6 Conclusion

To meet the organic principles, consumer credibility, and enhance farm resilience, it is evident that future organic pig production needs to reduce imported feed ingredients and be based on 100% locally grown feed. However, the amino acid profile and availability in Northern European protein crops such as peas, faba beans, lupins, rapeseeds, and others are less suitable for pigs than the imported organic soybean products (Van der Heide et al., 2023). With the use of 100% locally produced organic feeds, there is thus an imminent risk that future CP levels will increase even further, and N losses will be even higher than reported by Dorca-Preda et al. (2023).

Biorefining seems to be one feasible option to support local feed self-sufficiency through a high-quality protein source contributing considerably to organic pigs' amino acid requirement. Future research is needed, for example, to increase the protein recovery from grass and legumes, and to further investigate the interactions between green protein source and quality, pig performance, and meat quality.

The required daily access to roughage is an important characteristic of organic pig production. Besides animal health and welfare benefits, roughage like grass clover (ensiled or fresh) represents a potential contribution to the amino acid and energy requirements of pigs, especially gestating sows. However, more attention is needed on optimizing the palatability of the roughage to encourage voluntary intakes. There seems to be an unexploited potential for increasing uptakes by providing roughage sources freshly harvested in spring and summer seasons, but practical management strategies for incorporating fresh material into pig feeding systems need to be developed.

In free-range pig systems, direct foraging is a way to support feed self-sufficiency at farm level while supporting a more natural behavior of the pigs (Jakobsen et al., 2015b). While recent research shows a huge potential for meeting a large part of organic pigs' energy and amino acid requirements through grazed grass clover and lucerne in spring and summer seasons, and root crops in autumn and winter seasons, there has been little attention on how to improve forage crop availability, palatability, and quality through optimized crop and paddock management.

The optimization of nutrient supply from compound feed, roughage, and direct foraging must be balanced and tailored to organic pigs' actual nutrient requirements considering the higher need for energy for physical and thermoregulatory activity.

7 Where to look for further information

The following articles and reports provide a good overview of the subjects:

- Eskildsen, M. (2020). Nutrition of organic sows: Impact of energy and protein supply. Ph.D. thesis, Science and Technology, Aarhus University. 190 pp.
- Jørgensen, U., Kristensen, T., Jørgensen, J. R., Kongsted, A.G., De Notaris, C., Nielsen, C., Mortensen, E. Ø., Ambye-Jensen, M., Jensen, S. K., Stødkilde-Jørgensen, L. (2021). Green biorefining of grassland biomass. Advisory report no. 193 from DCA – Danish Centre for Food and Agriculture, Aarhus University, Denmark.
- Rivero, M. J., Rodríguez-Estévez, V., Petrosemoli, S., Carballo, C., Cooke, A. S., Kongsted, A. G. (2019). Forage consumption and its effects on the performance of growing swine – discussed in relation to European Wild Boar (Sus scrofa L) in semi-extensive systems: A review. *Animals* 9, 457.
- Van der Heide, M. E., Stødkilde, L., Værum Nørgaard, J., Studnitz, M. (2021). The Potential of Locally-Sourced European Protein Sources for Organic

Monogastric Production: A Review of Forage Crop Extracts, Seaweed, Starfish, Mussel, and Insects. *Sustainability* 13, 2303.

8 References

Andersen, H.-M. L., Mogensen, L. and Kristensen, T. (2021). *Klima- og Miljøpåvirkning ved produktion af grisekød - år 1990, 2005 og 2016: Rådgivningsrapport fra DCA - Nationalt Center for Fødevarer og Jordbrug, Aarhus Universitet.*

Andersen, H.-M. L., Kongsted, A. G., Mogensen, L. and Kristensen, T. (2022). Klima- og Miljøpåvirkning ved produktion af økologisk grisekød - år 2010 og 2020. Rådgivningsrapport fra DCA - Nationalt Center for Fødevarer og Jordbrug, Aarhus Universitet.

Bikker, P., Binnendijk, G., Vermeer, H. and Van der Peet-Schwering, C. (2014). Silage in diets for organic sows in gestation. In: Rahmann, G. and Aksoy, U. (Eds) Proceedings of the 4th ISOFAR Scientific Conference Building Organic Bridges at the Organic World Congress 2014, p. 819.

Buckner, L. J. (1996). Physical and mathematical modelling of the energy requirements of the outdoor sows. PhD thesis, The University of Aberdeen.

Buckner, L. J., Edwards, S. A. and Bruce, J. M. (1998). Behaviour and shelter use by outdoor sows. *Applied Animal Behaviour Science* 57(1), 69–80. https://doi.org/10.1016/S0168-1591(97)00111-1.

Carlson, D., Lærke, H. N., Poulsen, H. D. and Jørgensen, H. (1999). Roughages for growing pigs, with emphasis on chemical composition, ingestion and faecal digestibility. *Acta Agriculturae Scandinavica, Section A: Animal Science* 49(3), 129–136.

Carr, J. R. and Pearson, G. (1976). Photosensitisation, growth performance, and carcass measurements of pigs fed diets containing commercially prepared lucerne leaf-protein concentrate. *New Zealand Journal of Experimental Agriculture* 4(1), 45–50.

Close, W. H. and Poornan, P. K. (1993). Outdoor pigs - their nutrient requirements, appetite, and environmental responses. In: Garnsworth, P. C. and Wiseman, J. (Eds) *Recent Developments in Pig Nutrition 3*, pp. 353–378. Nottingham University Press, UK.

Damborg, V. K., Jensen, S. K., Weisbjerg, M. R., Adamsen, A. P. and Stødkilde, L. (2020). Screw-pressed fractions from green forages as animal feed: chemical composition and mass balances. *Animal Feed Science and Technology* 261, 114401.

Damborg, V. K., Krogh Jensen, S., Johansen, M., Ambye-Jensen, M. and Weisbjerg, M. R. (2019). Ensiled pulp from biorefining increased milk production in dairy cows compared with grass-clover silage. *Journal of Dairy Science* 102(10), 8883–8897.

Dorca-Preda, T., Kongsted, A. G., Andersen, H. M.-L., Kristensen, T., Theil, P. K., Knudsen, M. T. and Mogensen, L. (2023). Refining life cycle nutrient modeling in organic pig production: an analysis focusing on feeding strategies in organic Danish pig farming. *Livestock Science* 272. https://doi.org/10.1016/j.livsci.2023.105248.

Edwards, S. A. (2003). Intake of nutrients from pasture by pigs. *Proceedings of the Nutrition Society* 62(2), 257–265.

Edwards, S. (2022). Organic swine production and nutrition (chapter 24). In: Chiba, L. I. (Ed.) *Sustainable Swine Nutrition* (2nd edn.). John Wiley & Sons Ltd.

Eriksen, J., Petersen, S. O. and Sommer, S. G. (2002). The fate of nitrogen in outdoor pig production. *Agronomie* 22(7–8), 863–867. Available at: https://hal.archives-ouvertes.fr/hal-00885983/document.

Eriksen, J., Studnitz, M., Strudsholm, K., Kongsted, A. G. and Hermansen, J. E. (2006). Effect of nose ringing and stocking rate of pregnant and lactating outdoor sows on exploratory behaviour, grass cover and nutrient loss potential. *Livestock Science* 104(1-2), 91-102.

Eskildsen, M. (2020). Nutrition of organic sows: impact of energy and protein supply. PhD thesis, Science and Technology, Aarhus University, 190 pp.

Eskildsen, M., Hedemann, M. S., Theil, P. K. and Nørgaard, J. V. (2020a). Impact of substituting compound feed with increasing levels of fresh grass-clover on nitrogen metabolisms and plasma metabolites of sows. *Livestock Science* 242. https://doi.org/10.1016/j.livsci.2020.104269.

Eskildsen, M., Krogh, U., Nørgaard, J. V., Hedemann, M. S., Sørensen, M. T., Kongsted, A. G. and Theil, P. K. (2020b). Grass clover intake and effects of reduced dietary protein for organic sows during summer. *Livestock Science* 241, 104212.

Eskildsen, M., Krogh, U., Sørensen, M. T., Kongsted, A. G. and Theil, P. K. (2020c). Effect of reduced dietary protein level on energy metabolism, sow body composition and metabolites in plasma, milk and urine from gestating and lactating organic sows during temperate winter conditions. *Livestock Science* 240, 104088.

Eurostat (2023). Organic livestock. Update 31 October 2023. Statistics | Eurostat (europa.eu).

Fernández, J. A., Danielsen, V., Søegaard, K., Poulsen, H. D. and Jensen, S. K. (2006). Kløvergræs - afgræsset eller ensileret - kan dække mindst halvdelen af drægtige søers næringsbehov (In Danish). [Clover-grass –grazed – is able to cover at least half of the nutritional need of gestation sows]. Husdyrbrug nr. 72 Danmarks JordbrugsForskning, Ministeriet for Fødevarer, Landbrug og Fiskeri Danmarks JordbrugsForskning, Ministeriet for Fødevarer, Landbrug og Fiskeri 31.

Früh, B., Schlatter, B., Isensee, A., Maurer, V. and Willer, H. (2015). *Report on Organic Protein Availability and Demand in Europe*. Research Institute of Organic Agriculture (FiBL), Frick.

Gaffey, J., Rajuaria, G., McMahon, H., Ravindran, R., Dominguez, C., Jensen, M. A., Souza, M. F., Meers, E., Aragonés, M. M., Skunca, D. and Sanders, J. P. M. (2023). Green biorefinery systems for the production of climate-smart sustainable products from grasses, legumes and green crop residues. *Biotechnology Advances* 66, 108168.

Graves, H. B. (1984). Behaviour and ecology of wild and feral swine (*Sus scrofa*). *Journal of Animal Science* 58(2), 482–492.

Halberg, N., Hermansen, J. E., Kristensen, I. S., Eriksen, J., Tvedegaard, N. and Petersen, B. M. (2010). Impact of organic pig production systems on CO_2 emission, C sequestration and nitrate pollution. *Agronomy for Sustainable Development* 30(4), 721-731.

Horrell, R. I., A'Ness, P. J., Edwards, S. A. and Eddison, J. C. (2001). The use of nose-ring in pigs: consequences for rooting, other functional activities, and welfare. *Animal Welfare* 10(1), 3-22.

Hsu, A. and Allee, G. L. (1980). *Feeding Value of Alfalfa Leaf Protein Concentrate (ALPC) for Swine, Conference: Swine Day, Kansas State University*. Agricultural Experiment Station and Cooperative Extension Service, Manhattan, pp. 1-5.

Jacobsen, E. (2020). Fermentation of concentrated brown juice from a green biorefining platform. Production of high value carotenoids using recombinant oleaginous yeast - an experimental and industrial, techno-economic case study. MSc thesis, Department of Biological and Chemical Engineering, Aarhus University.

Jakobsen, M., Kongsted, A. G. and Hermansen, J. E. (2015a). Foraging behaviour, nutrient intake from pasture and performance of free-range growing pigs in relation to feed CP level in two organic cropping systems. *Animal* 9(12), 2006–2016. https://doi.org/10.1017/S1751731115001585.

Jakobsen, M., Preda, T., Kongsted, A. G. and Hermansen, J. E. (2015b). Increased foraging in outdoor organic pig production—modeling environmental consequences. *Foods* 4(4), 622.

Johannsen, J. C., Eskildsen, M., Kongsted, A. G. and Theil, P. K. (2022). Effect of reduced dietary protein on productivity and plasma, urine, and milk metabolites in organic sows during winter conditions. *Livestock Science* 263. https://doi.org/10.1016/j.livsci.2022.105019.

Jongbloed, A. W., van Diepen, J. T. M., Kemme, P. A. and Broz, J. (2004). Efficacy of microbial phytase on mineral digestibility in diets for gestating and lactating sows. *Livestock Production Science* 91(1–2), 143–155.

Juul, L., Kristensen, T., Theil, P. K., Therkildsen, M. and Kongsted, A. G. (2021). Effect of two different feeding strategies on energy intake from pasture, feed efficiency and growth performance of growing-finishing pigs in a mobile pasture system. *Livestock Science* 252, 104690.

Jørgensen, U., Kristensen, T., Jørgensen, J. R., Kongsted, A. G., De Notaris, C., Nielsen, C., Mortensen, E. Ø., Ambye-Jensen, M., Jensen, S. K. and Stødkilde-Jørgensen, L. (2021). *Green Biorefining of Grassland Biomass*. Advisory report no. 193 from DCA - Danish Centre for Food and Agriculture Aarhus University.

Jørgensen, U., Jensen, S. K. and Ambye-Jensen, M. (2022). Coupling the benefits of grassland crops and green biorefining to produce protein, materials and services for the green transition. *Grass and Forage Science* 77(4), 295–306. https://doi.org/10.1111/gfs.12594.

Kongsted, A. G., Andersen, M.-L., Salomon, E. and Kristensen, I. S. (2021). Pig integrated in cropping systems to support a sustainable and diversified organic meat production. In: Organic Animal Husbandry Systems – Ways to Improvement: Proceedings of the IAHA Pre-Conference on Organic Animal Husbandry linked to the 20th Organic World Congress in Rennes, France, pp. 105–111.

Kongsted, A. G., Horsted, K. and Hermansen, J. E. (2013). Free-range pigs foraging on Jerusalem artichokes (*Helianthus tuberosus* L.) – effect of feeding strategy on growth, feed conversion and animal behaviour. *Acta Agriculturae Scandinavica, Section A: Animal Science* 63(2) 76–83.

Kongsted, A. G. and Jakobsen, M. (2015). Effect of genotype and level of supplementary concentrate on foraging activity and vegetation cover in an organic free-range pig system. *Acta Agriculturae Scandinavica, Section A: Animal Science* 65(3-4), 139–147.

Kongsted, A. G., Larcher, J. and Larsen, V. A. (2000). Silage for lactating sows outdoors. In: *Ecological Animal Husbandry in the Nordic Countries*. Darcof report no. 2, pp. 125–130.

Kongsted, A. G., Nørgaard, J. V., Jensen, S. K., Lauridsen, C., Juul-Madsen, H. R., Norup, L. R., Engberg, R. M., Horsted, K. and Hermansen, J. E. (2015). Influence of genotype and feeding strategy on pig performance, plasma concentrations of micro nutrients, immune responses and faecal microbiota composition of growing-finishing pigs in a forage-based system. *Livestock Science* 178, 263–271.

Kyntäjä, S., Partanen, K., Siljander-Rasi, H. and Jalava, T. (2014). Tables of composition and nutritional values of organically produced feed materials for pigs and poultry. MTT report 164.

Larsen, S. U., Jørgensen, H., Bukh, C. and Schjoerring, J. K. (2019). Green biorefining: effect of nitrogen fertilization on protein yield, protein extractability and amino acid composition of tall fescue biomass. *Industrial Crops and Products* 130, 642-652.

Lund, V. (2006). Natural living - a precondition for animal welfare in organic farming. *Livestock Science* 100(2-3), 71-83.

Manevski, K., Jakobsen, M., Kongsted, A. G., Georgiadis, P., Labouriau, R., Hermansen, J. E. and Jørgensen, U. (2019). Effect of poplar trees on nitrogen and water balance in outdoor pig production - a case study in Denmark. *Science of the Total Environment* 646, 1448-1458.

Mogensen, L., Knudsen, M. T., Dorca-Preda, T., Nielsen, N. I., Kristensen, I. S. and Kristensen, T. (2018). Bæredygtighedsparametre for konventionelle fodermidler til kvæg - metode og tabelværdier. DCA rapport nr. 116.

Møller, A. H., Hammershøj, M., Dos Passos, N. H. M., Tanambell, H., Stødkilde, L., Ambye-Jensen, M., Danielsen, M., Jensen, S. K. and Dalsgaard, T. K. (2021). Biorefinery of green biomass - how to extract and evaluate high quality leaf protein for food? *Journal of Agricultural and Food Chemistry* 69(48), 14341-14357.

Petersen, J. (Ed.) (2010). Dansk, økologisk dyrkning af sojabønner til fødevare- og foderformål - temadag og markvandring 22. juni 2010. Intern rapport, Markbrug Nr. 27.

Pfeifer, S., Moakes, S., Salomon, E. and Kongsted, A. G. (2022). The role of diversity and circularity to enhance the resilience of organic pig producers in Europe. *Animal - Open Space* 1(1), 100009. ISSN 2772-6940. https://doi.org/10.1016/j.anopes.2022.100009.

Pietrzak, E. and Grela, E. R. (2015). The effects of adding lucerne protein concentrate to diets on the reproductive traits and blood metabolic profiles of sows and piglets. *Journal of Animal and Feed Sciences* 24(3), 216-225.

Reckmann, K., Blank, R., Traulsen, I. and Krieter, J. (2016). Comparative life cycle assessment (LCA) of pork using different protein sources in pig feed. *Archives Animal Breeding* 59(1), 27-36. http://doi.org/10.5194/aab-59-27-2016.

Regulation (EU) 2018/848 of the European Parliament and of the Council of 30 May 2018 on Organic Production and Labeling of Organic Products and Repealing Council Regulation (EC) No. 834/2007. https://eur-lex.europa.eu/eli/reg/2018/848/oj.

Renaudeau, D., Jensen, S. K., Ambye-Jensen, M., Adler, S., Bani, P., Juncker, E. and Stødkilde, L. (2022). Nutritional values of forage-legume-based silages and protein concentrates for growing pigs. *Animal* 16(7), 100572.

Rivera Ferre, M. G. R., Edwards, S. A., Mayes, R. W., Riddoch, I. and Hovell, F. D. D. (2001). The effect of season and level of concentrate on the voluntary intake and digestibility of herbage by outdoor sows. *Animal Science* 72(3), 501-510. https://doi.org/10.1017/S1357729800052024.

Rivero, M. J., Rodríguez-Estévez, V., Petrosemoli, S., Carballo, C., Cooke, A. S. and Kongsted, A. G. (2019). Forage consumption and its effects on the performance of growing swine - discussed in relation to European Wild Boar (*Sus scrofa* L) in semi-extensive systems: a review. *Animals* 9(7), 457. https://doi.org/10.3390/ani9070457.

Sandom, C. J., Hughes, J. and MacDonal, D. W. (2013). Rewilding the Scottish Highlands: do wild boar, *Sus scrofa*, use a suitable foraging strategy to be effective ecosystem engineers? *Restoration Ecology* 21(3), 336–343.

Sehested, J., Breinhild, K. K., Søegaard, K., Vognsen, L., Hansen, H. H., Fernandez, J. A., Danielsen, V. and Kristensen, V. F. (1999). Use of n-alkanes to estimate grass intake and digestibility in sows. In: Dove, H. and Coleman, S. W. (Eds) *Nutritional Ecology of Herbivores*. Satellite Symposium: Emerging Techniques for Studying the Nutrition of Free Ranging Herbivores, April 1999, Texas.

Solati, Z., Jørgensen, U., Eriksen, J. and Søegaard, K. (2018). Estimation of extractable protein in botanical fractions of legume and grass species. *Grass and Forage Science* 73(2), 572–581.

Sørensen, G. (1995). Helsædsensilage til løsgående drægtige søer. *Landbrug og Fødevarer, Sektor for Gris, Erfaring* nr. 9505.

Sørensen, G. (1997). Helsædsensilage til løsgående drægtige søer. *Landbrug og Fødevarer, Sektor for Gris, Meddelelse* nr. 373.

Stødkilde, L., Damborg, V. K., Jørgensen, H., Lærke, H. N. and Jensen, S. K. (2018). White clover fractions as protein source for monogastrics: dry matter digestibility and protein digestibility-corrected amino acid scores. *Journal of the Science of Food and Agriculture* 98(7), 2557–2563.

Stødkilde, L., Damborg, V. K., Jørgensen, H., Lærke, H. N. and Jensen, S. K. (2019). Digestibility of fractionated green biomass as protein source for monogastric animals. *Animal* 13(9), 1817–1825. https://doi.org/10.1017/S1751731119000156.

Stødkilde, L., Eriksen, J. and Jensen, S. (2020a). Biorefined grasses and legumes-effect of species and spring cutting date on extractable protein yield. Meeting the Future Demands for Grassland Production, p. 683.

Stødkilde, L., Ambye-Jensen, M. and Jensen, S. K. (2020b). Biorefined grass-clover protein composition and effect on organic broiler performance and meat fatty acid profile. *Journal of Animal Physiology and Animal Nutrition* 104(6), 1757–1767. https://doi.org/10.1111/jpn.13406.

Stødkilde, L., Ambye-Jensen, M. and Jensen, S. K. (2021a). Biorefined organic grass-clover protein concentrate for growing pigs: effect on growth performance and meat fatty acid profile. *Animal Feed Science and Technology* 276, 114943. https://doi.org/10.1016/j.anifeedsci.2021.114943.

Stødkilde, L., Lashkari, S., Eriksen, J. and Jensen, S. K. (2021b). Enhancing protein recovery in green biorefineries through selection of plant species and time of harvest. *Animal Feed Science and Technology* 278, 115016.

Stødkilde, L., Mogensen, L., Bache, J. K., Ambye-Jensen, M., Vinther, J. and Jensen, S. K. (2023). Local protein sources for growing-finishing pigs and their effects on pig performance, sensory quality and climate impact of the produced pork. *Livestock Science* 267, 105128. https://doi.org/10.1016/j.livsci.2022.105128.

Studnitz, M. (Ed.) (2019). Feeding monogastrics 100% organic and regionally produced feed. Knowledge synthesis. OK-net EcoFeed. *H2020 Project*. Available at: http://orgprints.org/34560.

Theil, P. K., Chwalibog, A. and Jørgensen, H. (2020). Energy for pigs: metabolism, requirement, utilisation and prediction of dietary content. In: *Textbook in Physiology*. Landbrug og Fødevarer, Sektor for Gris, Denmark.

Therkildsen, M., Jensen, S. K. and Kongsted, A. G. (2021). Crossbreed and feed allowance affect final meat and eating quality of pigs from a free-range system. *Livestock Science* 250, 104584.

Thers, H. and Eriksen, J. (2022). Annual protein yield and extractable protein potentials in three legumes and two grasses. *Journal of the Science of Food and Agriculture* 102(9), 3742–3751.

Thers, H., Stødkilde, L., Jensen, S. K. and Eriksen, J. (2021). Linking protein quality in biorefinery output to forage crop crude protein input via the Cornell net carbohydrate and protein system. *Applied Biochemistry and Biotechnology* 193(8), 2471-2482.

Tybirk, P., Sloth, N. M., Kjeldsen, N. and Shooter, L. (2019). *Danish Nutrient Requirement Standards (In Danish: Normer for Næringsstoffer)* (20th edn.). SEGES Pig Research Centre, Axelborg.

Tybirk, P., Strathe, A. B., Vils, E. and Sloth, N. M. (2006). The Danish feed evaluation system for pigs. Report 30. Danish Farming Consuling.

USDA (2012). *2011 Certified Organic Production Survey*. United States Department of Agriculture, National Agricultural Statistics Service, Washington.

USDA (2022). *Certified Organic Survey 2021 Summary*. United States Department of Agriculture, National Agricultural Statistics Service,Washington.

Van der Heide, M. E., Jørgensen, J. R. and Kongsted, A. G. (2023). Økologisk proteinfoder. *Rådgivningsnotat Fra DCA - National Center for Fødevarer og Jordbrug*. Nr. 2020-0125669, 42 s., jan. 12, 2023.

Van der Heide, M. E., Stødkilde, L., Værum Nørgaard, J. and Studnitz, M. (2021). The potential of locally-sourced European protein sources for organic monogastric production: a review of forage crop extracts, seaweed, starfish, mussel, and insects. *Sustainability* 13(4), 2303.

Van der Peet-Schwering, C. M. C., Binnendijk, G. P. and van Diepen, J. T. M. (2010). Verteerbaarheid en voederwaarde van diverse kwaliteiten kuilgras en van CCM bij biologische zeugen. Rapport 342, Wageningen UR Livestock Research, Lelystad.

Verhagen, J. M. F., Verstegen, M. W. A., Geuyen, T. P. A. and Kemp, B. (1986). Effect of environmental-temperature and feeding level on heat-production and lower critical-temperature of pregnant sows. *Journal of Animal Physiology and Animal Nutrition* 55(1-5), 246–256.

Weissensteiner, R., Baldinger, L., Hagmüller, W. and Zollitsch, W. (2018). Effects of two 100% organic diets differing in proportion of home-grown components and protein concentration on performance of lactating sows. *Livestock Science* 214, 211-218.

Whittemore, C. T. and Henderson, A. R. (1977). The digestibility of grass silage by sows. *Journal of the Science of Food and Agriculture* 28(6), 506–510.

Willer, H., Travnicek, J., Meier, C. and Schlatter, B. (Eds) (2021). *The World of Organic Agriculture. Statistics and Emerging Trends 2021*. Research Institute of Organic Agriculture FiBl, and IFOAM – Organics International, Bonn.

Wimmler, C., Vermeer, H. M., Leeb, C., Salomon, E. and Andersen, H. M.-L. (2022). Review: concrete outdoor runs for organic growing-finishing pigs – a legislative, ethological and environmental perspective. *Animal* 16(1). https://doi.org/10.1016/j.animal.2021.100435.

Zollitsch, W. (2007). Challenges in the nutrition of organic pigs. *Journal of the Science of Food and Agriculture* 87(15), 2747–2750.

Chapter 13

High protein corn fermentation products for swine derived from corn ethanol production

Peter E. V. Williams, FluidQuipTechnologies, USA

1 Introduction

Crops are a primary source of protein, and for many years, soybean meal (SBM) has been the protein of choice for livestock feed. However, the pleiotropic relationship between nitrogen and protein content has, over time, resulted in the reduction in protein content of crops such as soybean and corn (maize) as the economics of grain production has prioritized increased yield. This in turn has exacerbated the need for higher concentration protein supplements for feed formulation. In addition, there has also been a movement to reduce the reliance on soy products and replace them with more sustainable sources of protein and particularly vegetable proteins with the absence of anti-nutritional factors.

Decisions on how a feed is formulated are already being driven by the sustainability characteristics of individual feed ingredients. Sustainability (meeting energy and carbon emissions reduction targets) alongside

http://dx.doi.org/10.19103/AS.2024.0140.20

demographics (e.g. the age of those in the labour market) have been identified by the International Feed Industry Federation as key global issues facing the industry. Indexes of sustainability such as green house gas (GHG) and land use change (LUC) metrics are likely to be two new parameters that will become components of feed specification sheets in addition to the nutritional data that has been the norm.

To improve sustainability, processed animal proteins are being reintroduced into feed formulations and work is ongoing to produce a range of single-cell proteins, insect protein and microbial protein. Attention is also being paid to alternative crops high in protein. However, there are considerable challenges to growing new alternative crops on any scale. It takes several years to breed an elite variety to achieve maximum production in different geographic locations with different environmental conditions. Logistical factors such as acquiring appropriate machinery and inputs, as well as acquiring agronomic knowledge to successfully grow a new crop, all serve to favour the status quo. In addition, new crops require land and resources, and potentially compete with existing crop production, contributing to GHG emissions from agriculture as well as indirectly exacerbating climate change from changes in land use (through deforestation). In contrast, plant-based coproducts, secondary products generated during a manufacturing process, have a unique position in that they do not demand additional acreage and do not compete with human food consumption (Mottet et al., 2017).

2 Distillers dried grains with solubles

Corn is the major cereal crop grown in the USA and widely cultivated globally. Corn is traditionally considered an energy crop grown for the energy produced in the form of starch and oil but with a limited amount of protein. As well as being grown as a food crop, corn has also become an important feedstock for the sustainable biofuel industry, with over 100 million tons of corn annually processed into ethanol biofuel in the USA. A by-product of ethanol production is distillers dried grains with solubles (DDGS). This is the dried residue remaining after the starch fraction of corn is fermented with yeasts to produce ethanol. Following fermentation, the ethanol is removed by distillation and the remaining fermentation residues are dried to become DDGS. The growth in ethanol production from corn has resulted in a plentiful supply of DDGS as a co-product which has found a valuable market as a livestock feed. However, it is important to note that DDGS, as a by-product of the dry grind ethanol industry, was never designed as a livestock feed but emerged as a convenient means of marketing the residue from the ethanol production process. This creates a potential opportunity to further develop DDGS as a livestock feed product.

More broadly, the dry grind ethanol industry benefits from several important features as a source of animal feed. A dry grind ethanol plant has established logistics for the delivery and dispatch of grain feedstock and any feed products. Individual dry grind ethanol plants benefit from economies of scale, with large plants processing more than one million tons of grain per annum. Ethanol plants are also integrated bioprocessing facilities typically with integrated heat and power generation. These facilities are prime sites for the installation of processing technology to produce new products with higher nutritional value than traditional DDGS products.

3 Corn fermented protein

There are currently four different commercial processes to produce high concentration corn fermented protein (CFP) products (approximately >50% crude protein (CP)) (Fig. 1). By producing products with higher levels of protein compared with conventional DDGS, these technologies offer the potential to diversify and expand the revenue streams of the bioethanol plant in addition to providing new protein products for the feed industry.

The four processes are an evolution of the wet grind milling process and the separation of protein and fibre to produce separate high-value concentrated streams of protein and fibre. Mechanically separated CFP is produced from whole stillage when portions of fibre and oil are removed, concentrating residual grain proteins and yeast by methods involving centrifugation and washing. One version of CFP processing involves the separation of all fractions

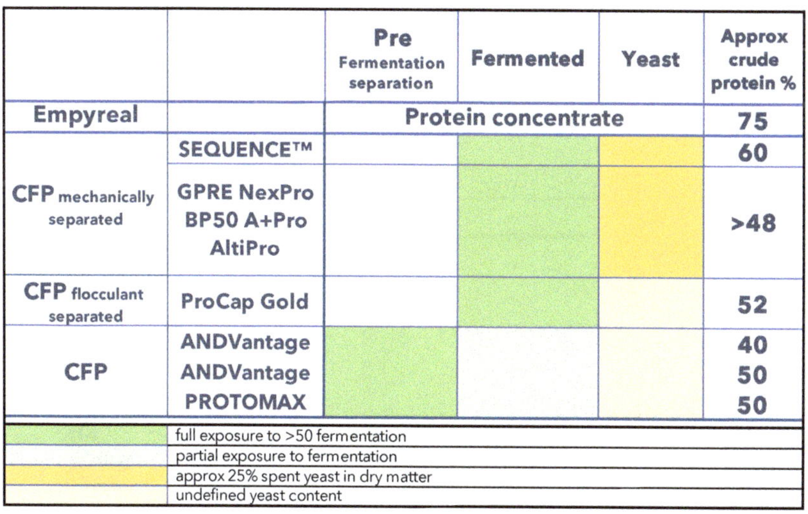

Figure 1 Technologies for producing a range of commercial high protein feed products.

post-fermentation with all components of the product exposed to fermentation: the Maximised Stillage Co-products (MSC™) process. A second version of CFP manufacture incorporates pre-fermentation grinding and separation of the whole grain. A third recovers the protein via flocculation and a fourth employs electrostatic forces to separate fibre from protein. Key differences between these processes are identified in Fig. 1.

The MSC™ process separates the protein from the whole stillage by mechanical centrifugation and alternate washing steps (Fig. 2). An important element to grasp is that the dry grind ethanol process starts with a 50–70 h fermentation of ground corn. The ground grain is exposed to moisture, raised temperature and enzymatic hydrolysis to facilitate the conversion of starch into ethanol. There has been little focus on the impact of the fermentation process on residual components of the grain, whether fermentation per se has a nutritional impact on the fibrous components found in whole stillage or the composition of protein components. Other than testing of phytase e.g. as a means of potentially increasing protein recovery, the potential value of fermentation in improving product functionality is unexplored. Xu et al. (2020) listed such potential improvements as improved protein digestibility, increased number of small peptides, increased energy digestibility, reduced fibre levels, improved nutritional value, bioavailability, weight gain and feed conversion rate. Work is in progress to investigate whether additional functional benefits can be obtained for CFP products via the use of fermentation.

In addition, little attention has been given to the fact that, prior to fermentation, there is a significant *in situ* generation of yeast to facilitate the fermentation process and that spent yeast at the end of the ethanol process is

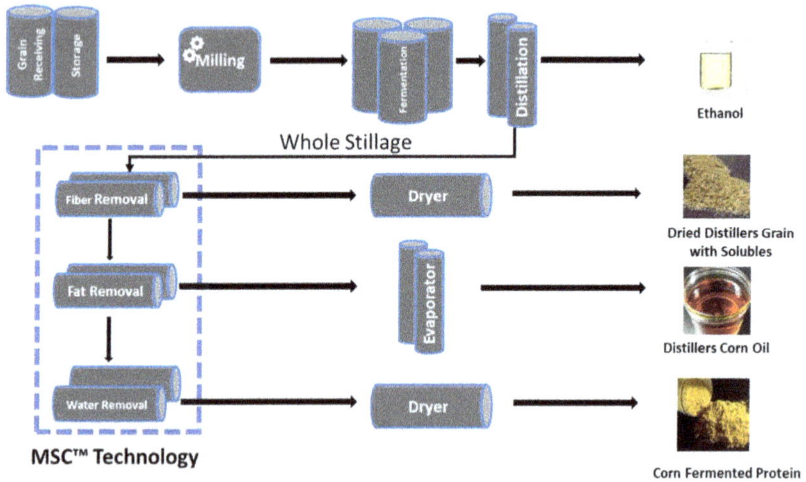

Figure 2 The Maximised Stillage Coproduct Process.

a valuable nutritional component of DDGS. Indeed, the spent yeast of DDGS has been calculated to be approximately 6% of DDGS dry matter (Han and Liu, 2010).

During fermentation, there is a significant net generation of protein from the yeast that is grown in the fermenter to facilitate fermentation. In addition to recovering corn protein, the MSC™ process also recovers a high proportion of spent yeast in particular cell wall components. Whole yeast cells and yeast cell components are recognized as valuable feed nutrient supplements (Shurson, 2018). This produces an exceptionally high-protein CFP product (up to 60% CP; ultra high Pro+, now given the trade name SEQUENCE) that has now been extensively and very successfully tested as a protein supplement for several different species including pigs.

A key challenge has been to measure the quantity of either yeast or yeast components in a dried material such as DDGS. The measurement of the yeast cell wall mannose is accepted by the yeast industry as a proxy measure for yeast cell material. Yeast generated in fermentation accompanies stillage in the ethanol process and passes through the distillation columns where the yeast cells are lysed, releasing the intracellular components including nucleotides. The components of yeast cell walls (mannan oligosaccharides and β-glucans) can be measured in dried CFP products. Using this proxy measure, the yeast cell wall material represents approximately 20–27% of the dry matter content of CFP. Approximately 9% of the protein in a 52% protein product is derived from yeast cell material. However, it is important to recognize that these spent yeast components are not equivalent to whole and viable yeast cells used as feed supplements though the immuno-stimulatory activity of yeast cell components may be valuable functional components of CFP.

4 Challenges in producing corn fermented protein

There are several challenges in producing a commercial CFP product. One potential safety issue is the use of antibiotics in the ethanol process. As a means of optimizing the efficiency of the yeast ethanol conversion step in the dry grind ethanol process, antibiotics have been used during fermentation to eliminate competitive bacterial activity. In the past, Virginiamycin, Erythromycin and Tylosin have been used routinely. Given that the stillage is exposed to high temperature during distillation there is no residual antibiotic activity in the co-products, but antibiotic residues are detectable. Plants producing CFP can use a number of alternative products to antibiotics and, with attention to improved cleaning in place (CIP), are able to operate without the use of antibiotics to control the fermentation process, allowing them to be compliant with No Antibiotics Ever (NAE) programmes for food and feed production.

Another key aspect of product development is its environmental impact. CFP is produced in a highly competitive, sustainable manner with low GHG emission values compared with a range of different protein products (Fig. 3). Using CFP in ration formulation as a partial replacement of SBM has been reported to reduce the GHG value of feed formulation in diets for turkey poults (Burton, 2021).

Key to commercial sustainability is the ability to produce at scale. Up to 500 k metric tons per annum of a specific formulated feed is a norm for commercial feed production. A batch of the formulated feed of up to 100 tons is made, then mixed in 10–12 tons mixer batches. Storage space for feed ingredients is limited. Generally large-scale integrators are limited to three to four feed bins on site, which limits the number of individual ingredients that can be stored. The availability of feed ingredients must merit bin space.

Key aspects of supply chain performance are Resilience, Redundancy and Reliability (the three R's). Changing a feed formulation, particularly for poultry, can be an expensive and time-consuming exercise. Resilience is key because feed producers will not work with a single source of supply, in case of a plant stoppage. It is essential to have one or more sites of manufacture of the product for there to be an alternative source of supply. For the same reason, there needs to be redundancy in volume production to account for increased demand. With resilience in the supply chain, there is an additional need for reliability (product consistency) across different suppliers. Producing a new novel feed protein ingredient is therefore not without significant challenge. CFP products are currently being produced from 17 different ethanol plants in the USA and

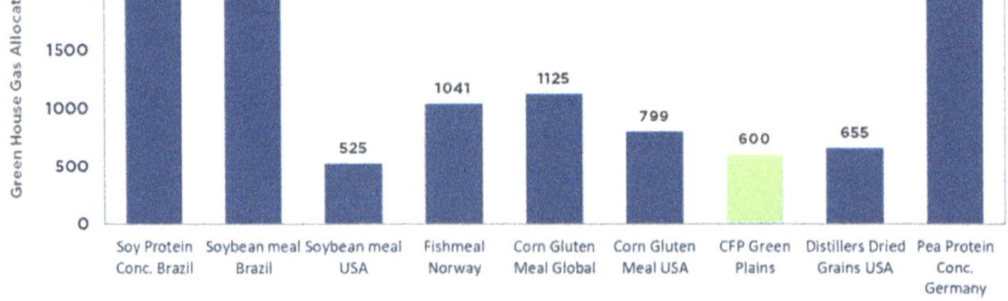

Figure 3 Greenhouse gas allocation values (kg CO_2 eq/ton protein) for a range of different protein products.

Brazil with an annual production capacity close to 1 million tons. Production of the product in Europe is imminent.

Since the first production of CFP 8 years ago standard measures of product composition including amino acid (AA) digestibility have been made using the precision-fed caecectomized rooster assay (PFR) (Parsons, 1985). The assay has been completed on nine separate occasions with samples of CFP obtained during the development of the product, from different ethanol plants as they came online over a period of approximately 6 years to monitor the consistency of the product over time and from different plants. The coefficient of variation of the *in vivo* mean standardized digestibility of the samples of CFP is 2.3%, demonstrating the excellent consistency in the nutritional *in vivo* value of CFP produced from different sites and over a period of approximately 6 years (Table 1).

5 Case study: standardized ileal digestibility of CFP for pigs

An experiment was conducted to compare the standardized ileal digestibility (SID) of CP and AA in high-protein corn fermented products and in the residual DDGSs in comparison with SBM. Nine pigs [initial body weight (BW): 30.56 ± 1.24 kg] were equipped with a T-cannula in the distal ileum (Stein et al., 2007) and were allotted to a 9 × 9 lLatin square design with nine diets and nine periods of seven days. All pigs were limit-fed 3.2 times the maintenance energy requirement (i.e. 197 kcal per kg $BW^{0.60}$; NRC, 2012). Two diets contained bakery meal or SBM as the sole source of AA. Six additional diets contained a mixture of bakery meal-CFP (i.e. at 5%, 10%, 15%, or 20%), bakery meal-ultra high pro+ (i.e. at 10%), or bakery meal-post-MSC DDGS (i.e. at 15%). Values for apparent ileal digestibility (AID), ileal endogenous losses and SID of CP and AA in each diet were calculated (Stein et al., 2007). The basal endogenous losses of CP and AA were calculated from pigs fed with nitrogen-free diet as previously described (Stein et al., 2007).

The AID of Arg, Leu, Thr, Val, Ala and Tyr was greater ($P < 0.05$) in CFP and ultra high pro+ compared with SBM and post-MSC DDGS (Table 2). However, no difference was observed on the SID of CP and AA (except Trp) in SBM, CFP and ultra high pro+. The SID of Lys, Trp, and Asp was less (P < 0.05) in Post-MSC DDGS compared with SBM, CFP, and Ultra High Pro+. The SID of His, Val, and Ala also tended to be less ($P < 0.10$) in post-MSC DDGS compared with SBM, CFP and ultra high pro+. Due to increased concentrations of CP and AA in ultra high pro+, concentrations of standardized ileal digestible CP and AA (except Arg, Lys, Trp and Asp) were greater ($P < 0.01$) in ultra high pro+ than in SBM, CFP and post-MSC DDGS (Table 3). The concentration of standardized ileal digestible Arg, Lys, Trp and Asp was greater ($P < 0.01$) in SBM than in

Table 1 Standardized ileal digestibility of amino acids in sample of corn fermented protein

	University Illinois samples 2017–2023																Univ Georgia		Mean
	CFP1		CFP2		CFP3		CFP4		CFP5		CFP6		CFP7		CFP8		CFP9	CFP9	
Amino acid	Digest value	SEM³	Digest value	SEM³	Digest value	SEM³	Digest value	SEM³	Digest value	SEM³	Digest value	SEM³	Digest value	SEM³	Digest value	SEM³	Digest value	Digest value	
Ala	91.5	0.8	92.6	0.6	94.7	0.3	91.8	0.9	91.1	0.8	90	1.6	91.7	0.4	91.6	0.8	89.5	88.6	
Arg	93.3	1	94.6	0.8	98.1	0.5	93.2	0.9	93.6	0.9	91.8	2.3	93.6	0.5	92.3	0.6	91.3	91.5	
Asp	87.7	1.1	87.8	1.3	91.3	0.8	86.1	1.4	84.6	1.3	85.7	1.9	87.2	0.6	87.9	2	85	84.2	
Cys	84.6	2.2	86.3	2.2	91.8	1.1	85.7	1.9	84.5	2.1	78.6	3.7	81.2	1.5	78.9	2.6	79.8	79.5	
Glu	93.1	0.7	93.7	0.6	96.1	0.2	93.1	0.8	92.2	0.7	91	1.7	92.8	0.5	92.9	1.3	91.5	90.8	
Gly																	68.9	52.5	
His	90.2	1.2	90.6	0.8	92.8	0.3	89.2	1.2	87.2	1.1	89.5	1.8	90.4	0.6	91.1	2.2	89.7	90.2	
Ile	90.8	0.8	91.9	0.7	93.2	0.2	90.5	1	89.9	0.9	87.3	2	89.3	0.4	89.3	1	87.6	86	
Leu	94.3	0.5	95.1	0.4	97.1	0.1	94.4	0.6	93.9	0.6	92.9	1.4	94.7	0.3	94	0.4	92.7	92.1	
Lys	83.4	1.5	83.2	1.1	90.5	0.8	81	2.3	80	1.6	85.5	2.4	88.9	0.9	89.2	2.4	80.5	90.7	
Met	91.5	0.4	93.3	0.5	96.9	0.1	92.3	0.9	92	0.6	92.8	1.7	93.9	0.4	93.9	0.7	91.6	90.5	
Phe	92.6	0.8	93.4	0.6	96.1	0.1	92.6	0.8	92.4	0.6	90.6	1.7	92.8	0.3	91.9	0.6	90.6	90.1	
Pro	92.4	0.9	92.4	1.1	95.4	0.2	92	1	90.9	0.7	88.5	1.9	91.1	0.7	89.7	1	90.8	90.2	
Ser	88.5	1.7	89.6	1.9	93	1	89.1	1.3	88.2	1.1	86.5	2.8	89.2	0.9	87.7	1.1	86.4	87.9	
Thr	86.1	1.5	87	1.6	89.9	1.2	85.7	1.6	83.5	1.1	85.5	2.6	87.1	1	87.2	2.3	84.2	84.1	
Trp	94.7	0.9	94.6	0.8	93.6	0.7	92.9	1.2	92.5	0.5	90.2	1.7	90.8	1.1	90.4	1.5	86.3	86.4	
Tyr	93.3	0.9	94.6	0.6	96.9	0.2	93.1	1	92.8	0.8	90.8	1.8	92.1	0.55	91.9	1.1	91.9	91.6	
Val	88.8	1.1	89.6	1.1	95.3	0.3	88.8	1.2	87.6	0.8	89	2	90.8	0.5	89.9	0.7	87.8	86	
Mean	**90.4**		**91.2**		**94.3**		**90.1**		**89.2**		**88.6**		**90.4**		**90**		**87**	**86.3**	**89.7**
																ex GLY	88.1	88.3	cv 2.3

CFP, ultra high pro+, and post-MSC DDGS. Post-MSC DDGS had the least ($P < 0.01$) concentration of standardized ileal digestible CP and AA among protein sources. The analysed CP and AA values in post-MSC DDGS concur with values reported for conventional DDGS and low-oil DDGS (Fastinger and Mahan, 2006; NRC, 2012; Espinosa et al., 2019). The analysed CP and AA values in CFP and ultra high pro+ are greater compared with values reported for high-protein DDGS (NRC, 2012; Espinosa et al., 2019), which is possibly due to different methods of production used to produce these high-protein corn fermented products. The analysed CP and AA values in CFP is in agreement with values reported by Acosta et al. (2021) for corn protein; however, ultra high pro+ had greater CP and AA compared with values reported by Acosta et al. (2021) for corn protein and values reported by Cristobal et al. (2020) for high- protein DDGS. The SID of AA in SBM was less than values reported by NRC (2012) and Lagos and Stein (2017) for US SBM but agrees with values reported by Espinosa et al. (2020).

The lack of difference observed in SID of CP and AA (except Trp) between SBM and high-protein corn fermented products indicates that these ingredients are of high quality and maybe used as a source of digestible AA in pig diets. The SID of CP and most AA in CFP and ultra high pro+ was not significantly different to that of SBM and was greater compared with values reported for other source of corn protein. The high-protein corn fermented products used in the present experiment were rich sources of CP and AA.

The coefficients of standardized AA digestibility measured in caecetomized cockerels (Table 1) (Parsons et al., 2023) (89–91%) shown for comparative purposes and standardized AA digestibility of AAs for pigs (85–89%) demonstrate that across species the ileal digestibility of these corn-based high protein products is high.

6 Case study: concentrations of digestible and metabolizable energy in CFP products fed to pigs

An experiment was conducted to evaluate the digestible energy (DE) and metabolizable energy (ME) content of high-protein corn fermented products and in post-MSC DDGS. Thirty-two pigs [initial BW: 20.51 ± 1.74 kg] were allotted to four diets with eight replicate pigs per diet using a randomized complete block design. Dietary treatments included a corn-based diet and three diets based on a mixture of corn and each source of high-protein corn fermented product/post-MSC DDGS. Pigs were limit fed at 3.2 times the energy requirement for maintenance (i.e. 197 kcal/kg × $BW^{0.60}$; NRC, 2012). The apparent total tract digestibility (ATTD) of GE was calculated for each diet and the DE and ME in each diet were calculated. The DE and ME in the corn diet were used to calculate the DE and ME in corn. These values were then used

Table 2 Apparent ileal digestibility (AID) (%) of crude protein (CP) and amino acids (AA) in soybean meal (SBM), corn fermented protein product (CFP), Ultra High Pro+, and Post-MSC DDGS1

Item, %	SBM	CFP5*	CFP10*	CFP15*	CFP20*	AV CFP	Ultra high pro+**	Post-MSC DDGS***	SEM	*P*-value
CP	89.3	83.3	83.0	85.1	84.5	84.0	87.7	74.2	5.3	0.554
Indispensable AA										
Arg	91.1	87.1	89.9	92.2	91.6	90.2	92.3	84.4	3.58	0.551
His	88.7	79.3	81.3	83.7	84.3	82.2	83.9	75.6^{z}	2.89	0.099
Ile	88.2	78.2	81.9	82.7	83.7	81.6	82.9	79.5	2.48	0.139
Leu	87.6	88.7	90.5	90.5	90.1	90.0	91.4	86.2	1.89	0.482
Lys	87.2	70.8	76.3	79	78.3	76.1	78.9	48.8^{b}	6.37	0.008
Met	88.2	88.9	88.8	88.6	88.4	88.7	90.4	88	1.93	0.966
Phe	88	84.9	86.2	87.9	88	86.8	88.4	85.2	2.08	0.691
Thr	84	80.8	82.9	82.6	83.1	82.4	83.7	71.9	4.35	0.513
Trp	93.2	75.8	70.9	75.5	78.5	75.2	80.8	58.7^{c}	5.46	0.008
Val	86.2	83.9	85.5	84.8	85	84.8	86.1	74.2^{y}	3.1	0.094
Ala	85.5	87.8	89.5	88.5	87.7	88.4	90.1	78.0^{y}	2.85	0.096
Asp	86.5	76.9	80.2	79.8	81.3	79.6	80.5	67.2^{b}	3.59	0.049
Cys	82.4	69.9	74	79.5	80.2	75.9	78	89.6	7.23	0.572
Glu	89.7	88.4	87.5	89	86.9	88.0	90.7	84.7	3	0.837

Gly	90.6	86	86.4	89.6	87.9	87.5	91.8	73.9	9.47	0.888
Ser	88.1	80.4	85.1	86.2	86.7	84.6	88.3	82.6	3.41	0.557
Tyr	88.7	90.3	87.9	90.7	89.7	89.7	91.9	84.6	2.48	0.476
	87.9	82.2	83.8	85.3	85.4	84.2	86.5	82.8		
Mean					**84.2**		86.5			

[1]Data are least squares means of nine observations per treatment.
[a–f]Means within a row lacking a common letter are different ($P < 0.05$).
[x–z]Means within a row lacking a common letter tend to be different ($P < 0.10$).
*Corn fermented protein included at 5%, 10%, 15% and 20% of the diet.
**Ultra high pro+ CFP produced with additional enzymatic processing during fermentation.
***Post MSC DDGS Residual DDGS after MSC process and CFP recovery.
[1]Data are least squares means of nine observations per treatment.
[a–d]Means within a row lacking a common letter are different ($P < 0.05$).

Table 3 Concentrations of standardized ileal digestible crude protein (CP) and amino acids (AAs) in soybean meal (SBM), corn fermented protein product (CFP), ultra high pro+ and post-MSC DDGS[1] fed to swine

Item g/kg	SBM	CFP5	CFP10	CFP15	CFP20	Ultra high pro	Post-MSC DDGS	SEM	*P*-value
CP	381.1[b]	410.6[b]	407.7[b]	419.4[b]	416.4[b]	501.0[a]	188.1[c]	25.54	<0.001
Indispensable AA									
Arg	27.8[a]	20.5[b]	21.2[b]	21.7[b]	21.6[b]	21.9[b]	10.2[c]	0.82	<0.001
His	9.9[c]	10.4[bc]	10.6[bc]	11.0[b]	11.0[b]	11.9[a]	5.3[d]	0.36	<0.001
Ile	18.6[b]	17.5[b]	18.4[b]	18.5[b]	18.8[b]	20.7[a]	8.0[c]	0.51	<0.001
Leu	29.3[c]	51.8[b]	52.8[b]	52.8[b]	52.6[b]	68.0[a]	26.9[c]	0.95	<0.001
Lys	23.8[a]	13.7[b]	14.7[b]	15.3[b]	15.2[b]	14.3[b]	4.1[c]	1.1	<0.001
Met	5.1[c]	9.4[b]	9.4[b]	9.4[b]	9.4[b]	11.8[a]	4.3[d]	0.19	<0.001
Phe	19.7[c]	22.7[b]	23.1[b]	23.5[b]	23.6[b]	27.9[a]	12.2[d]	0.51	<0.001
Thr	14.1[c]	15.9b[c]	16.3[ab]	16.3[ab]	16.4[ab]	17.9[a]	6.9[d]	0.73	<0.001
Trp	5.2[a]	2.8[b]	23.2[b]	23.1[b]	23.1[b]	25.4[a]	9.3[d]	0.2	<0.001
Val	18.3[c]	22.8[b]	23.2[b]	23.1[b]	23.1[b]	25.4[a]	9.3[d]	0.7	<0.001
Dispensable AA									
Ala	16.1[c]	31.4[b]	31.9[b]	31.6[b]	31.3[b]	38.7[a]	14.3[c]	0.86	<0.001
Asp	41.5[a]	28.8[c]	30.0[bc]	29.9[bc]	30.4[bc]	32.1[b]	10.6[d]	1.18	<0.001
Cys	4.5[c]	6.2[bc]	6.5[b]	7.0[ab]	7.1[ab]	8.3[a]	4.6[c]	0.68	<0.001
Glu	70.4[b]	71.7[b]	71.0[b]	72.2[b]	70.5[b]	91.7[a]	31.2[c]	2.33	<0.001
Gly	15.7[a]	17.3[a]	17.3[a]	18.0[a]	17.7[a]	19.2[a]	7.8[b]	1.97	<0.001
Ser	16.3[c]	17.9[bc]	19.0[b]	19.2[b]	19.3[b]	22.3[a]	9.5[d]	0.72	<0.001
Tyr	14.5[c]	19.14[b]	18.6[b]	19.1[b]	18.9[b]	23.2[a]	8.4[d]	0.4	<0.001

to calculate the contribution of corn to the corn-high-protein corn fermented product/Post MSC DDGS diets. The DE and ME in high-protein corn fermented products/post-MSC DDGS were calculated by difference (Widmer et al., 2007). Pigs fed the post-MSC DDGS diet had greater ($P < 0.01$) faecal GE loss compared with pigs fed the corn diet or diets containing the high-protein corn fermented products (Table 4). As a result, the ATTD of GE and concentrations of DE and ME in the post-MSC DDGS diet were less ($P < 0.01$) compared with the corn diet, CFP diet or ultra high pro+ diet. Concentrations of DE and ME in diets containing the high-protein corn fermented products were greater ($P < 0.01$) compared with the corn diet.

The ATTD of GE in post-MSC DDGS was less ($P < 0.01$) compared with that of corn, CFP or ultra high pro+. As a result, concentrations of DE and ME (as-fed basis and dry matter basis) in post-MSC DDGS were less ($P < 0.01$) compared with corn, CFP or ultra high pro+. The two high-protein corn fermented products had reduced ($P < 0.01$) ATTD of GE compared with corn; however, due to increased concentration of GE in CFP and ultra high pro+, greater ($P < 0.01$) values for DE and ME (as-fed basis) were observed for CFP and ultra high pro+ compared with corn. Likewise, the DE on a DM basis was also greater in CFP and ultra high pro+ than in corn.

The difference procedure was used to determine concentrations of DE and ME in the two high-protein corn fermented products and post-MSC DDGS, and a consequence of using the difference procedure is that reliable results for test ingredients will be obtained only if the DE and ME of the ingredient included in the corn basal diet are accurate. Values for DE and ME obtained for corn in the present experiment are in close agreement with previous data (NRC, 2012; Espinosa et al., 2019), which gives confidence that calculated values for DE and ME in CFP, ultra high pro+ and post-MSC DDGS are also accurate. Values for DE and ME in post-MSC DDGS used in the experiment were less than published data for conventional DDGS (NRC, 2012), which may be due to increased concentration of insoluble dietary fibre and reduced concentration of fat in post-MSC DDGS compared with conventional DDGS. However, the DE and ME in post-MSC DDGS is in agreement with values for DE and ME in low-oil DDGS (Curry et al., 2016). The observation that CFP and ultra high pro+ had greater concentrations of DE (DM basis) than corn is likely due to increased concentrations of CP and fat as a result of the mechanical separation of solids/protein from the bioethanol whole stillage after the distillation process. Concentrations of DE and ME in post-MSC DDGS were less compared with corn and the two high-protein corn fermented products. Corn had greater ATTD of GE compared with CFP and ultra high pro+; however, DE and ME (as fed basis) and DE on a DM basis were greater in CFP and ultra high pro+ than in corn. The ATTD of GE and concentrations of DE and ME were not different between the two high-protein corn fermented products.

Table 4 Concentration of digestible energy (DE) and metabolizable energy (ME) and apparent total tract digestibility (ATTD) of gross energy in experimental diets and in corn, CFP, ultra high pro+ and post-MSC DDGS[1] fed to swine

Item	Corn	CFP	Ultra high pro+	Post-MSC DDGS	SEM	*P*-value
Diets						
Gross energy intake, kcal/d	3,728	3,885	3,521	3,493	169	0.268
Gross energy in faeces, kcal/day	378[b]	440[b]	431[b]	535[a]	34	0.007
Gross energy in urine, kcal/day	48	83	71	59	14	0.061
ATTD of gross energy, %	89.9[a]	88.7[ab]	87.9[b]	84.8[c]	0.46	<0.001
DE, kcal/kg	3,289[b]	3,384[a]	3,394[a]	3,205[c]	17	<0.001
ME, kcal/kg	3,242[b]	3,300[a]	3,312[a]	3,142[c]	18	<0.001
Ingredients						
ATTD of gross energy, %	89.9[a]	81.6[b]	80.4[b]	61.8[c]	1.91	<0.001
DE, kcal/kg	3,398[b]	3,987[a]	4,057[a]	2,793[c]	59	<0.001
ME, kcal/kg	3,349[b]	3,707[a]	3,778[a]	2,640[c]	76	<0.001
DE, kcal/kg dry matter	3,914[b]	4,308[a]	4,374[a]	3,201[c]	68	<0.001
ME, kcal/kg dry matter	3,858[a]	4,005[a]	4,072[a]	3,025[b]	86	<0.001

[1]Data are means of eight observations per treatment.
Values within a row lacking a common superscript letter are statistically different ($P < 0.05$).

7 Case study: effects of adding phytase on availability of calcium and phosphorus in corn-fermented products

The 50+ h fermentation which is a critical component of the dry grind bioethanol process is an ideal environment in which to benefit from exogenous enzyme activity to improve nutrient availability in the final product. The standardized total tract digestibility (STTD) of phosphorus (P) in high-protein corn fermented products and in the residual distillers dried grains with solubles (DDGS) produced from a fermentation was evaluated with or without the addition of phytase either into the fermentation medium, or as a dietary supplement (Lopez et al., 2022). Three sources of high-protein corn-fermented products (i.e. CFP, ET-ultra high pro, and ET-ultra high pro+) and two sources of post-MSC DDGS were obtained (Table 5). Two sources of high-protein corn-fermented products (i.e. ET-ultra high pro and ET-ultra high pro+) and one source of post-MSC

Table 5 Analysed nutrient composition of ingredients

Item	CFP	ET[1]-ultra high pro	ET-ultra high pro+	Post-MSC DDGS	ET-post-MSC DDGS
Dry matter, %	91.39	92.55	92.77	87.75	87.26
Gross energy, kcal/kg	4,925	4,885	5,044	4,378	4,520
Ash, %	4.45	3.80	4.28	4.73	3.79
Ca, %	0.04	0.01	0.02	0.03	0.02
Total P, %	0.87	0.57	0.58	1.10	0.70
Phytic acid, %	2.14	<0.14	0.30	2.20	<0.14
Phytate-bound P,[2] %	0.60	<0.04	0.08	0.62	<0.04
Nonphytate P,[3] %	0.27	0.53	0.50	0.48	0.68
Myo-inositol, mg/100 g	29.70	65.50	58.60	77.40	150.00

[1]ET, enzyme-treated.
[2]Calculated as 28.2% of phytic acid (Tran and Sauvant, 2004).
[3]Calculated as total P – phytate-bound P.

DDGS (i.e. ET-post-MSC DDGS) were treated with microbial phytase during the fermentation process, whereas phytase treatment was not included for the other two corn co-products. Five diets were formulated by mixing each source of corn co-product with sucrose and cornstarch, and corn co-product was the only source of P in these diets. Five additional diets were also formulated similar to the previous two diets with the exception that 500 units of phytase per kilogram were added directly to the diets. A standard vitamin and minerals premix was included to meet the nutrient requirements for growing pigs (NRC, 2012).

Eighty pigs [initial BW: 14.38 ± 1.50 kg] were allotted to ten diets with eight replicate pigs per diet. All diets were fed in meal form. Pigs were limit fed at 3.2 times the energy requirement for maintenance (i.e. 197 kcal/kg × $^{BW0.60}$; NRC, 2012). The initial 5 days were considered the adaptation period to the diet, and faecal materials were collected from the feed provided during the following 4 days according to standard procedures using the marker to marker approach (Adeola, 2001).

The apparent total tract digestibility (ATTD) of Ca and P in each diet was calculated (Table 6). By correcting ATTD values for the basal endogenous loss of P (i.e. 190 mg per kg dry matter intake), values for the STTD of P in each diet were also calculated. Due to increased concentration of total P, pigs fed the diet containing CFP and post-MSC DDGS had greater ($P < 0.01$) P intake compared with that of pigs fed diets with other corn co-products. Inclusion of 500 FTU per kilogram of phytase in CFP and post-MSC DDGS diets reduced the

Table 6 Effects of microbial phytase on P balance, apparent total tract digestibility (ATTD), and standardized total tract digestibility (STTD) of P in three sources of high-protein corn fermented products and two sources of post-MSC DDGS fed to growing pigs[1]

Item	ADFI, g/day	P intake, g/day	P in faeces, %	P output, g/day	P absorption, g/day	ATTD of P, %	Basal EPL[2], mg/day	STTD of P[3], %
0 FTU[4]/kg								
CFP	525	1.92	2.13[a]	0.74[b]	1.19	62.07[d]	92.20	66.86[d]
ET-Ultra High Pro	588	1.49	0.68[d]	0.30[c]	1.18	79.74[ab]	102.52	86.63[a]
ET-Ultra High Pro+	499	1.33	0.58[d]	0.23[c]	1.10	82.95[a]	87.27	89.53[a]
Post-MSC DDGS	593	2.62	1.41[b]	1.01[a]	1.61	62.23[d]	101.52	66.10[d]
ET-Post-MSC DDGS	541	1.68	0.36[e]	0.34[c]	1.34	79.46[ab]	92.42	84.96[ab]
500 FTU/kg								
CFP	513	1.88	1.51[b]	0.57[b]	1.31	70.52[c]	90.00	75.30[c]
ET-Ultra High Pro	605	1.53	0.61[d]	0.27[c]	1.26	82.29[a]	105.50	89.19[a]
ET-Ultra High Pro+	524	1.38	0.58[d]	0.24[c]	1.15	82.74[a]	91.64	89.32[a]
Post-MSC DDGS	558	2.47	0.89[c]	0.61[b]	1.86	75.42[bc]	95.96	79.31[bc]
ET-Post-MSC DDGS	523	1.62	0.38[e]	0.30[c]	1.31	80.93[ab]	90.32	86.49[a]
SEM	34	0.12	0.06	0.06	0.09	2.28	5.86	2.28
P-values								
Corn co-product	0.006	<0.001	<0.001	<0.001	<0.001	<0.001	0.008	<0.001
Phytase	0.788	0.609	<0.001	0.003	0.028	0.001	0.862	0.001
CFP × Phytase	0.777	0.818	<0.001	0.012	0.309	0.024	0.802	0.025

[1]Data are least squares means of seven to eight observations per treatment.
[2]EPL = endogenous P loss. This value was estimated to be at 190 mg/kg DMI (dry matter intake). The daily basal EPL (mg/day) for each diet was calculated by multiplying the EPL (mg/kg DMI) by the daily DMI of each diet (Almeida and Stein, 2010).
[3]Values for STTD calculated by correcting ATTD values for the basal endogenous loss of P (NRC, 2012).
[4]FTU = phytase units.
[a–e]Means within a row lacking a common letter are different ($P < 0.05$).

concentration of P in faeces; however, phytase did not affect the concentration of P in faeces by pigs fed diets containing the ET corn co-products (interaction, $P < 0.01$). Due to increased P intake, P absorption was greater ($P < 0.01$) in pigs fed the diet containing CFP and post-MSC DDGS compared with that of pigs fed diets with other corn co-products regardless of phytase supplementation. Absorption of P also increased ($P < 0.05$) when diets were supplemented with phytase. The ATTD and STTD of P in the CFP and post-MSC DDGS diets increased upon phytase supplementation; however, phytase supplementation did not influence the ATTD and STTD of P in diets containing the enzyme-treated corn co-products (interaction, $P < 0.05$).

Dietary supplementation of phytase increased the ATTD and STTD of P in CFP and Post-MSC DDGS indicates that the exogenous phytase increased P absorption in the pigs (Adeola, 2001). The observed increase in Ca digestibility in experimental diets indicates that phytase was also effective in hydrolyzing the Ca-phytate complexes in the gastrointestinal tract of pigs (Selle et al., 2009). The observation that the ATTD and STTD of P in enzyme-treated corn co-products were greater compared with CFP and Post-MSC DDGS (without enzyme treatment) indicates that the use of phytase during the fermentation process was able to liberate some of the P in the phytic acid molecule. Indeed, this agreed with the observed reduction in the analysed concentration of phytate-bound P and a subsequent increase in the analysed concentration of myo-inositol in corn co-products that were treated with phytase during the fermentation process. The observation that dietary supplementation of phytase did not influence P digestibility in enzyme-treated corn co-products indicates that phytase treatment during the fermentation process is sufficient to maximize P utilization in high-protein corn fermented products and Post-MSC DDGS. Concentration of phytate-bound P was less and myo-inositol concentration was greater in enzyme-treated corn co-products compared with corn co-products without enzyme treatment. As a result, the ATTD and STTD of P in enzyme-treated corn co-products were greater compared with CFP and Post-MSC DDGS (without enzyme treatment). Dietary supplementation of phytase increased the ATTD of Ca in experimental diets and increased the ATTD and STTD of P in CFP and post-MSC DDGS. Addition of phytase to diets did not influence P digestibility in enzyme-treated corn co-products, which likely indicates that phytase treatment during the fermentation process is sufficient to maximize P utilization in corn co-products.

8 Case study: effects on performance of pigs of inclusion of corn-fermented protein in the diet

Formulations of creep feeds for nursery pigs are recognized to contain a wide variety of feed ingredients to stimulate intake and minimize the occurrence of diarrhoea in neonatal pigs. Highly digestible benign proteins, such as spray-dried animal plasma (SDAP), plant proteins such as enzyme-treated SBM or mixtures of high value protein plus additions of components of yeast, are common additions to nursery pig diets. Nutritionists are keen to minimize the number of products in nursery diets (phase 1 diets) that contain recognized anti-nutritional factors such as SBM due to the antigenic effects limiting early feed intake. One of the most appropriate uses of CFP co-products because of its high ME and digestible AA content and lack of anti-nutritional factors is in phase 1 and 2 diets for weaned pigs.

Trial 1: Martindale et al. (2018) determined the effects of increasing dietary levels (0%, 8%, 16%, and 24%) of CFP in nursery, phase 1 (0–14 days post-weaning) and phase 2 (14–28 days post-weaning), with a common corn-SBM diet fed during phase 3 (days 28–35 post-weaning) on growth performance of pigs weaned up to 21 days of age. No differences in average daily gain (ADG), average daily feed intake (ADFI) and gain:feed (G:F) were observed among dietary treatments during phase 1, but pigs fed the 24% CFP diet during phase 2 had reduced ADG and ADFI compared with pigs fed the control (0% CFP) diet. There were no effects on growth performance during phase 3 and the overall 35-day trial. The researchers concluded that adding up to 16% CFP to phase 1 and phase 2 nursery diets does not negatively affect growth performance.

Trial 2: Acosta et al. (2021) evaluated the impact on growth performance and faecal scores when feeding diets containing various amounts of CFP to partially replace SDAP and enzyme-treated SBM (ES) for 21 days post- weaning during phase 1 (days 1–7) and phase 2 (days 8–21). Feeding the control phase 1 diet containing 5% ES and 2.5% SDAP for the first 7 days after weaning tended to increase ADG (0.12 kg/d) and G:F (0.78) compared with feeding diets containing 4.5% ES + 5% CFP and 10% CFP (0.08 kg/days; 0.57 respectively), but there was no significant difference in ADFI among dietary treatments. However in phase 2, the amount of CFP (0–10%) fed from day 8–21 post-weaning had no effect on ADG, ADFI and G:F. All pigs were fed a common corn–SBM diet during phase 3 (day 22–35 post-weaning) and there was no differences in subsequent growth. The results indicate that if 2.5% SDAP is included in the diet adding 5% CFP to phase 1 diets for weaned pigs can partially replace spray-dried porcine plasma providing equivalent performance but this was not the case with ES. Adding 10% CFP to phase 1 nursery diets appeared to be excessive to support optimal growth performance without addition of SDAP or ES. However, adding 10% CFP to phase 2 diets supported acceptable growth performance comparable to feeding the control diet containing 7.5% ES.

Trial 3: The effect of CFP at 7.5% and 15% inclusion rate (replacing soy- derived ingredients on an equivalent CP basis) in piglet weaner diets, formulated to meet current nutrient requirements, was evaluated on piglet growth performance when fed a single diet for 28 days post-weaning. Three treatments included a high-specification control diet containing whey (15%)

and soy (SBM (10%) and Alphasoy 530 (5%)) as the primary protein sources, and two similar diets but with CFP added at 7.5% or 15% at the expense of the soya-derived protein ingredients on an equivalent CP basis. All diets were formulated to be isoenergetic and equivalent in terms of nutrient composition. A total of 240 mixed sex pigs ([Large White × Landrace] × Danish Duroc) were weaned onto the experimental diets at an initial weight of 8.24 ± 1.19 kg (mean ± SD) at 26.4 ± 0.56 days of age. Pig health, faecal consistency and cleanliness scores were recorded daily. Pig performance was in line with expectations for the facility, with growth rates averaging 376 (± 56.4) g/pig/day over the course of the study. Health, faecal and cleanliness scores were uniformly good throughout the study, with no differences observed between treatments and there were no pig mortalities during the experiment.

There were no differences in pig growth rates or feed conversion ratios between the three treatments during the first 13 days of the study although there was a tendency for a reduction in ADFI in response to feeding CFP at both levels. After 28 days on trial pigs receiving the diets containing CFP were 1.27 and 2.01 kg lighter than control-fed pigs, respectively ($P < 0.01$). This difference in BW was attributed to a reduction in feed intake in response to CFP inclusion over the course of the study, with those fed CFP 7.5% consuming 11% less feed than the control pigs, and those fed CFP 15% consuming 16% less feed. However there was no difference in pig FCR between the three treatments throughout the trial, indicating that CFP is utilized as efficiently as AlphaSoy 530 and SBM. There was no effect of replacing the soy derived proteins with CFP on piglet health or feed conversion efficiency after 28 days of feeding, indicating that this yeast/grain-based protein ingredient can be utilized as efficiently as a conventional soy derived product. The results indicate that CFP can be utilized as an alternative protein supplement in piglet diets up to 15% inclusion for the first 13 days post-weaning, without compromizing piglet health or growth performance. However, upon extending the feeding duration of CFP to 28 days post-weaning, depressions in piglet growth rates were observed at both the 7.5% and 15% inclusion levels. This effect was attributed to reductions in piglet feed intake occurring primarily throughout the final two weeks of the study, which suggest that there was an effect of CFP on diet palatability. The results suggest that in young pigs there is a component of CFP that has a negative impact on feed intake. It is reported that excess Leucine competes with tryptophan (Trp) transport through the blood into the brain which reduces serotonin synthesis and consequently reduces ADFI (Kwon et al., 2019; Yang et al., 2019). It was postulated that an imbalance of AAs caused by the inclusion of corn protein may be responsible for the effect on palatability.

Trial 4: The earlier experiments demonstrated that whilst CFP can be used to replace a range of different protein sources in weaner diets for pigs, there is a limit to CFP inclusion above which there is appetite depression. Diets with a high inclusion levels of high-protein distillers dried grain with solubles (HP-DDGS) or CFP may contain an excess of dietary Leucine (Leu), and this may be responsible for the negative effects on feed intake and reduced growth performance. It was hypothesized that the negative effect of using CFP in diets for weanling pigs could be avoided if the AA imbalance could be corrected by addition of crystalline sources of Valine (Val), Tryptophan (Try) and (or) Isoleucine (Ile) (Stein personal communication). A total of 320 weanling pigs (BW: 6.11 ± 0.66 kg) were randomly allotted to one of 10 dietary treatments in a completely randomized design. A two-phase feeding programme was used, days 1 to 14 (phase 1) and days 15 to 28 (phase 2). Three basal diets were formulated for each phase, a corn-SBM diet and 2 test diets based on corn plus 10% CFP or corn plus 20% CFP. Seven additional diets were formulated by adding Val, Ile, Trp, Val and Ile, Val and Trp, Ile and Trp, or Val, Ile, and Trp to the basal diet with 20% CFP. Phase 1 diets were fed from day 1 to 14 post-weaning and phase 2 diets were from day 15 to 28. ADFI, ADG and average G:F were calculated for each phase and for the overall experiment. Faecal scores were recorded every other day. There was no effect of the treatments on the health of the piglets. Results indicated that inclusion of 10% or 20% CFP in diets reduced ($P < 0.05$) final BW on day 28, ADG and ADFI in phase 2 and for the entire experimental period and G:F for the entire experiment similar to results obtained in trial 3. However, pigs fed SBM and CFP supplemented with the 3 AAs had a greater (+1.72 kg live weight; $P < 0.05$) final BW and ADG compared with pigs fed the other diets. On day 28, pigs fed the diet with 20% CFP and only Val, Val and Trp, or Val, Trp, and Ile had reduced ($P < 0.01$) blood urea N compared with pigs fed the corn-SBM diet or the other CFP-based diets. The fact that high concentrations of leucine negatively affects the utilization and metabolism of valine and isoleucine (Ile), causing a nutritional challenge is recognized (Harris et al., 2004; Cemin et al., 2019a,b; Kwon et al., 2019; Yang et al., 2019) and CFP contains high levels of leucine. In addition, excess Leu competes with tryptophan (Trp) transport through the blood into the brain which reduces serotonin synthesis and consequently reduces ADFI (Kwon et al., 2019; Yang et al., 2019). These results indicate that CFP can be used as an alternative vegetable protein source and included in phase 1 and phase 2 nursery diets at levels up to 16% as long as attention is paid to balancing the AA content of the formulation with additional Val, Trp and Ile.

9 Conclusion

There are currently 17 plants producing CFP in the USA, 2 in South America and 1 in construction in Europe with a combined production capacity of well over 1 million tons of CFP per annum, providing resilience and redundancy in the supply chain. Although the dry grind ethanol process is a global source of medium- quality protein, the opportunity to turn corn into a multi-purpose energy and protein crop is a major breakthrough. The trials carried out to evaluate the use of CFP as an alternative protein supplement in diets for broilers and turkey poults demonstrate that CFP can be used in formulation as an alternative protein.

In order to complete these trials the standardized digestibility and true metabolizable energy were measured in the target species. In all the early growth experiments, the maximum inclusion rate of CFP was restricted to not exceed 10% of dietary inclusion. Since these experiments were exploratory experiments to test the use of the product in diets for poultry and CFP contains approximately 20–27% yeast cell wall material, this was a precautionary measure to not exceed the yeast cell wall dietary inclusion level of 5%. CFP can be used as a dietary replacement for SBM and with appropriate balancing of essential AAs will achieve performance at a minimum equal to the performance of the control diets. However, there is evidence to suggest that either the content of spent yeast material or the combination of spent yeast with a fermented carbohydrate fraction may be beneficial in terms of gut health.

A trial with turkey poults is particularly interesting (Williams, 2024). Overall growth performance of poults across all treatments aligned well with industry standards (Aviagen, 2014), indicating that the diets conformed well to industry standards. However, in terms of improved nitrogen end energy retention, there is evidence to suggest that there may be additional benefits to the partial replacement of SBM with CFP.

CFP contains a functional component of approximately 24% (on a DM basis) spent yeast content. Several nutraceutical components of yeast are reported to have beneficial effects in poultry and may indirectly improve growth performance. β-1,3-glucans, the major component of yeast cell walls, have prebiotic effects due to their ability to bind toxins and pathogens (Vetvicka and Oliveira, 2014). Yeasts contain mannan oligosaccharides which have an overall positive effect on animal performance (Spring et al., 2015) via improving intestinal architecture, physical gut tissue turnover, a change in microbiota or reduction in immune stimulation (Baurhoo et al., 2007). The effects on ceacal VFA concentrations reported in a broiler trial are evidence that the fermented fibre component of CFP has the potential to increase hind gut fermentation although to date there is no evidence to indicate an improvement in the histology of the gut.

Identifying new sources of high-quality vegetable protein that are available for use in animal feed is a priority. High quality vegetable proteins are extensively used for food and can therefore demand a price premium. Identifying a source of high concentration vegetable protein such as CFP that is not available for use in food since it is derived from a non-food industrial process is an opportunity for the feed industry to source an alternative cost competitive product. While SBM is recognized as an excellent protein source, residual trypsin inhibitor, lipoxygenases and antigenic proteins retain a degree of activity post-processing which slightly but significantly reduce dietary protein digestion (Clarke and Wiseman, 2005). Thus, some of the performance improvement in the current study could be related to the reduction in anti-nutritional effects associated with partial replacement of SBM with CFP. However, the results demonstrate that CFP can be used as a dietary replacement for SBM and with appropriate balancing of essential AAs will achieve performance at a minimum equal to the performance of the control diets.

The current programme covering the use of CFP in pigs is part of an extensive programme on the use of the protein in diets for poultry, aquaculture, companion animals and ruminants in addition to poultry. The results mainly reflect the beneficial effects reported in poultry. The development of CFP is an excellent example of how the introduction of tested processing technology with existing commodity feed materials can significantly improve the value a low-quality feed material and significantly increase the flexibility and value of feed materials used by animal nutritionists in feed formulation. The composition of CFP with a high protein content and potential functional characteristics based on the content of spent yeast and fermented fibre, point towards the product being an ideal vegetable protein supplement for swine and in particular for use in the nursery phase 0–21 days of age. There have been few if any new and significant developments in the vegetable protein space for animal feed in the past 20 years. CFP is the first major development in terms of a new commercially viable high protein feed for all classes of livestock and companion animals.

10 References

Acosta, J. P., Espinosa, C. D., Jaworski, N. W. and Stein, H. H. 2021. Corn protein has greater concentrations of digestible amino acids and energy than low-oil corn distillers dried grains with solubles when fed to pigs nut does not affect performance of weanling pigs. *J. Anim. Sci.* 99(7):1–12. doi: 10.1093/jas/skab175.

Adeola, O. 2001. Digestion and balance techniques in pigs. In Lewis, A. J. and Southern, L. L. (eds). *Swine Nutrition*, Second edition, CRC Press, Washington DC, USA.

Aviagen 2014. Ross 308 broiler: nutrition specifications. Available at: http://en.aviagen.com/assets/Tech_Center/Ross_Broiler/Ross308BroilerNutritionSpecs2014-EN.pdf, accessed 21 December 2016.

Baurhoo, B., Phillip, L. and Ruiz-Feria, C. A. 2007. Effects of purified lignin and mannan oligosaccharides on intestinal integrity and microbial populations in the ceca and litter of broiler chickens. Poult. Sci. 86(6):1070–1078 doi: 10.1093/ps/86.6.1070.

Burton, E. 2021. Use of an ethanol bio-refinery product as a soy bean alternative in diets for fast-growing meat production species. *Sustainability* 13(19):11019.

Cemin, H. S., Tokach, M. D., Dritz, S. S., Woodworth, J. C., DeRouchey, J. M. and Goodband, R. D. 2019a. Meta-regression analysis to predict the influence of branched-chain and large neutral amino acids on growth performance of pigs. *J. Anim. Sci.* 97(6):2505-2514. doi: 10.1093/jas/skz118.

Cemin, H. S., Tokach, M. D., Woodworth, J. C., Dritz, S. S., DeRouchey, J. M. and Goodband, R. D. 2019b. Branched-chain amino acid interactions in growing pig diets. *Transl. Anim. Sci.* 3(4):1246-1253. doi: 10.1093/tas/txz087.

Clarke, E. and Wiseman, J. 2005. Effects of variability on trypsin inhibitor content of soy bean meals on true and apparent ileal digestibility of amino acids and pancreas size in broiler chicks. *Anim. Feed Sci. Technol.* doi:10.1016/j.anifeedsci.2005.02.012.

Cristobal, M., Acosta, J. P., Lee, S. A. and Stein H. H. 2020. A new source of high-protein distillers dried grains with solubles (DDGS) has greater digestibility of amino acids and energy, but less digestibility of phosphorus, than de-oiled DDGS when fed to growing pigs. *J. Anim. Sci.* 98(7): 1-9.

Curry, S. M., Rojas, O. J. and Stein, H. H. 2016. Concentration of digestible and metabolizable energy and digestibility of energy and nutrients by growing pigs in distillers dried grains with solubles produced in and around Illinois. *The Professional Animal Scientist* 32(5):687-694.

Espinosa, C. D., Fry, R. S., Kocher, M. E. and Stein, H. H. 2019. Effects of copper hydroxychloride and distillers dried grains with solubles on intestinal microbial concentration and apparent ileal and total tract digestibility of energy and nutrients by growing pigs. *J. Anim. Sci.* 97(12):4904-4911. doi: 10.1093/jas/skz340.

Espinosa, C. D., Oliveira, M. S. F., Lagos, L. V., Weeden, T. L., Mercado, A. J. and Stein, H. H. 2020. Nutritional value of a new source of fermented soybean meal fed to growing pigs. *J. Anim. Sci.* 98(12):1-9.

Fastinger, N. D. and Mahan, D. C. 2006. Determination of the ileal amino acid and energy digestibilities of corn distillers dried grains with solubles using grower-finisher pigs. *J. Anim. Sci.* 84(7):1722-1728. doi: 10.2527/jas.2005-308.

Harris, R. A., Joshi, M. and Jeoung, N. H. 2004. Mechanism responsible for regulation of branched-chain amino acid catabolism. *Biochem. Res. Commun.* 313:391-396. doi: 10/1093/jas/skz259.

Han, J. and Liu, K. 2010. Changes in composition and amino acid profile during dry grind ethanol processing from corn and estimation of yeast contribution towards DDGS proteins. *J. Agric. Food Chem.* 58(6):3430-3437.

Kwon, W. B., Touchette, K. J., Simongiovanni, A., Syriopoulos, K., Wessels, A. and Stein, H. H. 2019. Excess dietary leucine in diets for growing pigs reduces growth performance, biological value of protein, protein retention, and serotonin synthesis. *J. Anim. Sci.* 97(10):4282-4292. doi: 10.1093/jas/skz259.

Lagos, L. V. and Stein, H. H. 2017. Chemical composition and amino acid digestibility of soybean meal produced in the United States, China, Argentina, Brazil or India. *J. Anim. Sci.* 95:1626-1636.

Lopez, D. A., Lee, S. A. and Stein, H. H. 2022. Effects of microbial phytase on standardized total tract digestibility of phosphorus in feed phosphates fed to growing pigs. *J. Anim. Sci.* 100(12):skac350.

Martindale, A., Trenhaile-Grannemann, M., Barnett, S., Miller, P. and Burkey, T. 2018. Growth performance of weaned pigs fed high-protein corn product. *J. Anim. Sci.* 96(Suppl. 3):295.

Mottet, A., de Haan, C., Falcuucci, A., Tempio, G., Opio, C. and Gerber, P. 2017. Livestock: on our plates or eating at our table? A new analysis of the feed/food debate. *Glob. Food Sec.* 14:1-8. doi: 10.1016/j.gfs.2017.01.001.

NRC 2012. *Nutrient Requirements of Swine* (11th rev. edn). Washington (DC): The National Academies Press.

Parsons, B. W., Utterback, P. L., Parsons, C. M. and Emmert, J. L. 2023. Standardized amino acid digestibility and true metabolizable energy for several increased protein ethanol co-products using back-end fractionation systems. *Poult. Sci.* 102(2):102329.

Parsons, C. M. 1985. Influence of caecectomy on digestibility of amino acids by roosters fed distillers dried grains with solubles. *J. Agric. Sci.* 104(2):469–472.

Rho, Y., Zhu, C., Kiarie, E. and de Lange, C. F. M. 2017. Standardized ileal digestible amino acids and digestible energy contents in high-protein distiller's dried grains with solubles fed to growing pigs. *J. Anim. Sci.* 95(8):3591–3597. doi: 10.2527/jas.2017.1553.

Scholey, D., Alkhtib, A., Williams, P. and Burton, E. 2023. Corn fermented protein, an alternative protein to soybean meal to support growth in young turkey poults. *J. Appl. Anim. Nutr.* doi: 10.3920/JAAN2023.0002.

Selle, P. H., Cowieson, A. J. and Ravindran, V. 2009. Consequences of calcium interactions with phytate and phytase for poultry and pigs. *Livest. Sci.* 124:126–141.

Shurson, G. C. 2018. Yeast and yeast derivatives in feed additives and ingredients: sources, characteristics, animal responses and quantification methods. *Anim. Feed Sci. Technol.* 235:60–76 . doi: 10.1016/j.anifeedsci.2017.11.010.

Spring, P., Wenk, C., Connolly, A. and Kiers, A. 2015. A review of 733 published trials on Bio-Moss, a mannan oligosaccharide and Actigen, a second generation mannose rich fraction, on farm and companion animals. *J. Appl. Anim. Nutr.* 3:1–11.

Stein, H. H., Sève, B., Fuller, M. F., Moughan, P. J. and de Lange, C. F. M. 2007. Invited review: amino acid bioavailability and digestibility in pig feed ingredients: terminology and application. *J. Anim. Sci.* 85(1):172–180. doi: 10.2527/jas.2005-742.

Vetvicka, V. and Oliveira, C. 2014. β(1-3)(1-6)-D-glucans modulate immune status in pigs: potential importance for efficiency of commercial farming. *Ann. Transl. Med.* 2(2):16. doi: 10.3978/j.issn.2305-5839.2014.01.04.

Widmer, M. R., McGinnis, L. M. and Stein, H. H. 2007. Energy, phosphorus, and amino acid digestibility of high-protein distillers dried grains and corn germ fed to growing pigs. *J. Anim. Sci.* 85(11):2994–3003. doi: 10.2527/jas.2006-840.

Williams, P. E. V. 2024. High protein corn fermentation products for poultry derived from corn ethanol production. In Applegate, T. (ed.). *Advances in Poultry Nutrition*, Burleigh Dodds Science Publishing, Cambridge, UK.

Xu, B., Li, Z., Wang, C., Fu, J., Zhang, Y., Wang, Y. and Lu, Z. 2020. Effects of fermented feed supplementation on pig growth performance: a meta-analysis. Anim. Feed Sci. Technol. 259:114315. doi: 10.1016/j.anifeedsci.2019.114315.

Yang, Z., Urriola, P. E., Hilbrands, A. M., Johnston, L. J. and Shurson, G. C. 2019. Growth performance of nursery pigs fed diets containing increasing levels of a novel high-protein corn distillers dried grains with solubles. *Transl. Anim. Sci.* 3(1):350–358. doi: 10.1093/tas/txy101.

Index

www.ingramcontent.com/pod-product-compliance
Ingram Content Group UK Ltd.
Pitfield, Milton Keynes, MK11 3LW, UK
UKHW021825100926
8556IPUK00003B/164

9 781801 466943